CHEMISTRY RESEARCH AND APPLICATIONS

CURRENT APPLICATIONS OF CHEMOMETRICS

Chemistry Research and Applications

Additional books in this series can be found on Nova's website under the Series tab.

Additional e-books in this series can be found on Nova's website under the e-book tab.

CHEMISTRY RESEARCH AND APPLICATIONS

CURRENT APPLICATIONS OF CHEMOMETRICS

MOHAMMADREZA KHANMOHAMMADI
EDITOR

nova publishers
New York

Library of Congress Cataloging-in-Publication Data

Current applications of chemometrics / [edited by] Mohammadreza Khanmohammadi (Chemistry Department, Faculty of Science, IKIU, Qazvin, Iran).
pages cm. -- (Chemistry research and applications)
Includes bibliographical references and index.
ISBN 978-1-63463-117-4 (hardcover)
1. Chemometrics. I. Khanmohammadi, Mohammadreza, editor.
QD75.4.C45C87 2014
543.01'5195--dc23

2014036535

Published by Nova Science Publishers, Inc. † New York

CONTENTS

PREFACE

Known as a scientific domain in which the scientist would try to extract information from chemical systems by data-driven means, Chemometrics is a fast spreading field, being applied to solve both descriptive and predictive problems in experimental life sciences, especially in chemistry. It is defined as a highly interfacial discipline, which employs multivariate statistics, applied mathematics, and computer science via using methods frequently employed in core data-analytic, in order to address problems in chemistry, biochemistry, medicine, biology and chemical engineering. Initiated by analysts, now the discipline is widened by other chemistry discipline researches and even those from medical and biological areas. Chemometric techniques are particularly heavily used in analytical chemistry and metabolomics, and the theoretical development of chemometric methods of analysis also continues to advance the state of the art in analytical instrumentation and methodology improvements. It is an application driven discipline, and thus while the standard chemometric methodologies are very widely used industrially, academic groups are dedicated to the continued development of chemometric theory, method and application development. My research interest since 2 decades ago was to connect the bench scale applications of chemometric methodologies to industrial level, in order to introduce its superb benefits and capabilities to be generalized. Thus in all my research activities, everywhere possible, chemical data has been processed by chemometric approaches, especially while the spectrometric analytical procedure were conducted. This edited collection would try to review current applications of chemometrics for different aims e.g., multivariate calibrations, pattern recognitions, curve resolution, etc. It has been tried to organize the project, considering the interest of chemometrics researchers for both method evaluation and technical applications. A wide range of distinguished researches from all around the world have prepared their contribution in the frame of Current Applications of Chemometrics and on behalf of all colleagues who have collaborated in this edited collection, I hope it would be interesting and useful for academic and industrial researches. Finally, I would like to express my sincere gratitude to all the colleagues who accepted my invitation and also to my friends and co-workers in the SPECHEMO research group.

Mohammadreza Khanmohammadi
Chemistry Department, Faculty of Science, IKIU
P.O. Box 34149-1-6818, Qazvin, Iran
Tel-Fax: +98 281 3780040
Email: m.khanmohammadi@sci.ikiu.ac.ir
June 2014, Qazvin, Iran

AUTHOR INFORMATION

Abdollahi, Hamid

Institute for Advanced Studies in Basic Sciences, Zanjan, Iran

Ahmadi, Golnar

Institute for Advanced Studies in Basic Sciences, Zanjan, Iran

Alcaráz, Mirta R.

Universidad Nacional del Litoral-CONICET, Santa Fe, Argentina

Bagheri Garmarudi, Amir

Imam Khomeini International University, Qazvin, Iran

Balasaheb Aher, Rahul

Jadavpur University, Kolkata, India

Baumgartner, Christian

University for Health Sciences, Medical Informatics and Technology, Austria

Ambure, Pravin

Jadavpur University, Kolkata, India

Brasca, Romina

Universidad Nacional del Litoral-CONICET, Santa Fe, Argentina

Breit, Marc

University for Health Sciences, Medical Informatics and Technology, Austria

Filzmoser, Peter

Vienna University of Technology, Vienna, Austria

Cámara, María S.

Universidad Nacional del Litoral-CONICET, Santa Fe, Argentina

Garrigues, Salvador

University of Valencia, Valencia, Spain

Culzoni, María J.

Universidad Nacional del Litoral-CONICET, Santa Fe, Argentina

Ghasemi, Keyvan

Imam Khomeini International University, Qazvin, Iran

de la Guardia, Miguel

University of Valencia, Valencia, Spain

Goicoechea, Héctor C.

Universidad Nacional del Litoral-CONICET, Santa Fe, Argentina

Heberger, Karoly

Hungarian Academy of Sciences, Budapest, Hungary

Mozaffari, Faezeh

Imam Khomeini International University, Qazvin, Iran

Khanmohammadi, Mohammadreza

Imam Khomeini International University, Qazvin, Iran

Neymeyr, Klaus

Universität Rostock & Leibniz-Institut für Katalyse, Rostock, Germany.

Khorani, Mona

Imam Khomeini International University, Qazvin, Iran

Parto Dezfouli, Mohammad Ali

Tehran University of Medical Sciences, Tehran, Iran

Kuligowski, Julia

Health Research Institute Hospital La Fe, Valencia, Spain

Perez-Guaita, David

University of Valencia, Valencia, Spain

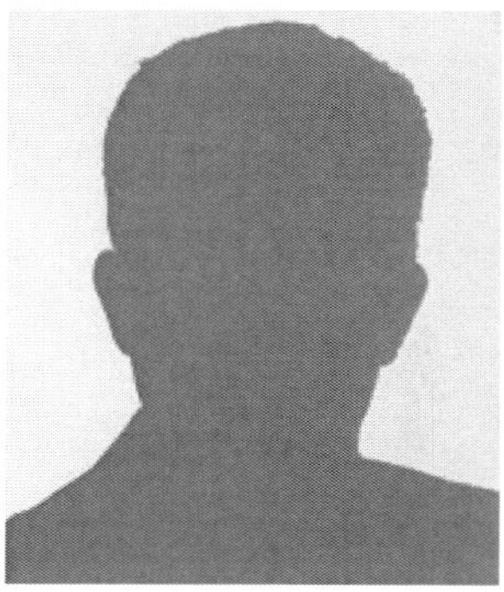

Pérez-Rambla, C.

University Hospital Doctor Peset Aleixandre, Valencia, Spain

Sancho-Andreu, M.

University Hospital Doctor Peset Aleixandre, Valencia, Spain

Quintás, Guillermo

Leitat Technological Center, Valencia, Spain

Sârbu, Costel

Babeş-Bolyai University, Cluj-Napoca, Romania

Roy, Kunal

Jadavpur University, Kolkata, India

Sawall, Mathias

Universität Rostock, Rostock, Germany

Saligheh Rad, Hamidreza

Tehran University of Medical Sciences, Tehran, Iran

Schenone, Agustina V.

Universidad Nacional del Litoral-CONICET, Santa Fe, Argentina

Teglia, Carla M.

Universidad Nacional del Litoral-CONICET, Santa Fe, Argentina

Weinberger, Klaus M.

University for Health Sciences, Medical Informatics and Technology (UMIT), Austria sAnalytiCo Ltd, Belfast, United Kingdom

Varmuza, Kurt

Vienna University of Technology, Vienna, Austria

Ventura-Gayet, J.

University Hospital Doctor Peset Aleixandre, Valencia, Spain

Vera-Candioti, Luciana

Universidad Nacional del Litoral-CONICET, Santa Fe, Argentina

In: Current Applications of Chemometrics
Editor: Mohammadreza Khanmohammadi

ISBN: 978-1-63463-117-4

Chapter 1

HOW TO DETERMINE THE ADEQUATENESS OF MULTIPLE LINEAR REGRESSION AND QSAR MODELS?

Károly Héberger[*]
Research Centre for Natural Sciences, Hungarian Academy of Sciences,
Budapest, Hungary

ABSTRACT

Performance parameters (coefficient of determination - R^2, cross-validated correlation coefficients - Q^2(external), R^2(leave-one-out) R^2(leave-many-out), R^2(bootstrap), etc.) are frequently used to test the predictive performance of models. It is relatively easy to check whether the performance parameters for linear fits stem from the same distribution or not. Tóth's test (J. Comput. Aided Mol. Des. (2013) 27:837–844), difference test (a variant of *t*-test) and Williams' *t*-test, show whether the difference between correlation coefficient (R) and cross-validated correlation coefficient (Q) are significantly different at a given probability level. An ordering procedure for comparison of models is able to rank the models as compared to the experimental values, and provides an easy way to test the adequacy of the splits for training and prediction sets. Ranking of performance parameters can be completed similarly. The variants of *t*-tests and the ranking procedure are recommended to be applied generally: they are simple and provide practical ways to test the adequacy of models.

INTRODUCTION

Performance parameters as coefficient of determination, R^2; cross-validated correlation coefficients leave-one-out, leave-many-out (Q^2, R^2_{LOO}, R^2_{LMO}, etc.) are frequently used to validate models.

[*] Corresponding author: Károly Héberger. Research Centre for Natural Sciences, Hungarian Academy of Sciences, H-1117 Budapest XI., Magyar Tudósok krt 2, Hungary. Phone: +36 1 382 65 09.

There is a general belief in the QSAR field that the difference between correlation coefficient (*R*) and cross-validated correlation coefficient (*q* or *Q*)) cannot be too high. A high value of the leave-one-out cross-validated statistical characteristic ($q^2 > 0.5$) is generally considered as a proof of the high predictive ability of the model. It has already been shown [1] that this assumption is usually incorrect. Some authors make categories high, medium/ acceptable and weak correlation [2], but these groups are not usable without knowing the degree of freedom. Similarly, how small the mentioned difference should be (how large might be), whereas the fit remains acceptable is not known and generally governed by feelings of the modelers. Recently, it was proven in our laboratory [3] that the ratio

$$\frac{PRESS/dfPRESS}{RSS/dfRSS} = \frac{1-Q^2}{1-R^2} \approx F - distributed \tag{1}$$

where:

$$PRESS = \sum_{i=1}^{n} \left(\hat{y}_{i/i} - \bar{y}\right)^2 \tag{2}$$

$\hat{y}_{i/i}$ denotes the value calculated for the *i*-th experiment leaving out the *i*-th experiment in the parameter estimation of the model;
and

$$RSS = \sum_{i=1}^{n} (\hat{y}_i - \bar{y})^2 \tag{3}$$

whereas

$$TSS = \sum_{i=1}^{n} (y_i - \bar{y})^2 \tag{4}$$

and the cross-validated correlation coefficient and the coefficient of determination are defined in Equation 5.

$$Q^2 = 1 - \frac{PRESS}{RSS} \quad \text{and} \quad R^2 = 1 - \frac{RSS}{TSS}\text{, respectively.} \tag{5}$$

The basic assumption in the above test is that the ratio of two variances sampling from the same normal distribution follows *F* distribution with the corresponding degree of freedom. Both RSS and PRESS are sum of squares with *dfRSS* and *dfPRESS* degrees of freedom.

The rapid calculation of PRESS/RSS ratio and the above F test are fast methods to pre-estimate the presence of outliers, i.e., the internal predictive character of a model. If a model fails the test, it is expedient to change the training data.

Editors and reviewers can check the submitted models easily if a model fails the above variance ratio test the model has little generalization ability, if at all [3].

As a logical continuation another test is introduced here for checking whether the training and test (prediction) sets stem from the same distribution. Todeschini and coworkers have compared three different calculation methods for performance indicators of external validation [4]: The methods were that of Schüürmann *et al.* [5]:

$$Q_{F1}^2 = 1 - \frac{\sum_{i=1}^{nEXT} (\hat{y}_i - y_i)^2}{\sum_{i=1}^{nEXT} (y_i - \bar{y}_{TRAIN})^2} \tag{6}$$

that of Sheenan et al. [6]:

$$Q_{F2}^2 = 1 - \frac{\sum_{i=1}^{nEXT} (\hat{y}_i - y_i)^2}{\sum_{i=1}^{nEXT} (y_i - \bar{y}_{EXT})^2} \tag{7}$$

and the “hybrid” recommendation of Todeschini et al. [7]:

$$Q_{F3}^2 = 1 - \frac{\left[\sum_{i=1}^{nEXT} (\hat{y}_i - y_i)^2\right]/nEXT}{\left[\sum_{i=1}^{nEXT} (y_i - \bar{y}_{TRAIN})^2\right]/nTRAIN} \tag{8}$$

It turned out that all three functions produce bias-free estimation, but Q_{F3}^2 has far the least variability [4]. We will see later that the superiority of Q_{F3}^2 is not as unambiguous as thought at the first glance.

This work present a known statistical test, applied in a unique and creative way on the above performance parameters. The test is called difference test and has not been used for the given purpose till know (to my knowledge). In the second part of the paper a ranking procedure for checking the model consistency is shown if some assumptions are fulfilled. In case the calculated (modeled) values are also available not only the performance parameters, then, the models can be ranked together with the experimental values [8-10]. The ordering on the training, test and both sets allows us to compare the representative character of the splits, as well.

THEORY

Difference Test [11]

The significance of the difference between two correlation coefficients ($\boldsymbol{r}_1$ and $\boldsymbol{r}_2$) is computed as follows: the test statistics is

$$t = d/sd \tag{9}$$

where d is the difference between the two Fisher z-transformed [12] correlation coefficients:

$$d = \frac{1}{2}\left[\ln\left(\frac{1+r_1}{1-r_1}\right)\right] - \frac{1}{2}\left[\ln\left(\frac{1+r_2}{1-r_2}\right)\right] \tag{10}$$

and sd is the standard error of the difference between the two normalized (Fisher z-transformed) correlation coefficients:

$$sd = \sqrt{(n_1 + n_2 - 5)/(n_1 - 3)(n_2 - 3)} \tag{11}$$

where n_1 and n_2 are the two sample sizes for r_1 and r_2 respectively. (It means that the test cannot be applied if n_1 or $n_2 < 4$). The test statistics (Equation (1)) is evaluated against the t-distribution with degree of freedom

$$n_1 + n_2 - 4.$$

The Statistica program package [11] calculates the two sided p-values. If the probability is smaller than a predefined probability level, say 5 %, then the null-hypothesis is rejected: H_0 is the two correlation coefficients are not significantly different, its rejection means they are significantly different. The test concerns for simple correlation coefficients, but it can be extended for the multivariate case. The above test is well-known and textbook material.

However, the main idea is to substitute one of the correlation coefficients with the cross-validated correlation coefficient (Q). The test statistic is the same as above (Equations (9-11) just instead of r_2 stands Q. (and instead of r_2 stands r).

This is equivalent with the assumption that the correlation coefficient, and the cross-validated correlation coefficient, stem from the same distribution. The two coefficients are independent, i.e., their sampling from the same distribution has no common objects. With other words the test is applicable for external validation only.

Two Nonindependent Correlations with a Variable in Common [13]

A frequent situation commonly appears when we would like to test about significant difference of correlation coefficients using solely three vectors, say y, x_1 and x_2. We test the correlation coefficients of (y,x_1) and (y,x_2) on the condition that x_1 and x_2 are also correlated. The test is commonly known as Williams' t-test and the test statistic is as follows.

$$t_{n-3} = (r_{y,x1} - r_{y,x2})\sqrt{\frac{(n-1)(1 + r_{x1,x2})}{2\left(\frac{n-1}{n-3}\right)|R| + \frac{(r_{y,x1} + r_{y,x2})^2}{4}(1 - r_{x1,x2})^3}} \tag{12}$$

where

$$|R| = 1 - r_{y,x1}^2 - r_{1y,x2}^2 - r_{x1,x2}^2 + 2r_{y,x1}r_{y,x2}r_{x1,x2} \tag{13}$$

The test is especially suitable for discrimination of x_1 and x_2, if they are correlated with the same y vector [14]. Two models can also be compared, if calculated (predicted) y-s are used (commonly denoted as $\hat{y}_1$ and $\hat{y}_2$). A computer code, which realizes a multivariable selection (and ordering) of x variables, can be downloaded for free from the following link:

http://aki.ttk.mta.hu/gpcm

The ordering possibilities i.e., the algorithm is described in reference [15] in detail. The test is applicable for leave-one-out cross-validation (LOO CV): y corresponds to experimental values, x_1 to the modeled ones and x_2 for the values calculated n times leaving out the i-th experiment one by one (denoted by $\hat{y}_{i/i}$ in Equation (2)).

Ranking Procedure

Recently a novel procedure [8-10], called sum of ranking difference (SRD), is introduced for model comparison. It can be applied for performance parameters, as well. In this case, however, more data sets are necessary for ordering. Different models can also be compared (together with experimental values). The SRD procedure is based on sum of ranking differences, which provides an easy ordering. The average can preferably be accepted as reference for ranking (consensus). The smaller the SRD value for a given model the better the model. The models are ranked for the training, prediction sets and eventually for the overall data sets. If the ordering is different for the various sets, the split cannot be considered as representative; the models are not optimal for the training or prediction sets. If we compare performance parameters for the training and for the prediction sets, the results of difference tests (above) should be in conformity with the results of the ranking procedure. All details of the ranking procedure can be found in references [8-10]. An excel macro is written and is downloadable freely together with sample input- and output files from:

http://aki.ttk.mta.hu/srd

Results and Discussion

Case Study No 1

Recently Haoliang Yuan *et al.* have published a paper about the structural optimization of c-Met inhibitor scaffold 5H-benzo[4,5]cyclohepta[1, 2-b]pyridin-5-one [16]. Their fragment based approach produced a detailed 3D-QSAR study: a comparison of CoMFA and CoMSIA models with experimental values (Table 3 in reference [16]).

The data of Table 3 [16] are suitable for testing using the difference test:

Training set: the correlation coefficient for experimental values and modeled by CoMFA is 0.94201, whereas the same for experimental values and modeled by CoMSIA is 0.97476 (n=44, training set). The two sided test gives a 0.0737 limit probability, i.e., the two models (CoMFA and CoMSIA) are not significantly different at the 5 % level, but IS significant at the 10 % level.

Prediction set: the correlation coefficient for experimental values and modeled by CoMFA is 0.90073, whereas the same for experimental values and modeled by CoMSIA is 0.96146 (n=44, prediction set). The two sided difference test gives a 0.3560 limit probability, i.e., the two models (CoMFA and CoMSIA) are not significantly different at the 5 % level, neither at the 10 % level. As the difference test cannot give a definite answer about the equivalency of the two models, other approaches should be considered for a decision. First the average of the experimental and the two calculated values are considered and used for Williams' t-test, as we know for sure that the errors (both biases and random ones) cancel each other at least partially (maximum likelihood principle). Table 1 shows the dramatic difference of the Williams' t-test results for the training and prediction sets. Calculated values by CoMSIA are significantly better correlated to the average than the calculated values by CoMFA (on the condition that modeled values by CoMFA and CoMSIA are also correlated).

Even the experimental values are significantly better correlated to the average than the calculated values by CoMFA, whereas none of the correlations are significant for the prediction set. The significant differences in Table 1 should be reflected if using the ranking procedure. The experimental and predicted inhibitor activities (summarized in Table 3 of reference [16]) are suitable for an ordering between CoMFA, CoMSIA predicted and experimental values.

Table 1. Case study No 1 (Inhibitor activity) Pearson correlation coefficients for the training and prediction sets

Training set[a]		Prediction set[b]	
r(exp,CoMFA)	0.94201	r(exp,CoMFA)	0.90073
r(Ave,exp)	0.98755	r(Ave,exp)	0.98491
r(Ave,CoMFA)	0.97750	r(Ave,CoMFA)	0.95328
Test statistic	2.49[a]	Test statistic	2.15[b]
p (theor.)	**0.01677**[c]	p (theor.)	0.063577
r(exp,CoMSIA)	0.97476	r(exp,CoMSIA)	0.96146
r(Ave,exp)	0.98755	r(Ave,exp)	0.98491
r(Ave,CoMSIA)	0.98960	r(Ave,CoMSIA)	0.97570
Test statistic	0.577[a]	Test statistic	0.676[b]
p (theor.)	0.567223	p (theor.)	0.518376
r(CoMFA,CoMSIA)	0.94827	r(CoMFA,CoMSIA)	0.88430
r(Ave,CoMFA)	0.97750	r(Ave,CoMFA)	0.95328
r(Ave,CoMSIA)	0.98960	r(Ave,CoMSIA)	0.97570
Test statistic	3.07[a]	Test statistic	1.15[b]
p (theor.)	**0.00379**[c]	p (theor.)	0.282789

[a]The test statistic is to be compared with the critical value from table 2.0195; n=44.
[b]Test statistic is to be compared with the critical value from table 2.3060; n=11.
[c]significant at the predefined p = 0.05 level, indicated by bold.

The training set (44) and the prediction set (11) provide a reliable basis for ranking. Indeed, the ordering is far from being random as can be seen from the Figures 1 and 2. Sum of ranking differences orders the modeling methods and experimental values for the training and prediction sets alike as shown in Figures 1 and 2.

Ordering of modeling methods and experimental values using sum of ranking differences for the training set (n=44). Row-minimums were used as a benchmark for ranking. Scaled SRD values (between zero and hundred) are plotted on x axis and left y axis (the smaller the better). Right y axis shows the relative frequencies (only for black Gaussian curve). Parameters of the fit are m=66.79 s=6.48. Probability levels 5 % (XX1), Median (Med), and 95 % (XX19) are also given.

It can be seen immediately that the rankings for training and prediction sets are not the same. The order of lines for CoMFA and CoMSIA are reversed from the training and prediction sets. However, both rankings are located in between zero and the random ranking. The random ordering is shown by Gaussian and Gaussian-like black curves.

As SRD the smaller the better, we can state for sure that the experimental values rationalize the information present in the data better than any of the modeling methods CoMFA or CoMSIA. The row average was accepted as the reference for ranking similarly to the results of Williams' t-test (see above).

It is reassuring that the CoMFA and CoMSIA models are not random ones but (much) better models. The ordering is different for the training and prediction set: the SRD values are better (somewhat smaller) for the training set than for the prediction set.

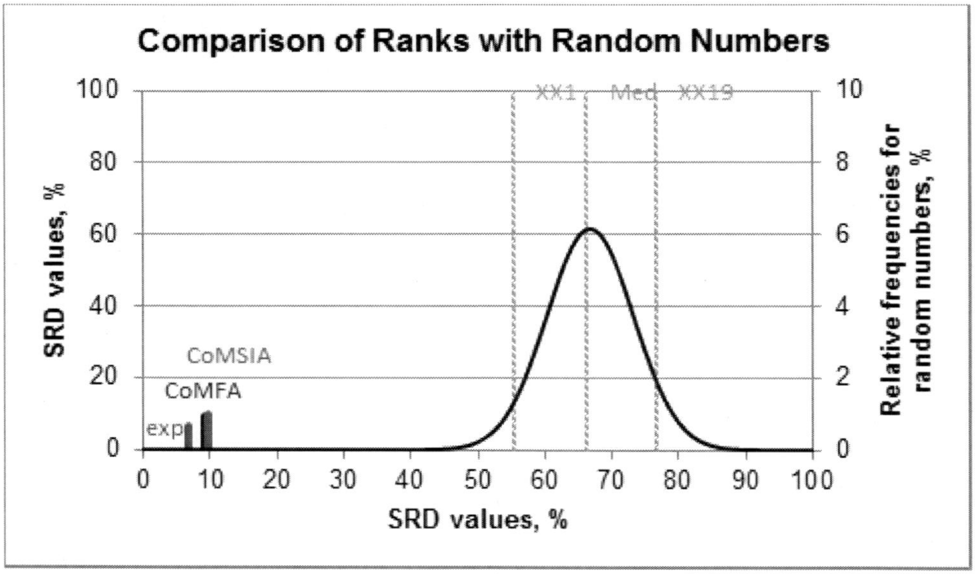

Figure 1. Ordering of modeling methods and experimental values using sum of ranking differences for the training set (n=44). Row-minimums were used as a benchmark for ranking. Scaled SRD values (between zero and hundred) are plotted on x axis and left y axis (the smaller the better). Right y axis shows the relative frequencies (only for black Gaussian curve). Parameters of the fit are m=66.79 s=6.48. Probability levels 5 % (XX1), Median (Med), and 95 % (XX19) are also given.

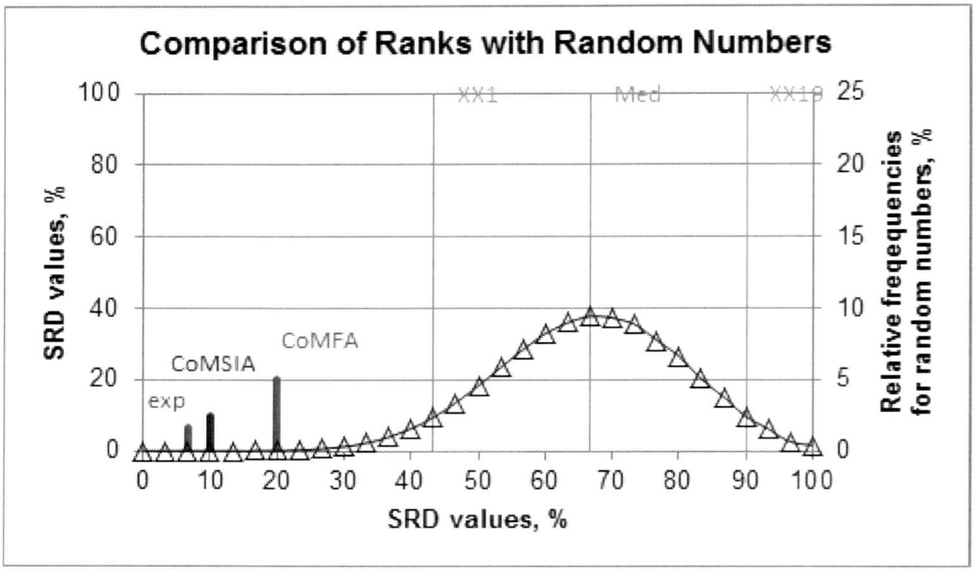

Figure 2. Ordering of modeling methods and experimental values using sum of ranking differences for the prediction set (n=11). Row-minimums were used as a benchmark for ranking. Scaled SRD values (between zero and hundred) are plotted on x axis and left y axis (the smaller the better). Right y axis shows the relative frequencies (only for black Gaussian-type curve). The exact theoretical distribution is given (triangles). Probability levels 5 % (XX1), Median (Med), and 95 % (XX19) are also given.

The ordering is reversed for the prediction and training sets: for the prediction set the better one is the CoMSIA model and the worse one are CoMFA model; for the training set the case is just the opposite CoMFA have a bit better SRD values than CoMSIA, whereas experimental values were the best for both sets.

A sevenfold cross-validation for the training set and a leave-one-out cross-validation for the prediction set generate uncertainties for the SRD values.

A combination of SRD values and ANOVA provides a unique and unambiguous way of decomposing the effects and determine the best combination of factors [17].

Herewith the effect of two factors can be examined: (i) data sets – two levels (training and prediction) and (ii) modeling – three levels (CoMFA, CoMSIA and experimental), while all SRD values are united into one variable. It is interesting to know, which factors are significant and how they influence the SRD values.

A constant term (intercept), the two factors and their interaction term are all (highly) significant. The variances for factors and the interaction term are not homogeneous, as shown by the highly significant Hartley, Cochran and Bartlett tests as well as by Levene test. The prediction set has significantly higher SRD values (recall the smaller SRD the better). Experimental values exert the smallest SRD values, CoMSIA models the second smallest and CoMFA is the highest. Figure 3 plots the most informative weighted means.

Even laymen can judge that the ANOVA results of training and prediction sets are completely different: CoMFA and CoMSIA models are ranked differently. All groups are significantly different except one: the experimental values for the training and test sets produce the same – smallest – mean SRD values, but their variances are significant even in this case, as well. The smallest significance was observed between mean SRD values for CoMFA and CoMSIA in case of the training set: $p = 0.039623$.

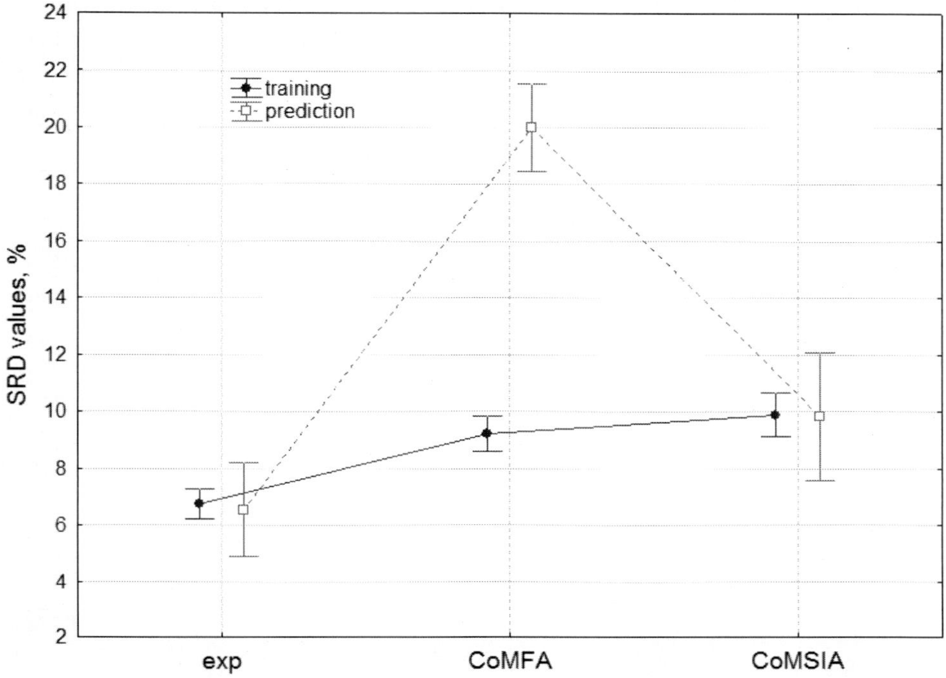

Figure 3. Variance analysis for sum of ranking differences: weighted means (sigma-restricted parameterization, effective hypothesis decomposition). Vertical bars denote 0.95 confidence intervals.

The figure strengthens our earlier inference: the experimental values rationalize the information in the data better than the models; the models behave on the training and prediction sets differently, i.e., the model split cannot be considered as representative.

Case Study No 2

Todeschini et al. [7] have compared different ways to evaluate the predictive ability of QSAR models. Three performance indicators for external validation were called predictive squared correlation coefficient Q^2 and defined as given in Equations (6-8). The indicators behave differently as the allocations of the points are different. Their Tables 5 and 7 summarizes performance parameters; the data are suitable for a detailed statistical analysis.

Tóth's test (given by Equation (1)) shows, whether the correlation coefficients (R) and cross-validated correlation coefficients (Q) are derived from the same or different distribution. As an illustration the first line of Table 5 of reference [7] is utilized: $R^2 = 0.972$; $Q^2(\mathrm{F1}) = 0.873$; $Q^2(\mathrm{F2}) = 0.843$; $Q^2(\mathrm{F3}) = 0.877$. As a consequence $Q^2(\mathrm{F1})$ is significantly different from R^2 and several serious outliers are suspected *c.f.* Figure 1 in reference [3]. The other two indicators ($Q^2(\mathrm{F2})$, $Q^2(\mathrm{F3})$) seem to be acceptable.

Merging the data of Table 5 and Table 7 several difference tests can be conducted: there is no significant differences between R^2 and $Q^2(\mathrm{F1})$ $p = 0.1170$; between R^2 and $Q^2(\mathrm{F2})$ $p = 0.6569$; and between R^2 and $Q^2(\mathrm{F3})$ $p = 0.4010$.

Table 2. Case study No 2; Comparison of various external validation options, Equations (6-8) Pearson correlation coefficients (r) and their significance according to the Williams' *t*-test

$r(R^2,Q^2(F1))$	-0.28253	$r(Q^2(F1),Q^2(F2))$	0.24936
$r(Ave,R^2)$	-0.39074	$r(Ave, Q^2(F1))$	0.46365
$r(Ave,Q^2(F1))$	0.46365	$r(Ave,Q^2(F2))$	0.96722
Test statistic	0.21[a]	Test statistic	3.78[a]
p (theor.)	0.836793	p (theor.)	**0.0043191**[b]
$r(R^2,Q^2(F2))$	-0.44894	$r(Q^2(F1),Q^2(F3))$	-0.35871
$r(Ave,R^2)$	-0.39074	$r(Ave, Q^2(F1))$	0.46365
$r(Ave,Q^2(F2))$	0.96722	$r(Ave,Q^2(F3))$	0.63177
Test statistic	5.17[a]	Test statistic	0.47937[a]
p (theor.)	**0.0005892**[b]	p (theor.)	0.6431031
$r(R^2,Q^2(F3))$	-0.07516	$r(Q^2(F2),Q^2(F3))$	0.75431
$r(Ave,R^2)$	-0.39074	$r(Ave, Q^2(F2))$	0.96722
$r(Ave,Q^2(F3))$	0.63177	$r(Ave,Q^2(F3))$	0.63177
Test statistic	0.71[a]	Test statistic	6.34004[a]
p (theor.)	0.498172	p (theor.)	**0.0001344**[b]

[a] The test statistic is to be compared with the critical value from table 2.26215; n=12.

[b] significant at the predefined $p = 0.05$ level, indicated by bold.

Therefore a more sensitive test should be applied; the results of Williams' *t*-test are summarized in Table 2. Here, again, the consensus approach, the mean values were used for data fusion. In fact, we do not know which performance indicator has the lowest random error and bias. However, the only reasonable assumption is that all expresses the performance with some error. The maximum likelihood principle suggests using the average value as the best estimate. Then, Q^2(F2) is significantly correlated with the mean value of all performance parameters; moreover, it is superior to both other indicators (Q^2(F1) and Q^2(F3)). All other comparisons are not significant (c.f. Table 2). This finding is in conformity with Schüürmann et al.'s suggestion [5]; they also argued in favor of a function given by Equation (2).

The information in the united Tables 5 and 7 can also be utilized for ranking of performance parameters. Again, the average of all performance indicators (row-wise) was used as a benchmark for ranking. There is no doubt $Q^2(F2)$ has the smallest (the best) SRD value i.e., its ranking is the closest one to the reference ranking, even $Q^2(F1)$ is out of the error limit (first two dotted lines on the figure). On the other hand, $Q^2(F3)$ and R^2 are comparable with the random ranking. With other words they are indistinguishable from chance ordering (Figure 4). The fact that R^2 is the worst performance parameter is in conformity with the expectations and widely accepted beliefs. The SRD ranking results support the independent calculations by Williams' *t*-test and not in contradiction with the results of Tóth's test.

SRD ranking reinforces our investigations above: Williams' *t*-test and SRD methodology questions the superiority of $Q^2(F3)$ seriously. A leave-one-out cross-validation is able to assign uncertainty values to the SRD ranking provided in Figure 4.

A box and whisker plot shows unambiguously the distinction of the ranking: no overlapping is observed; pairwise Wilcoxon's matched pair tests confirm all positions in the ranking: They are all significantly different (Figure 5).

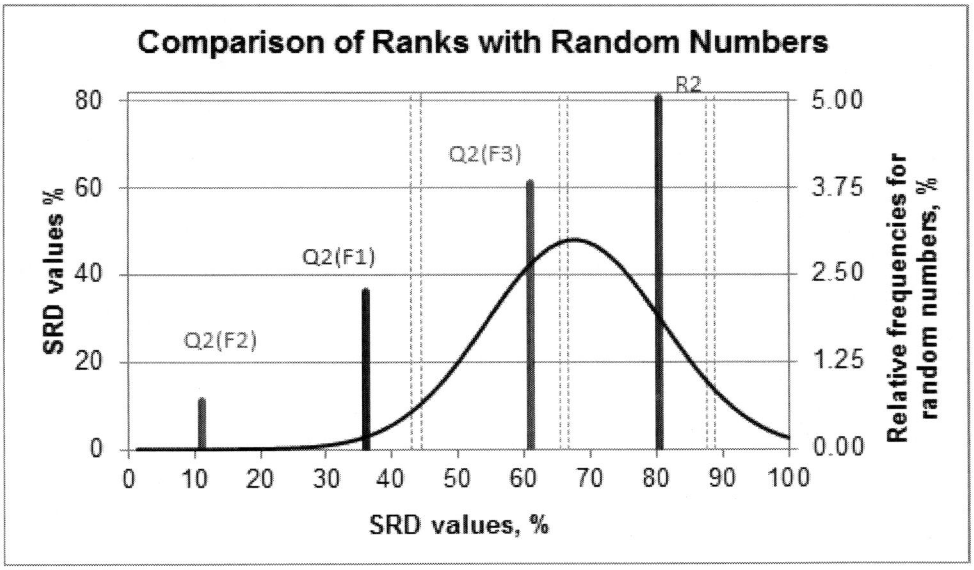

Figure 4. Ordering of squared correlation coefficient (R) and predictive squared correlation coefficient (Q) values using sum of ranking differences (n=12). Row-minimums were used as a benchmark for ranking. Scaled SRD values (between zero and hundred) are plotted on x axis and left y axis (the smaller the better). Right y axis shows the relative frequencies (only for black Gaussian fitted curve). Probability levels are also given by double dotted lines: 5 % Median, and 95 %, respectively.

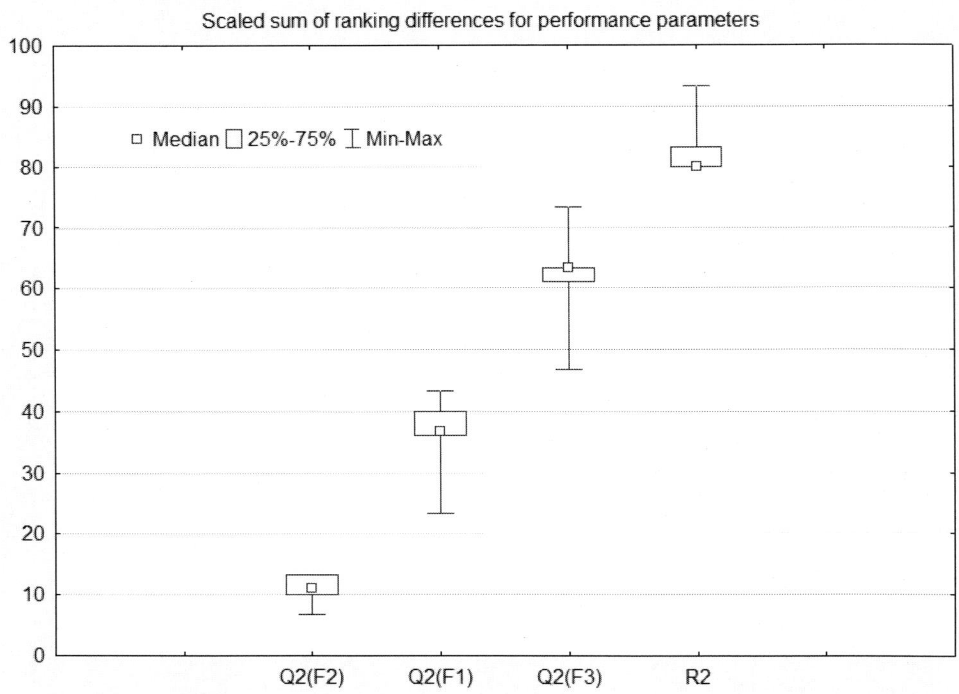

Figure 5. Box and whisker plot for leave-one-out cross-validation of SRD ranking.

Conclusion

The difference test for correlation coefficients is too conservative; it signalizes differences in case of correlation coefficients (R) and externally cross-validated correlation coefficients (Q) only if the difference is indeed large.

Williams' t-test is suitable to prove differences in correlation coefficients (R) and predictive squared correlation coefficient (Q) (leave-one-out cross-validated correlation coefficients and correlation coefficients of external validation). In general, Williams' t-test is more sensitive than the difference test.

SRD methodology is able to testify whether the training - prediction set split is consistent or not.

Any of four methods (three statistical tests and a ranking procedure) may reveal whether the performance parameters for validation are significantly different from each other or not as shown by two case studies.

It is suggested to test the difference (ratio) between correlation coefficient and cross-validated correlation coefficients with the above methods before publication.

Acknowledgment

The author is indebted to the financial support from OTKA, contract No K112547.

References

[1] Golbraikh, A., Tropsha, A. Beware of q2! *J. Molec. Graph. Model.* 2002, 20, 269-276.
[2] Tropsha, A., Gramatica, P., Gombar, V. K. *QSAR Comb. Sci.* 2003, 22, 69-77.
[3] Tóth, G., Bodai, Zs., Héberger, K. *J. Comput. Aided Mol. Des.* 2013, 27, 837-844.
[4] Consonni, V., Ballabio, D., Todeschini, R. *J. Chemometrics* 2010, 24, 194-201.
[5] Schüürmann, G., Ebert, R.-U., Chen, J., Wang, B., Kühne, R. *J. Chem. Inf. Model.* 2008, 48, 2140-2145.
[6] Shi, L. M., Fang, H., Tomg, W., Wu, J., Perkins, R., Blair, R. M., Branham, W. S. Dial, S. L., Moland, C. L., Sheenan, D. M. *J. Chem. Inf. Comput. Sci.* 2001, 41, 186-195.
[7] Consonni, V., Ballabio, D., Todeschini, R. *J. Chem. Inf. Model.* 2009, 49, 1669-1678.
[8] Héberger, K. *Trends Anal. Chem.* 2010, 29, 101-109.
[9] Héberger, K., Kollar-Hunek, K. J. *Chemometr.* 2010, 25, 151-158.
[10] Kollár-Hunek, K., Heberger, K., *Chemometr. Intell. Lab. Syst.*, 2013, 127, 139-146.
[11] StatSoft, Inc. (2004). STATISTICA (data analysis software system), version 6.0, Tulsa, Oklahoma, US, www.statsoft.com.
[12] Fisher, R. A. *Metron* 1921, 1, 3-32.
[13] Weaver, B., Wuensch, K. L. *Behav. Res.* 2013, 45, 880-895.
[14] Rajkó, R., Héberger, K. *Chemometr. Intell. Lab. Syst.*, 2001, 57, 1-14.
[15] Héberger, K., Rajkó, R. *J. Chemometr.* 2002, 16, 436-443.

[16] Yuan, H., Tai, W., Hu, Sh., Liu, H., Zhang, Y., Yao, S., Ran, T., Lu, Sh., Ke, Zh., Xiong, X., Xu, J., Chen, Y., Lu, T., *J. Comput. Aided Mol. Des*. 2013, 27, 897-915.

[17] Héberger, K., Kolarevic, S., Kracun-Kolarevic, M., Sunjog, K., Gacic, Z., Kljajic, Z., Mitric, M., Vukovic-Gacic, B. *Mutat. Res.: Genet. Toxicol. Environ. Mutagen*. 2014, 771, 15-22. http://dx.doi.org/10.1016/j.mrgentox.2014.04.028.

In: Current Applications of Chemometrics
Editor: Mohammadreza Khanmohammadi

ISBN: 978-1-63463-117-4

Chapter 2

Repeated Double Cross Validation (rdCV) - A Strategy for Optimizing Empirical Multivariate Models, and for Comparing Their Prediction Performances

Kurt Varmuza * ***and Peter Filzmoser***
Vienna University of Technology, Department of Statistics and Probability Theory, Austria

Abstract

Repeated double cross validation (rdCV) is a resampling strategy for the development of multivariate models in calibration or classification. The basic ideas of rdCV are described, and application examples are discussed. rdCV consists of three nested loops. The outmost loop performs repetitions with different random sequences of the objects. The second loop performs CV by splitting the objects into calibration and test sets. The third (most inner) loop uses a calibration set for estimating an optimal model complexity by CV. Optimization of model complexity is strictly separated from the estimation of the model performance. Model performance is solely derived from test set objects. The repetition loop allows an estimation of the variability of the used performance measure and thus makes comparisons of models more reliable. rdCV is combined with PLS regression, DPLS classification, and KNN classification. Worked out examples with data sets from analytical chemistry demonstrate the use of rdCV for (1) determination of the ethanol concentration in fermentation samples using NIR data; (2) comparison of variable selection methods; (3) classification of synthetic organic pigments (relevant for artists' paints) using IR data; (4) classification of AMES mutagenicity (QSAR). Results are mostly presented in plots directly interpretable by the user. Software for rdCV has been developed in the R programming environment.

* Corresponding author: Kurt Varmuza. E-mail: kvarmuza@email.tuwien.ac.at.

INTRODUCTION

Prominent tasks in chemometrics and chemoinformatics are the creation of empirical multivariate models for calibration and classification. A multivariate calibration model defines a relationship between a set of numerical x-variables and a numerical continuous property y; a multivariate classification models between a set of x-variables and a categorical (discrete) property y (encoded class memberships of the objects). Mostly, the goal of such a model is the application to new cases to predict the unknown y from given x-variables. Essential topics for the creation of powerful models and successful applications are (1) Good estimation of the optimum complexity of the model; (2) a cautious estimation of the prediction performance for new cases; and (3) a reasonable estimation of the applicability domain of the model (this topic is not treated here). The model complexity is related to the degree of freedom (the number of independent parameters, more specifically, e.g., the number of PLS components). An optimum complexity is essential for a good compromise between underfitting (a too simple model) and overfitting (the model is well adapted to the calibration data but does not possess sufficient generalization necessary for new cases).

Empirical models are usually derived from a data set from n reference objects (cases), each characterized by m x-variables (features), forming a matrix $\boldsymbol{X}(n \times m)$, and a corresponding $\boldsymbol{y}$-vector (containing the known properties of the n objects). Model building is an inductive approach, using a specific data sample, and resulting in a generalized model which should be capable to be successfully applied to new cases. Inductive models are inherently uncertain, and therefore great effort is necessary for a reasonable estimation of the model performance and the applicability range. In many practical situations, the number of reference objects is rather small and this fact limits the significance of model parameters. An appropriate theory for the relationships between the x-variables and the desired property y is often not available and the existence of a relationship is solely a hypothesis. Various strategies have been suggested for handling these problems; most often applied in chemometrics are strategies based on cross validation (CV), others use the bootstrap concept.

This chapter focuses on CV and presents a combination of a procedure that was suggested 2009 [1, 2] under the name *repeated double cross validation* (rdCV). rdCV (and bootstrap techniques) are so called resampling strategies (repeated sampling within the same sample) which are necessary for small data sets (say 20 to 200 objects) in order to obtain a sufficient large number of predictions. For this purpose the n objects are repeatedly split into three sets: A training set for making models with different complexities, a validation set for estimating the optimum complexity, and a test set for getting predictions by simulating new cases. Training set and validation set together form the calibration set. Various strategies (for CV or bootstrap) have been suggested and are applied depending on data size (n), and accepted computational effort, but also on availability in commercial software. A too simple approach gives misleading, often too optimistic, result for the prediction performance. rdCV estimates an optimization parameter (usually the optimum model complexity) together with its variation, and rigorously estimates the prediction performance together with its variation. Variation concerns different random splits into the three mentioned sets.

The strategy is applicable, e.g., for calibration and for classification, and requires at least about 15 objects. Optimization of the model complexity is separated from the estimation of the model performance.

rdCV is implemented in a group of user-friendly functions in the R programming environment [3], based on the R package *chemometrics* [4, 5]. Typical computation time (n = 100, m = 20) is a few seconds (3.4 GHz processor).

Strategies similar to rdCV have been published for classification as cross model validation (2CV) [6, 7], and for bootstrap techniques [8], as well as for principal component analysis as double full cross validation [9]. Critical contributions to CV have frequently been published, e.g., [10-12]; and recently (2014) an almost identical strategy, named repeated grid-search (nested) V-fold cross-validation, has been described [13].

The outline of the next parts of this chapter is as follows: In the Methods' Section the basic principles of cross validation, double cross validation, and repeated double cross validation are described. In the Applications' Section, rdCV is applied to multivariate calibration, using PLS regression with an example from bioanalytical chemistry (including a comparison of variable selection methods). Then, rdCV is applied to multivariate classification, using KNN classification, and DPLS classification.

METHODS

Cross Validation

The most used resampling strategy in chemometrics for optimization of a model parameter (e.g., the model complexity) is cross validation (CV) [2, 14-16]. The idea is to split the objects into groups, using one for making models by varying the parameter to be optimized, and the other to validate the models for finding the optimum value. Simple CV can be described as follows. The considered set of n objects is split into s segments (groups, parts, folds) of approximately equal size. The split is usually random, but alternatively may be (if reasonable) in the mode "123123" (object 1 into segment 1, object 2 into segment 2, and so on), or in the mode "111222333" (first n/s objects - rounded to integer - into segment 1, and so on). More sophisticated pseudo-random procedures - related to experimental design and sampling theory - are rarely applied. E.g., the Kennard-Stone algorithm [17, 18] covers the x-space of the calibration set as uniformly as possible; stratified CV [13, 19] tries to have the same region (distribution) of y-values in each segment.

The number of segments is typically between 2 and 10. Figure 1 demonstrates simple CV with three segments. Two segments form a training set, the left out segment is the validation set. From the training set, models (e.g., for calibration of a property y) are calculated with different values for the parameter to be optimized (e.g., the number of PLS components, A_1, A_2, ..., A_D). The models are applied to the objects of the validation set, giving predicted properties, $\hat{y}$, for each object in the validation set, and each tested value of A. The procedure is repeated with each of the segments being a validation set. From the obtained matrix $\hat{\boldsymbol{Y}}_{CV}$ ($n \times D$) an error measure is calculated column-wise (e.g., the mean of the squared errors, MSE_{CV}). Typically, MSE_{CV} varies with A. In a simple approach the value of A with the smallest MSE_{CV} (global minimum) is used as an optimum A, A_{OPT}. Alternatively, heuristic algorithms are applied for searching a value for a more parsimonious model to avoid overfitting; e.g., the one standard error rule [1, 20] uses a statistical concept (but requires at least four segments, and about at least three objects per segment).

Only applying a single CV may have several pitfalls: (1) Estimation of optimum model complexity and estimation of prediction performance are not separated and the performance measure obtained is usually too optimistic [21]. (2) The result may heavily depend on the (random) split of the objects into segments; the resulting single performance measure may be an outlier, and does not allow a reasonable comparison of model performances. Repeating the CV with different random splits is recommended to estimate the variation of the performance measure (and the variation of the optimum model complexity).

A special case of simple CV uses as many segments as objects ($s = n$), called leave-one-out CV or full CV. Full CV often gives too optimistic results, especially if pair-wise similar objects are in the data set (e.g., duplicate measurements). However, if the number of objects is very small (say $n < 15$) this strategy may be the only applicable one; the resulting performance has to be considered as a (crude) estimation and must not be over-interpreted.

Double Cross Validation

In a single CV the performance measure is guided by the optimum parameter, A_{OPT}, (model complexity) of the model. A_{OPT} is best for the used data but not necessarily enough generalized for new cases. In double cross validation (dCV) the optimization is separated from the model assessment. The dCV strategy can be realized by two nested loops - two nested CVs (Figure 2). In the outer CV all considered n objects are split into s_{TEST} segments; typically $s_{TEST} = 2$ to 10 (for split strategies see previous subsection.). Based on the CV idea, one segment constitutes a test set, which is not used for optimization of the model complexity. The other segments form a calibration set. First, the optimum complexity for the current calibration set is estimated - and the method of choice for this purpose is a single CV as described before, considered as the inner CV. We denote the number of segments in the inner CV (working with a calibration set) by s_{CALIB}. Second, the obtained A_{OPT} is stored, e.g., in a vector $\boldsymbol{A}_{CV}$ ($s_{TEST} \times 1$). Note that the outer CV produces s_{TEST} different (partly overlapping) calibration sets, and the resulting values for A_{OPT} are in general not equal. Actually, only at the end of the outer CV a final (or several final) values for the optimum model complexity are selected from the obtained s_{TEST} values. From a practical point of view it is therefore useful to make models from the whole current calibration set for all values of the optimization parameter (A_1, A_2, ..., A_D) and to apply the models to the current test set.

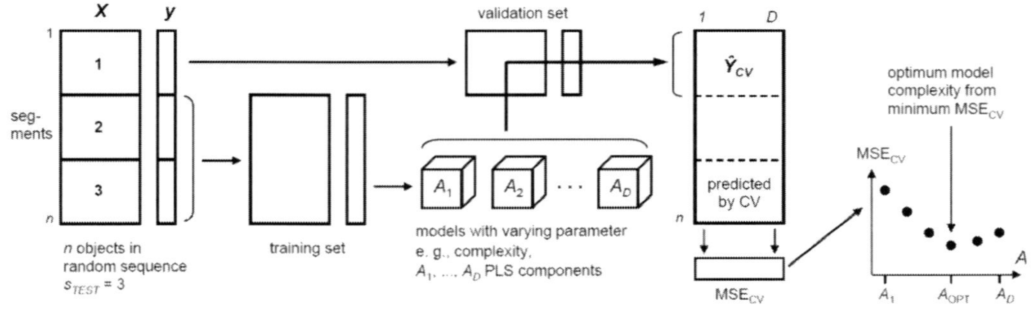

Figure 1. Simple CV with three segments for optimization of the model complexity (e.g., number of PLS components).

After the first loop of the outer CV we have a predicted value for each object of the current test set using models of varying complexity and made from the current calibration set. These predicted values are "tests set predicted" results - we need for a careful estimation of the model performance for new cases.

After all s_{TEST} loops of the outer CV, the next step searches a final optimum model complexity, A_{FINAL}. A convenient method is to choose the most frequent value among the s_{TEST} results. Alternatively, more than one value can be considered, e.g., when working with a consensus approach (using several models in parallel, and combining the prediction results). The distribution of the obtained values for A_{OPT} is instructive, as it often demonstrates that no unique optimum exists.

Now having a final model complexity, A_{FINAL}, the n test set predicted values for this complexity are used to calculate a final prediction performance. This performance measure refers to test set conditions and simulates new cases; it is calculated from one prediction for each of the n objects. Note that s_{TEST} different calibration sets have been used (all with complexity A_{FINAL}). A final model can be made from all n objects, using complexity A_{FINAL}; a good estimation of the performance of this model for new cases is the final result from rdCV. The still present drawback is the fact that only a single number for the model performance has been estimated without knowing its variability; this problem can be overcome by repeated double cross validation (rdCV).

Repeated Double Cross Validation (rdCV)

The results from CV depend on the (usually random) split of the objects into segments, and therefore a single CV or dCV may be very misleading. Furthermore, a comparison of the performances of models requires an estimation of the variabilities of the compared measures. Repeated double cross validation (rdCV) repeats dCV several times with different random splits of the objects into segments. The number of repetitions, n_{REP}, is typically 10 to 100.

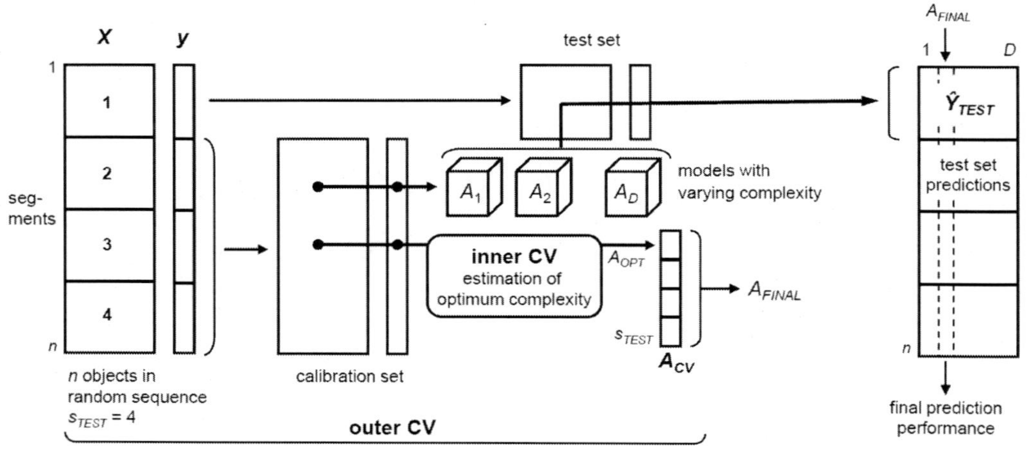

Figure 2. Double CV with four segments for the outer CV (creation of test sets); the optimum complexity (e.g., number of PLS components) is estimated for each calibration set in an inner CV.

The obtained n_{REP} values of the prediction performance give a more complete picture than a single value; the median (or mean) can be used as a single final performance measure; for the comparison of models boxplots are a convenient visualization of the distributions. The number of estimations of the optimum complexity in rdCV is $n_{REP} \times s_{TEST}$.

Figure 3 summarizes the output data from rdCV, some results and diagnostic plots. Basic result is a 3-dimensional array $\hat{\boldsymbol{Y}}_{rdCV}$ with the predicted y-properties (or predicted class membership codes). The dimensions of this array are the object numbers, 1, ... , n; the tested values for the model complexity, A_1, ... , A_D; and the repetitions, 1, ... , n_{REP}. For the further evaluation the slice for the identified final optimum model complexity, A_{FINAL}, is most relevant. A corresponding array with the residuals, y - $\hat{y}$, is relevant for regression models. The distribution of the residuals is often similar to a normal distribution, and then plus/minus twice the standard deviation of the residuals define a 95% tolerance interval for predictions; if the distribution significantly differs from a normal distribution, the quantiles 0.025 and 0.975 can be used instead. A widely used measure for the prediction performance of regression models is the standard error of prediction (SEP), defined as the standard deviation of the residuals. The SEP can be calculated separately for the repetitions, giving an estimation of the variability (due to different random splits of the objects into CV segments). The repetitive predictions are also instructive in the plots $\hat{y}$ versus y, and residuals versus y, as they indicate problematic objects. Further plots - showing the dependence of rdCV results (SEP, residuals) from the model complexity - have only diagnostic value, and indicate the (in)stability of the evaluation, but, of course must not be used to adjust the final model complexity.

The main parameters of rdCV are: s_{TEST}, the number of segments in the outer CV (split into calibration and test sets); s_{CALIB}, the number of segments in the inner CV (split of the calibration set into training and validation sets for estimating an optimum model complexity); n_{REP}, the number of repetitions.

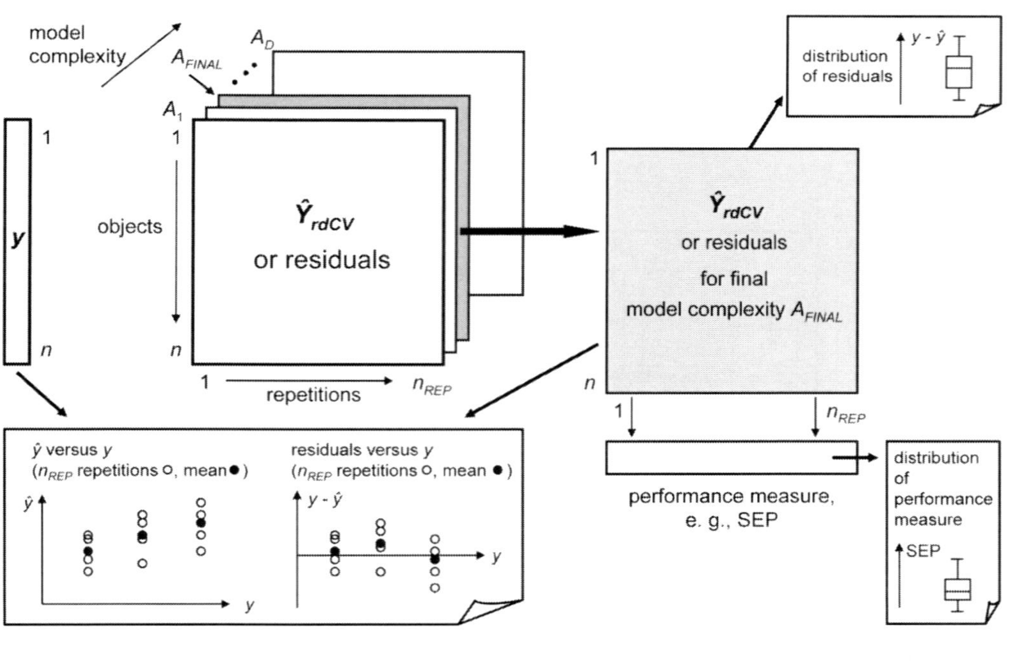

Figure 3. Repeated double CV (rdCV): Resulting evaluation data and a selection of plots.

The number of segments is usually chosen by considering the number of objects, n. If n is small (say <30) care has to be taken for sufficiently large calibrations sets, and therefore s_{TEST} will be large, even having the maximum value n. The larger s_{TEST}, the more resembles the process a leave-one-out CV; however, the random split in the inner CV keeps some variability in the estimation of A_{OPT}. For a large number of objects (say $n > 2000$), s_{TEST} may be kept small to save computing time. The restrictions for the number of segments in the inner CV, s_{CALIB}, are given by the used one standard error method [1, 20] for estimating the optimum model complexity, A_{OPT}, for each calibration set separately. As mentioned before. the error measure MSE is calculated for each segment requiring a reasonable number of objects per segments (in the inner CV), say ≥ 3. From the resulting s_{CALIB} MSE values the standard error (standard deviation of means) is calculated which requires $s_{CALIB} \geq 4$. So we have two contrary effects: a large s_{CALIB} increases the number of values used in the calculation of the standard error (an advantage), but decreases the number of values in each segment (a disadvantage). The minimum number of objects in a calibration set therefore is about 12 (3 time 4), and consequently the minimum total number of objects for rdCV is about 15 (with $s_{TEST} = 5$). In general, $s_{TEST} = 4$, and $s_{CALIB} = 7$ seems to be a good compromise between computational effort and stability of the validation results [1].

The computing time is proportional to the number of repetitions, n_{REP}. For a large number of objects the variations of the rdCV results (e.g., SEP) are rather small, and 10 to 30 repetitions may be sufficient. In general, 50 to 100 repetitions have been recommended [1]. For a small number of objects the probability of identical random sequences increases but is not a practical problem.

APPLICATIONS

Multivariate Calibration with PLS-rdCV

Partial least-squares (PLS) regression is a frequently applied method in chemometrics and chemoinformatics for linear multivariate calibration models; the various PLS methods are well documented elsewhere [2, 18, 22]. The complexity of a PLS model is controlled by the number of PLS components, A, (orthogonal latent variables with maximum covariance with y, defined as linear combinations of all x-variables).

The larger A, the more complex is the model and the bigger is the chance of overfitting. The rdCV strategy gives a set of estimations for the optimum number of PLS components, from which a final value, A_{FINAL}, is selected (Figure 2).

The performance measure of the model used here is the standard error of prediction, SEP, defined as the standard deviation of prediction errors (residuals), $e_i = y_i - \hat{y}_i$, with y_i for the given (experimental, "true") value, and $\hat{y}_i$ the predicted (modeled) value for a property of an object i. Note that in rdCV, all $\hat{y}_i$ are test set predicted results, and n_{REP} predictions are available for each object (for A_{FINAL} PLS components). The single parameter SEP of the distribution of the residuals is usually appropriate because the distribution is mostly similar to a normal distribution with a mean ("bias") near zero. SEP is a user-oriented performance criterion, as it can be given in the units of y. rdCV gives a set of values for SEP demonstrating the variability for different splits of the objects into the sets for CV.

The data set *ethanol* (see Appendix) is used to demonstrate some aspects of rdCV. The x-data contain n = 166 objects (fermentation samples) with m = 235 variables (first derivatives of NIR absorbances). The property y to be modeled is the ethanol concentration (21.7 - 88.1 g/L). The rdCV parameters applied are: s_{TEST} = 4; s_{CALIB} = 7; n_{REP} = 30; the number of PLS components considered 1 - 20. The computing time was 6 seconds (3.4 GHz processor). Figure 4a shows the distributions of the prediction errors for the 30 repetitions (thin gray lines) and their mean (thick black line). 95% of the errors (means) are between -2.81 and 3.80 (dashed vertical lines). We obtain $s_{TEST} \times n_{REP}$ = 120 estimations of the optimum number of PLS components, A_{OPT}, between 10 and 20 (Figure 4b). The highest frequencies (24 from 120) have A = 13 and A = 14. Considering the parsimonious principle for model generation, A_{FINAL} = 13 has been selected. The result demonstrates that usually more than one value for the optimum number of PLS components arises, and that a single CV may give a misleading, outlying value of this essential parameter of a PLS model.

Figure 5a compares experimental y with the predicted y (in gray the 30 repetitions, in black their mean). A zoomed area is shown in Figure 5b indicating that the spread (in this example) is similar for all objects, however, some tend to have outliers. The distribution of the 30 SEP values obtained for A_{FINAL} = 13 is shown in Figure 6a; the median is about 1.6 (g/L ethanol), however, single values are between 1.5 and 1.8 - a rather small range for this obviously good relationship between x-variables and y. Figure 6b shows the dependence of SEP from the number of PLS components. From the used rdCV procedure A_{FINAL} =13 PLS components were found to be optimal. Note that the optimum complexity must not be derived or adjusted from the results shown in Figure 6b, because this would introduce a model optimization based on test set results.

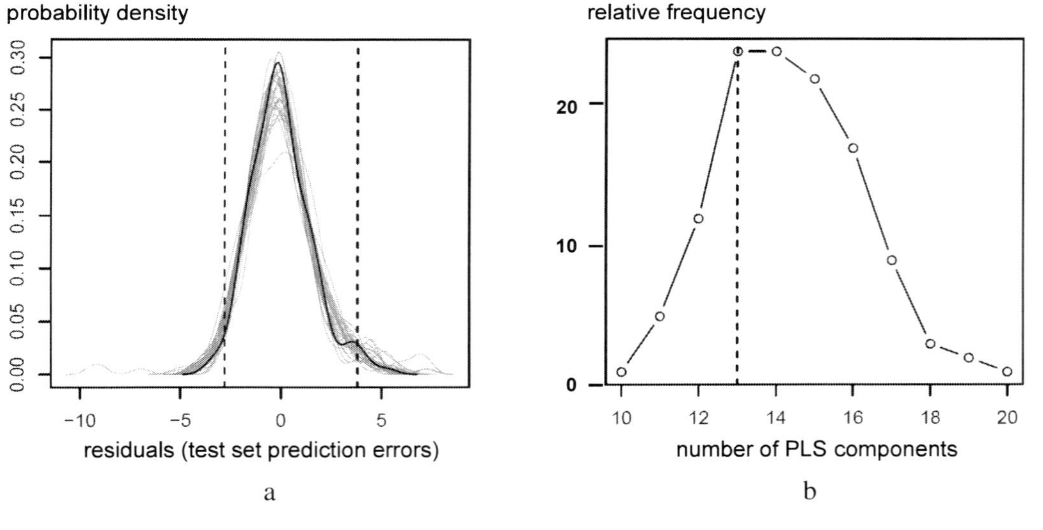

Figure 4. PLS-rdCV results for ethanol data set. (a) Distributions of the prediction errors from 30 repetitions (thin gray lines) and their mean (thick black line). (b) Relative frequencies of the 120 estimations of the optimum number of PLS components; final result is 13.

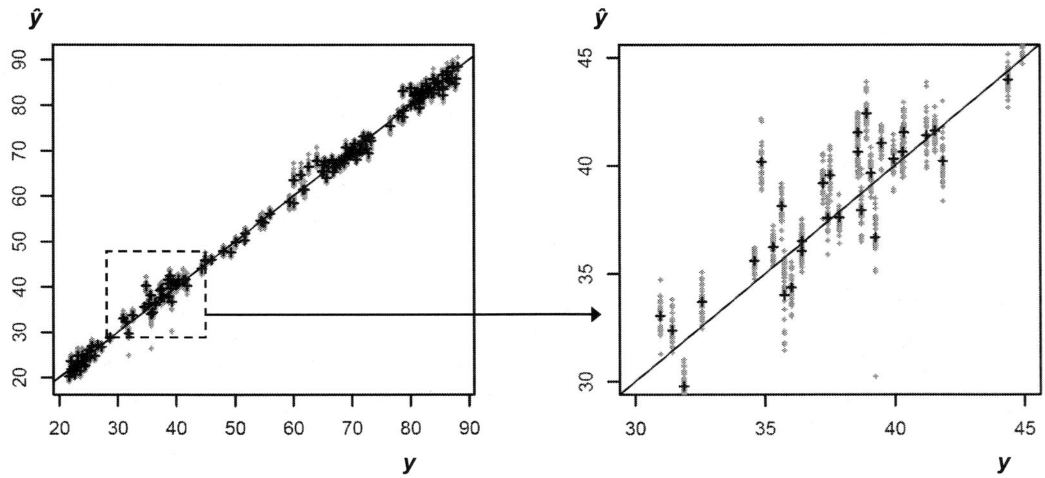

a b

Figure 5. PLS-rdCV results for the ethanol data set. (a) Test-set predicted y (in gray for the 30 repetitions, in black the mean) versus y. (b) Zoom.

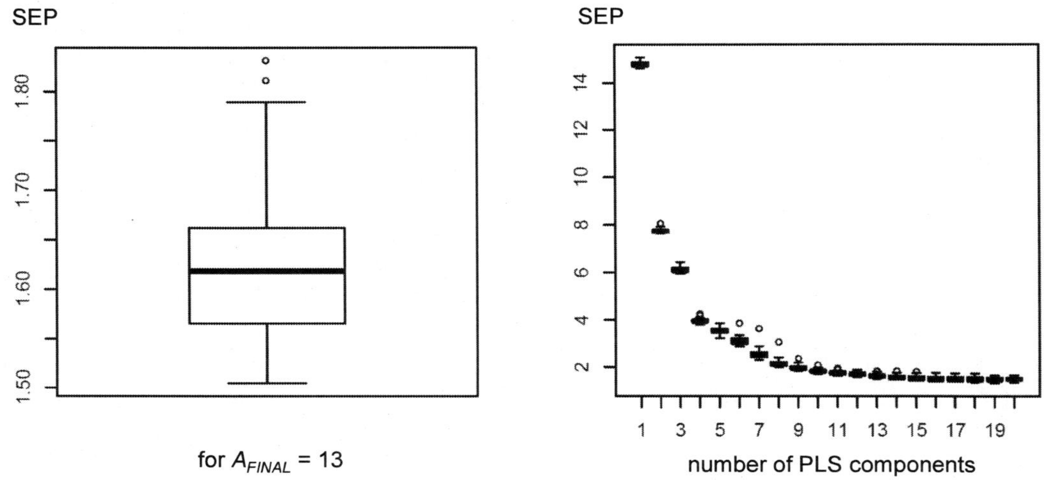

a b

Figure 6. PLS-rdCV results for the ethanol data set. (a) Distribution of the 30 SEP values obtained for the final model complexity with 13 PLS components. (b) Distribution of the 30 SEP values for all investigated numbers of PLS components.

Comparison of Variable Selection Methods by rdCV

Data sets in chemometrics may have up to some thousands of variables (e.g., molecular descriptors [23, 24]), therefore a reduction of the number of variables is often useful. Although PLS regression can handle many ($m > n$), and correlating variables, a reduction may enhance the prediction performance. Furthermore, models with a small number of variables are easier to interpret.

However, variable selection is a controversially discussed topic, for at least the following reasons: (1) For practical problems (say more than 30 variables) only heuristic approaches can be used, that have no guarantee to find a very good solution. (2) Within the variable selection mostly fit criteria are applied to measure the performance because test set approaches (like rdCV) are too laborious. (3) Performance results from variable selection are inappropriately considered as the final performance.

Another undecided problem is the question whether variable selection can or should be done with all objects, or must be done with subsets of the objects within CV. The latter approach sounds very serious, however, the practical problems connected with it are not solved, and reservations against the first approach are often caused by mixing variable selection and evaluation of the model performance. We advocate here for using all objects in variable selection and consider this step completely independent from model assessment by rdCV. The strategy (Figure 7) is to apply several methods of variable selection (and varying their parameters). Preliminary results may consist of several variable subsets, only considered as suggestions, that have to be evaluated by rdCV. The resulting boxplots for SEP allow a clear and visual inspection of the different powers of the variable subsets [25].

For a demonstration of this strategy the data set *ethanol* is used and three different selection methods are applied to the original 235 variables. The methods tested are: (1) Selection of a preset number of variables (m = 20, 40, 60) with highest absolute Pearson correlation coefficient with y. (2) Stepwise variable selection in forward/backward mode using the criterion BIC (Bayes information criterion) [2, 25]. (3) Sequential replacement method [26-28] as implemented in the function *regsubsets*() in the R-library *leaps*, using the criterion BIC. The rdCV parameter applied are: $s_{TEST} = 4$; $s_{CALIB} = 7$; $n_{REP} = 30$; the number of PLS components considered is 1 - min(m, 20).

Figure 8 shows the distributions of the 30 SEP values for the various data sets. Selection of 20 - 60 variables with highest absolute Pearson correlation coefficient only slightly improves the model performance. The variation in the results again stresses the importance of repetitions with different random splits of the objects.

Stepwise selection also gives only a minor improvement, however, only requires 6 variables. Significantly best results have been obtained with 20 variables selected by the sequential replacement method.

Multivariate Classification with KNN-rdCV and DPLS-rdCV

The most used and most powerful methods for classification in chemometrics [2, 16] are discriminant partial least-squares (DPLS) classification, k-nearest neighbor (KNN) classification, and support vector machines (SVM). Like in multivariate calibration, these methods require an optimization of the complexity of the models, and - independently - a cautious estimation of the classification performance for new cases.

Optimizing the complexity of a DPLS classifier means estimation of an optimum number of PLS components - similar to PLS calibration models as described above.

In KNN classification, the number of considered neighbors (k) is the complexity parameter which influences the smoothness (complexity) of the decision surface between the object classes. Note that a small k gives a decision boundary that well fits the calibration data (for an example see Figure 5.13 in [2]).

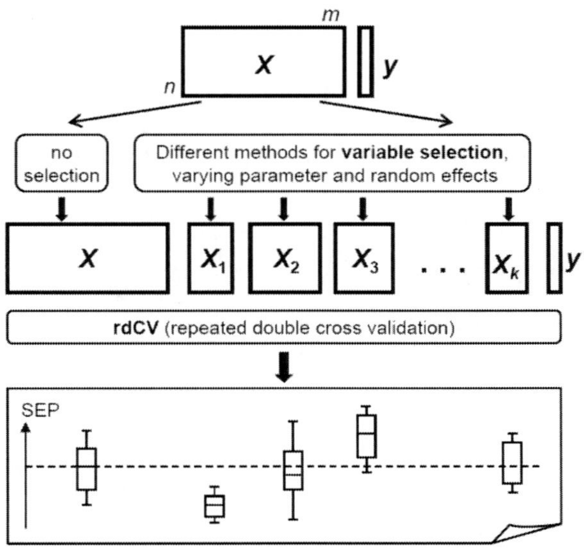

Figure 7. Strategy for variable selection combined with rdCV.

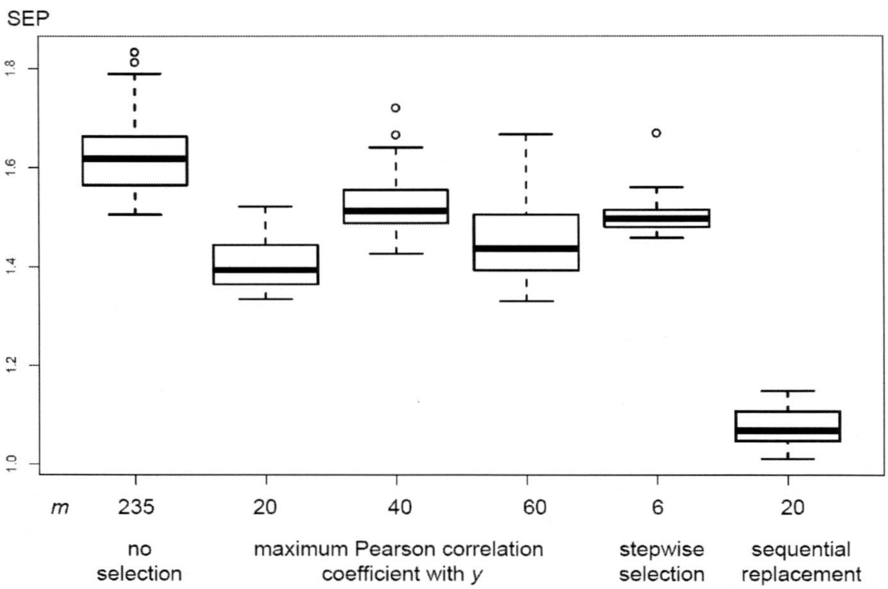

Figure 8. rdCV results for ethanol data set. Comparison of variable selection methods; *m*, number of variables.

The smaller *k*, the more complex is the model and the bigger is the chance for an undesired overfitting. The rdCV strategy gives a set of estimations for the optimum number of neighbors, from which a final value, k_{FINAL}, is selected.

The performance of a classifier is measured here by predictive abilities, P_g, defined as the fraction of correctly assigned objects of a class (group) g ($g = 1, \ldots, G$, with G for the number of classes). An appropriate single number for the total performance, considering all classes equally, is the arithmetic mean, P_{MEAN}, of all P_g.

The sometimes used overall predictive ability, calculated by n_{CORR}/n , with n the number of all tested objects, and n_{CORR} the number of all correctly assigned objects, depends on the relative frequencies of the classes; therefore it may be misleading and is not used. Note that in rdCV, all predictive abilities are obtained from test set objects, and n_{REP} predictions are available for each object (for A_{FINAL} PLS components in DPLS, or k_{FINAL} neighbors in KNN). rdCV gives a set of values, e.g., for P_{MEAN}, demonstrating the variability for different splits of the objects into the sets for CV.

The data set *pigments* (see Appendix) is used to demonstrate some aspects of KNN-rdCV. The x-data contain $n = 227$ objects (synthetic organic pigments) with $m = 526$ variables (IR transmission data). The objects belong to $G = 8$ mutually exclusive classes (11 to 49 objects per class). The rdCV parameter applied are: $s_{TEST} = 4$; $s_{CALIB} = 7$; $n_{REP} = 50$; for KNN $k = 1, \ldots, 5$ neighbors have been considered.

We obtain $s_{TEST} \times n_{REP} = 200$ estimations of the optimum number of neighbors, k_{OPT}. The highest frequency of 91.5% is for $k = 1$; $k = 2, 3, 4$ have 5.5, 2.5, and 0.5%, respectively. Thus, $k_{FINAL} = 1$ is easily selected.

Figure 9a shows the distributions of the predictive abilities, P_g, for the 8 classes separately, and for the mean P_{MEAN} - as obtained in 50 repetitions with different random splits into segments. P_{MEAN} has a narrow distribution with an average of 0.95. Classes 3, 4, and 7 show almost 100% correct class assignments. Class 5 appears worst with a mean of $P_5 = 0.88$ and a rather large spread. Some outlying results with $P < 0.8$ warn from making only a single CV run without repetitions. Figure 9b shows P_{MEAN} as a function of k; this diagnostic plot demonstrates the stability of the results, and confirms that the best result is obtained by considering only the first neighbor; however, these results must not be used for an adjustment of k_{FINAL}.

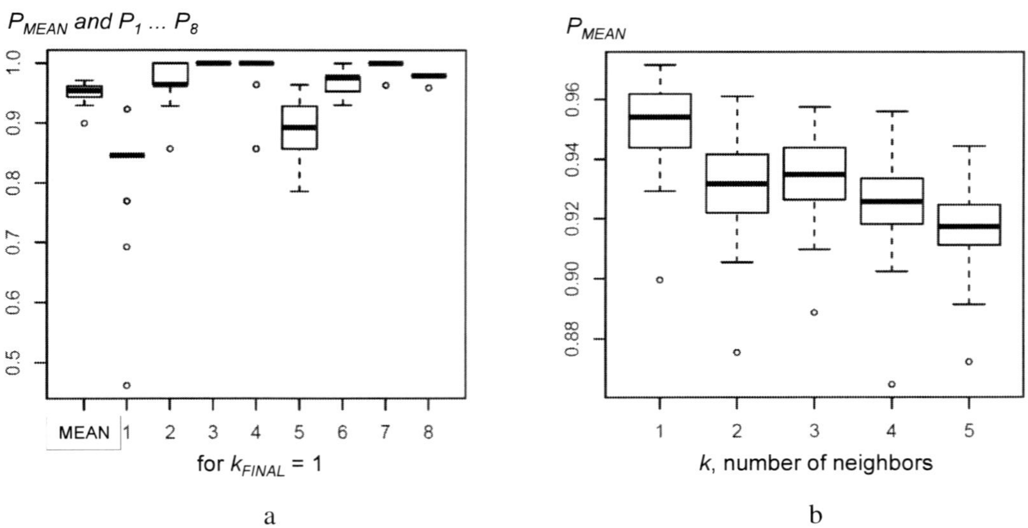

Figure 9. KNN-rdCV results for the pigments data set. (a) Predictive abilities for the classes 1 to 8, and their mean (from 50 repetitions and the final number of neighbors, k_{FINAL}. (b) Distribution of P_{MEAN} for all investigated numbers of neighbors.

The data set *mutagenicity* (see Appendix) is used as a QSAR example for binary classification with DPLS-rdCV. The *x*-data contain n = 6458 objects (chemical structures) with m = 1440 variables (molecular descriptors). The objects belong to G = 2 mutually exclusive classes, with 2970 not mutagenic compounds (class 1), and 3488 mutagenic compounds (class 2). The rdCV parameters applied are: $s_{TEST} = 2$; $s_{CALIB} = 7$; $n_{REP} = 30$; the number of PLS components considered 1 - 30. We obtain $s_{TEST} \times n_{REP} = 60$ estimations of the optimum number of PLS components, A_{OPT}, with results between 20 and 29 (Figure 10a). The highest frequency has $A = 25$ which is used as A_{FINAL}.

In Figure 10b the resulting distributions of the predictive abilities are shown for this model complexity. The large number of objects gives only small variations of the performance measures; the mutagenic compounds are better recognized (ca 78% correct) than the not mutagenic compounds (ca 68% correct).

The rather low total prediction performance, P_{MEAN}, of only ca 73% is not surprising for this very challenging classification problem and the simple approach, namely a linear classifier applied to chemical structures with high diversity.

Finally, the rdCV strategy has been applied to estimate the prediction performance of DPLS classification for some subgroups of these very diverse structures. A comprehensive study of local models is out of the scope of this study, and only the applicability of rdCV is demonstrated.

(1) First, subgroups with defined ranges of the molecular mass are investigated. The molecular masses of all 6458 chemical structures are between 26 and 1550 (median 229); the tested four groups have the ranges 30 - 100, 101 - 200, 201 - 300, and 301 -600, see Figure 11. The compound set with low molecular masses (30 - 100) shows considerably lower prediction performance than the other subgroups, in particular caused by only random class assignments of the mutagenic compounds.

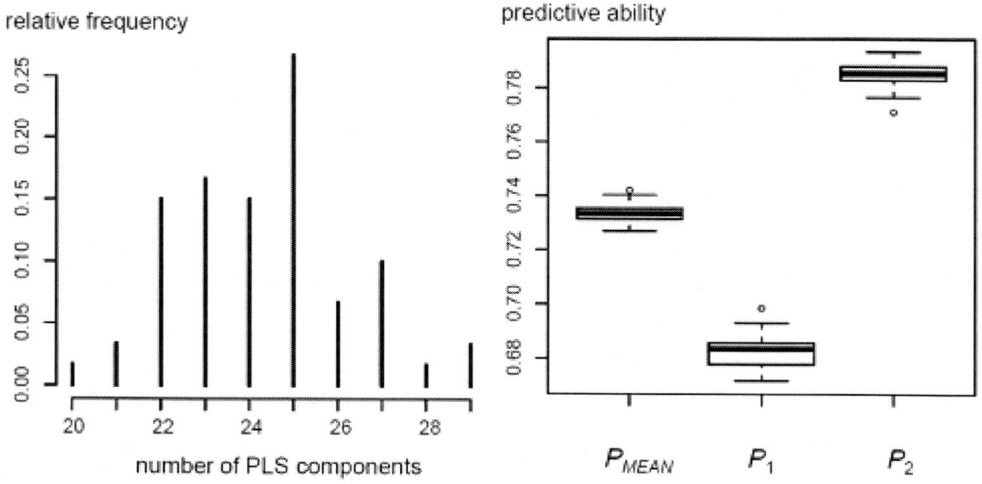

a b

Figure 10. DPLS-rdCV results for the mutagenicity data set. (a) Histogram with relative frequencies obtained for the optimum number of PLS components (total 60 values). (b) Predictive abilities P_1 for class 1 (not mutagenic), P_2 for class 2 (mutagenic), and P_{MEAN}, the mean of both, as obtained from 30 repetitions.

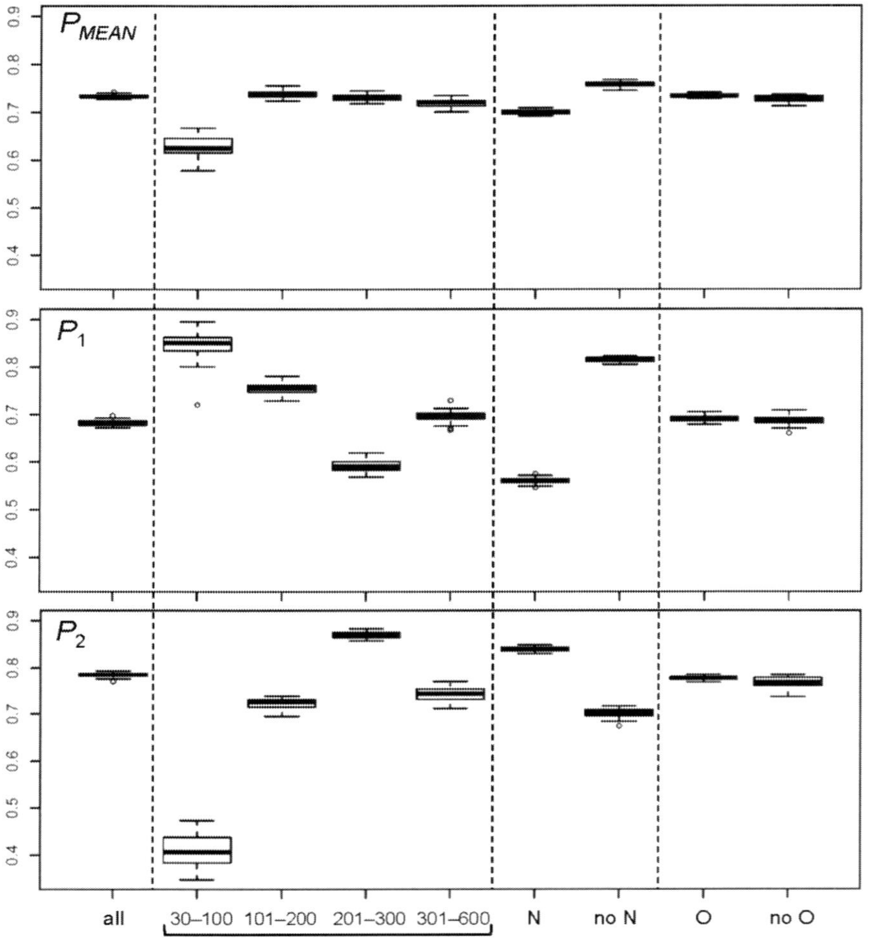

	all	30–100	101–200	201–300	301–600	N	no N	O	no O
n	6458	323	2178	2441	1345	4086	2372	5049	1409
n_1	2970	211	1097	874	682	1622	1348	2345	625
n_2	3488	112	1081	1567	663	2464	1024	2704	784

Figure 11. DPLS-rdCV results for the mutagenicity data set for selected groups of compounds. The rdCV results are compared for "all" compounds, compounds in four ranges of molecular mass (30-100, 101-200, 201-300, and 301-600), and compounds containing nitrogen (N), no nitrogen (no N), containing oxygen (O) and no oxygen (no O). n, number of compounds; n_1, number of not mutagenic compounds (class 1); n_2, number of mutagenic compounds (class 2); for predictive abilities, P, see Figure 10.

Consequently, compounds with molecular masses below 100 should be excluded from DPLS classification with these data.

(2) Second, subsets of compounds have been investigated, containing a certain element or not. In this demonstration only nitrogen and oxygen have been considered (more detailed studies will apply chemical substructure searches for defining subsets). rdCV results show that the prediction performances for these compound subsets are not very different from those for all compounds.

CONCLUSION

Repeated double cross validation (rdCV) is a general and basic strategy for the development and assessment of multivariate models for calibration or classification. Such heuristic models require an optimization of the model complexity, which is defined by a method specific parameter, e.g., the number of PLS components, or the number of neighbors in KNN classification. rdCV provides a set of estimations of this parameter from which a final single value or a set of values is selected. Estimation of the optimum model complexity is separated from the evaluation of the model performance for new cases by applying a double cross validation (dCV). By repeating the dCV with different random splits of the objects into calibration and test sets, several estimations of the performance measure are obtained, and its variability can be characterized. This variability may be high for small data sets. Knowing the variability of the performance measure (at least for different splits of the data during CV) is necessary for a reasonable comparison of models.

APPENDIX

(1) Data Sets

Data set *ethanol* [29, 30]. The n = 166 objects are centrifuged samples of fermentation mashes, related to bioethanol production from different feedstock (corn, rye, wheat). The m = 235 variables are first derivatives (Savitzky-Golay, 7 points, 2nd order) of NIR absorbances in the wavelength range 1115 to 2285 nm, step 5 nm. The property y is the ethanol concentration (21.7 - 88.1 g/L) as determined by HPLC.

Data set *pigments* [31]. For the identification of pure synthetic organic pigments FTIR spectroscopy has been established as a powerful analytical technique. The n = 227 objects are synthetic organic pigments used for artists' paints. The selected 8 classes of pigments are defined according to the usual classification system of organic colorants based on their chemical constitution. The m = 526 variables are IR transmissions (%) in the spectral range 4000 - 580 cm^{-1}.

Class no.	Pigment type	No. of samples
1	Monoazo yellow pigments	13
2	Disazo yellow pigments	28
3	s-Naphthol pigments	11
4	Naphthol_AS pigments	28
5	s-Naphthol Lake pigments	28
6	BONS Lake pigments	43
7	Benzimidazolone pigments	27
8	Phthalocyanine pigments	49

Data set *mutagenicity*. Data are from n = 6458 chemical structures with the mutagenicity determined by AMES tests as collected and provided by Katja Hansen (University of Technology, Berlin, Germany, 2009) [32].

Class 1 contains 2970 not mutagenic compounds, class 2 contains 3488 mutagenic compounds. Each structure has been characterized by m = 1440 molecular descriptors calculated by software Dragon 6.0 [24] from approximated 3D structures and all H-atoms explicitly given (calculated by software Corina [33]). This data set for binary classification has been described and used previously [30, 34].

(2) Software

Software used in this work has been written in the R programming environment [3], in particular using the R packages *chemometrics* [4, 5], and *pls* [35].

ACKNOWLEDGMENTS

We thank Bettina Liebmann for the NIR data from fermentation samples; Anke Schäning and Manfred Schreiner (Academy of Fine Arts Vienna, Austria) for the IR data from pigment samples; and Katja Hansen (former at University of Technology, Berlin, Germany) for the chemical structure data with AMES mutagenicity.

REFERENCES

[1] Filzmoser, P., Liebmann, B., Varmuza, K. *J. Chemometrics* 23 (2009) 160-171.

[2] Varmuza, K., Filzmoser, P. *Introduction to multivariate statistical analysis in chemometrics.* CRC Press, Boca Raton, FL, US, 2009.

[3] *R: A language and environment for statistical computing*. R Development Core Team, Foundation for Statistical Computing, www.r-project.org, Vienna, Austria, 2014.

[4] Filzmoser, P., Varmuza, K. *R package chemometrics.* http://cran.at.r-project.org/web/packages/chemometrics/index.html, Vienna, Austria, 2010.

[5] Garcia, H., Filzmoser, P. *Multivariate statistical analysis using the R package chemometrics (R-vignette).* http://cran.at.rproject.org/web/packages/chemometrics/vignettes/chemometrics-vignette.pdf, Vienna, Austria, 2010.

[6] Smit, S., Hoefsloot, H. C. J., Smilde, A. K. *J. Chromatogr. B* 866 (2008) 77-88.

[7] Westerhuis, J. A., Hoefsloot, H. C. J., Smit, S., Vis, D. J., Smilde, A. K., van Velzen, E. J. J., van Duijnhoven, J. P. M., van Dorsten, F. A. *Metabolomics* 4 (2008) 81-89.

[8] Dixon, S. J., Xu, Y., Brereton, R. G., Soini, H. A., Novotny, M. V., Oberzaucher, E., Grammer, K., Penn, D. J. *Chemometr. Intell. Lab. Syst.* 87 (2007) 161-172.

[9] Forina, M., Lanteri, S., Boggia, R., Bertran, E. *Quimica Analitica* (Barcelona, Spain) 12 (1993) 128-135.

[10] Esbensen, K. H., Geladi, P. *J. Chemometr.* 24 (2010) 168-187.

[11] Martens, H. A., Dardenne, P. *Chemometr. Intell. Lab. Syst.* 44 (1998) 99-121.

[12] Martins, J. P. A., Teofilo, R. F., Ferreira, M. M. *J. Chemometr.* 24 (2010) 320-332.

[13] Krstajic, D., Buturovic, L. J., Leahy, D. E., Thomas, S. *J. Chemoinformatics* 6:10 (2014) 1-15.

[14] Brereton, R. G. *Chemometrics - Applications of mathematics and statistics to laboratory systems.* Ellis Horwood, New York, NY, US, 1990.

[15] Massart, D. L., Vandeginste, B. G. M., Buydens, L. C. M., De Jong, S., Smeyers-Verbeke, J. *Handbook of chemometrics and qualimetrics: Part A*. Elsevier, Amsterdam, The Netherlands, 1997.

[16] Naes, T., Isaksson, T., Fearn, T., Davies, T. *A user-friendly guide to multivariate calibration and classification*. NIR Publications, Chichester, United Kingdom, 2004.

[17] Kennard, R. W., Stone, L. A. *Technometrics* 11 (1969) 137-148.

[18] Vandeginste, B. G. M., Massart, D. L., Buydens, L. C. M., De Jong, S., Smeyers-Verbeke, J. *Handbook of chemometrics and qualimetrics: Part B*. Elsevier, Amsterdam, The Netherlands, 1998.

[19] Breiman, L., Spector, P. *Int. Stat. Rev.* 60 (1992) 291-319.

[20] Hastie, T., Tibshirani, R. J., Friedman, J. *The elements of statistical learning - Data mining, inference, and prediction*, edn. 2nd. Springer, New York, 2009.

[21] Baumann, K. *Trends Anal. Chem.* 22 (2003) 395-406.

[22] Wold, H., Sjoestroem, M., Eriksson, L. *Chemom. Intell. Lab. Syst.* 58 (2001) 109-130.

[23] *ADRIANA: Software for the encoding of molecular structures,* version 2.2.4. Molecular Networks GmbH, www. molecular-networks.com, Erlangen, Germany, 2011.

[24] *Dragon: Software for molecular descriptor calculation*, version 6.0, by Todeschini, R., Consonni, V., Mauri, A., Pavan, M. Talete srl, www.talete.mi.it, Milan, Italy, 2010.

[25] Varmuza, K., Filzmoser, P., Dehmer, M. *Computational and Structural Biotechnology Journal* 5:e201302007 (2013) 1-10. Open access: http://dx.doi.org/10.5936/csbj. 201302007; data and software: www.lcm.tuwien.ac.at/R/.

[26] Grisoni, F., Cassotti, M., Todeschini, R. *J. Chemometr.* 28 (2014) 249-259.

[27] Mercader, A. C., Castro, E. A. In: Dehmer, M., Varmuza, K., Bonchev, D. (Ed.), *Statistical Modelling of Molecular Descriptors in QSAR/QSPR*, Wiley-VCH, Weinheim, Germany, 2012, p. 149-174.

[28] Miller, A. *Subset selection in regression*, 2nd ed. Chapman Hall, CRC Press, Boca Raton, FL, 2002.

[29] Liebmann, B., Friedl, A., Varmuza, K. *Anal. Chim. Acta* 642 (2009) 171-178.

[30] Varmuza, K., Filzmoser, P., Liebmann, B. *J. Chemom.* 24 (2010) 209-217.

[31] Schäning, A., Varmuza, K., Schreiner, M. *e-Preservation Science* 6 (2009) 75-80.

[32] Hansen, K., Mika, S., Schroeter, T., Sutter, A., ter Laak, A., Steger-Hartmann, T., Heinrich, N., Müller, K. R. *J. Chem. Inf. Model.* 49 (2009) 2077-2081.

[33] *Corina: Software for the generation of high-quality three-dimensional molecular models,* by Sadowski, J., Schwab, C. H., Gasteiger, J. Molecular Networks GmbH, www.molecular-networks.com, Erlangen, Germany, 2004.

[34] Varmuza, K., Filzmoser, P., Liebmann, B., Dehmer, M. *Chemom. Intell. Lab. Syst.* 117 (2012) 31-41.

[35] Mevik, B. H., Wehrens, R. *J. Stat. Software* 18 (2007) 1-24.

In: Current Applications of Chemometrics
Editor: Mohammadreza Khanmohammadi

ISBN: 978-1-63463-117-4

Chapter 3

FUZZY CLUSTERING OF ENVIRONMENTAL DATA

Costel Sârbu*
Faculty of Chemistry and Chemical Engineering,
Babeş-Bolyai University, Cluj-Napoca, Romania

ABSTRACT

Cluster analysis is a large field, both within fuzzy sets and beyond it. Many algorithms have been developed to obtain hard clusters from a given data set. Among those, the *c-means algorithms* are probably the most widely used. *Hard c-means* execute a sharp classification, in which each object is either assigned to a class or not. The membership to a class of objects (samples) therefore amounts to either 1 or 0. The application of *Fuzzy sets* in a *classification function* causes this class membership to become a relative one and consequently a sample can belong to several classes at the same time but with different degrees. The c-means algorithms are *prototype-based procedures*, which minimize the total of the distances between the prototypes and the samples by the construction of a target function. *Fuzzy Generalized n-Means* is easy and well improved tool, which have been applied in many fields of science and technique. In this chapter, different Fuzzy clustering algorithms applied to various environmental data described by different characteristics (parameters) allow an objective interpretation of their structure concerning their similarities and differences, respectively.

The very informative fuzzy approach namely, *Fuzzy Cross-Clustering Algorithm*, (FHCsC) allows the qualitative and quantitative identification of the characteristics (parameters) responsible for the observed similarities and dissimilarities between various environmental samples (water, soil) and related data.

In addition, the fuzzy hierarchical characteristics clustering (FHiCC) and fuzzy horizontal characteristics clustering (FHoCC) procedures revealed the similarity or differences between considered parameters and finding out fuzzy partitions and outliers.

* Corresponding author: Costel Sârbu. Faculty of Chemistry and Chemical Engineering, Babeş-Bolyai University, Arany Janos Str., No 11, RO-400028, Cluj-Napoca, Romania. E-mail: csarbu@chem.ubbcluj.ro.

INTRODUCTION

Imprecision in data and information gathered from measurements and observations about our universe is either *statistical* or *nonstatistical.*

This latter type of uncertainty is called *fuzziness*. We all assimilate and use fuzzy data, vague rules, and imprecise information, just as we are able to make decisions about situations that seems to be governed by an element of chance. Accordingly, computational models of real systems should also be able to recognize, represent, manipulate, interpret, and use both fuzzy and statistical uncertainties. *Statistical models* deal with random events and outcomes; *fuzzy models* attempt to capture and quantify nonrandom imprecision.

The mathematics of *fuzzy sets theory* was originated by L. A. Zadeh in 1965. It deals with the uncertainty and fuzziness arising from interrelated, reasoning, cognition, and perception. This type of uncertainty is characterized by structure that lack sharp boundaries.

The fuzzy approach provides a way to translate a linguistic model of the human thinking process into a mathematical framework for developing the computer algorithms for computerized decision-making processes.

A fuzzy set is a generalized set to which objects can belongs with various degrees (grades) of memberships over the interval [0,1]. *Fuzzy systems* are processes that are too complex to be modeled by using conventional mathematical methods (*hard models*).

In general, fuzziness describes objects or processes that are not amenable to precise definition or precise measurement. Thus, fuzzy processes can be defined as processes that are vaguely defined and have some uncertainty in their description.

The data arising from fuzzy systems are in general, *soft*, with no precise boundaries. Fuzziness of this type can often be used to represent situations in the real world better than the rigorous definitions of *crisp set theory*. For example, water is to some extent an acid, germanium is to some extent a metal, and pharmaceutical drugs are effective to some extent. Fuzzy sets can be particularly useful when dealing with imprecise statements, ill-defined data, incomplete observations or inexact reasoning.

There is often fuzziness in our everyday world: varying signal heights in spectra from the same substance, or varying patterns in QSAR pattern recognition studies.

Fuzzy logic is a tool that permits us to manipulate and work with uncertain data. Fuzzy logic differs from Boolean or classical logic in that it is many-valued rather than two-valued.

In classical logic, statements are either *true* or *false* - which corresponds to a one or zero membership function - whereas fuzzy logic the truth values are graded and lie somewhere between true and false.

Many rules have been developed for dealing with fuzzy logic statements which enable the statements to be processed and inferences to be reached from the statements. The outcome of fuzzy logic processing is usually a fuzzy set. So that the final answer is one that has practical significance, a *defuzzification* of this fuzzy set is performed to yield a concrete single number that can be used directly.

Such procedures are now regularly performed in a host different appliances including, cement kiln control, navigation system for automatic cars, a predictive fuzzy logic controller for automatic operation of trains, laboratory water level controllers, controllers for robot arc-welders, feature-definition controllers for robot vision, graphics controllers for automatic police sketchers, and more.

HARD AND FUZZY CLUSTERING

A basic problem that arises in a wide variety of fields, including environmental chemistry, is the so-called *clustering problem* [1-3]. Classification is useful, since it allows meaningful generalizations to be made about large quantities of data by recognizing among them a few basic patterns. It plays a key role in searching for structures in data.

Each of these structures is called *cluster* or *class*. A class is a group of individuals (e.g., soil samples or pixels of a image) which resemble each other more strongly, in terms of particular properties, than they resemble members of other classes [4]. There are various ways in which classification may be carried out. Generally, two types of algorithm are distinguished, these being *hierarchical* and *non-hierarchical* or relocation clustering.

Both methods require the calculation of a *(dis)similarity matrix*. This (dis)similarity which is really a measure of the proximity of the pair of objects (points) in the *p-dimensional characteristic space,* defined by the *p* properties measured for each individual, is usually expressed in terms of either the *Euclidean* or the *Mahalanobis distance* between the two points [1-10]. Either the number of clusters to be generated can be specified in advance (*semi-supervised learning approach*), or it may be optimized by the algorithm itself according to certain criteria (*unsupervised learning approach*).

There are two opposing approaches to hierarchical clustering, these being *agglomerative* and *divisive procedures*.

In the *agglomerative hierarchical methods*, we begin with *n* items and end with a single cluster containing all *n* items. A *divisive hierarchical method* starts with a single cluster of *n* items and divides it into two groups. At each step thereafter, one of the groups is divided into two subgroups. The ultimate result of a divisive algorithm is *n* clusters of one item each. The results can be shown in a *dendrogram*. Divisive methods suffer from the same potential drawback as the agglomerative methods—namely, once a partition is made, an item cannot be moved into another group it does not belong to at the time of the partitioning.

However, if larger clusters are of interest, then the divisive approach may sometimes be preferred over the agglomerative approach, in which the larger clusters are reached only after a large number of joinings of smaller groups.

Divisive algorithms are generally of two classes: *monothetic* and *polythetic*. In a *monothetic* approach, the division of a group into two subgroups is based on a single variable, whereas, the *polythetic* approach uses all *p* variables to make the split.

If the variables are binary (quantitative variables can be converted to binary variables), the monothetic approach can easily be applied. Division into two groups is based on presence or absence of an attribute.

In *classical cluster analysis* (hard clustering) each object must be assigned to exactly one cluster. This is a source of ambiguity and error in cases of outliers or overlapping clusters and affords a loss of information. This kind of vagueness and uncertainty can, however, be taken into account by using the *theory of fuzzy sets* as one have already been mentioning above [11]. A *fuzzy set* or a fuzzy subset is a collection of ill-defined and not-distinct objects with un-sharp boundaries in which the transition from membership to non-membership in a subset of a reference set is gradual rather than abrupt. The theory of fuzzy set is basically a theory of graded concepts. It is an extreme generalization of *ordinary set theory* and is basically designed to handle the concept of partial truth.

A central concept of fuzzy set theory is that it is permissible for an element to belong partly to a fuzzy set. It provides an adequate conceptual framework as well as a mathematical tool to model the real world problems which are often obscure and indistinct.

Most *fuzzy clustering algorithms* are *objective function* based [12, 14]: they determine an optimal classification by minimizing an objective function. In objective function based clustering usually each cluster is represented by a *cluster prototype* or *cluster support*. This prototype consists of a cluster center (whose name already indicates its meaning) and may be some additional information about the size and the shape of the cluster. However, the cluster center is computed by the clustering algorithm and may or may not appear in the dataset. The *degrees of membership* to which a given data point belongs to the different clusters are computed from the distances of the data point to the cluster centers.

The closer a data point lies to the center of a cluster, the higher is its degree of membership to this cluster. Hence the problem to divide a data set into c clusters can be stated as the task to minimize the distances of the data points to the cluster centers, since, of course, we want to maximize the degrees of membership.

Several fuzzy clustering algorithms can be distinguished depending on the additional size and shape information contained in the cluster prototypes, the way in which the distances are determined, and the restrictions that are placed on the membership degrees. Here we focus on the G*eneralized Fuzzy c-Means Algorithms*, as an extension of the well-known *Fuzzy c-Means Algorithms* [14, 15], which use only cluster centers and a Euclidean distance function. *The Gustafson-Kessel Algorithm* is also discussed and applied.

The fuzzy set and corresponding *fuzzy logic* formula enhance the robustness of the fuzzy set clustering for *pattern recognition* [16-18]. Fuzzy set and fuzzy logic are principal constituents of *soft computing*. In contrast to traditional *hard computing*, soft computing is tolerant to imprecision, uncertainty and partial truth. Soft computing models utilize fuzzy logic, artificial neural networks, approximation reasoning, possibility theory, fuzzy clustering, and so on, to discover relationships in dynamic, non-linear, complex, and uncertain environments. These techniques often borrow the mechanics of cognitive processes and laws of nature to provide us with a better understanding of and solution to real world problems including, of course, chemistry, environment and related fields [19, 20].

FUZZY C-MEANS ALGORITHM

Fuzzy c-means (FCM) algorithm, also known as Fuzzy ISODATA, was introduced by Bezdek [14, 15] as an extension to Dunn's [21] and Ruspini's algorithm [22]. The FCM-based algorithms are the most widely used fuzzy clustering algorithms in practice. Let, $X = \{x_1, x_2, ..., x_n\}$, where $x_i \in R^p$ present a given set of feature data. The aim of the FCM algorithm is to find an optimal fuzzy c-partition and corresponding prototypes minimizing the objective (dissimilarity) function formulated usually as

$$J(P,L) = \sum_{j=1}^{c} \sum_{i=1}^{n} (A_{ij})^m \left\| x_i - L_j \right\|^2 \qquad (1)$$

where $L = \{L_1, L_2, ..., L_c\}$ are the cluster centers and $P = (A_{ij})_{nxc}$ is a fuzzy partition matrix, in which each member indicates the degree of membership between the data vector and the cluster j. The values of matrix P should satisfy the following conditions

$$A_{ij} \in [0,1], \forall i = 1, 2, ..., n, \forall j = 1, 2, ..., c \tag{2}$$

$$\sum_{j=1}^{c} A_{ij} = 1, \forall i = 1, 2, ..., n \tag{3}$$

The weighting exponent m (so-called "fuzzifier") is any real number in [1, ∞], which determines the fuzziness of the clusters; for m→1 the A_{ij} approach 0 or 1, for m→∝ the memberships tend to be "totally fuzzy" A_{ij}→1/c. The most commonly used distance norm is the Euclidean distance, $d_{ij} = \|x_i - L_j\|$, although other distance norms, expressing the similarity between any measured data and the prototype, could produce better results [23]. Minimization of the objective function *J(P,L)* is a nonlinear optimization problem, which can be minimized with respect to P (assuming the prototypes L to be constant), a fuzzy c-partition of the data set, and to L (assuming the memberships P to be constant), a set of c prototypes $(2 \le c < n)$, by applying the following iterative algorithm:

Step 1: Initialize the membership matrix P with random values so that the conditions (2) and (3) are satisfied. Choose appropriate exponent m and the termination criteria.

Step 2: Calculate the cluster centers L according to the equation:

$$L_j = \frac{\sum_{i=1}^{n} (A_{ij})^m x_i}{\sum_{i=1}^{n} (A_{ij})^m}, \forall j = 1, 2, ..., c \tag{4}$$

Step 3: Calculate the new distance norms:

$$d_{ij} = \|x_i - L_j\|, \forall i = 1,2,...,n, \forall j = 1, 2, ..., c \tag{5}$$

Step 4: Update the fuzzy partition matrix P:

If $d_{ij} > 0$ (indicating that $x_i \neq L_j$)

$$A_{ij} = \frac{1}{\sum_{k=1}^{c} \left(\frac{d_{ij}}{d_{ik}} \right)^{\frac{2}{m-1}}} \tag{6}$$

Else

$A_{ij} = 1$

Step 5: *If* the termination criterion has been met, stop
Else go to Step 2

A suitable termination criterion could be to evaluate the objective function (Equation 1) and to see whether it is below a certain tolerance value or if its improvement compared to the previous iteration is below a certain threshold between 0 and 1 [24]. Also the maximum number of iteration cycles can be used as a termination criterion.

GENERALIZED FUZZY C-MEANS ALGORITHM

In this section we will recall the so-called *Generalized Fuzzy c-Means Algorithm* (GFCM) [25-28]. This algorithm is a generalization of the Fuzzy c-Means Algorithm discussed above.

In the most general way, a fuzzy clustering algorithm with objective function can be formulated as follows: let $X = \{ x^1, ..., x^n\} \subset R^p$ be a finite set of feature vectors, where n is the number of objects (measurements) and p is the number of the original variables, $x^j = [x_1^j, x_2^j, ..., x_p^j]^T$ and $L = (L^1, L^2, ..., L^c)$ be a c-tuple of prototypes (supports) each of which characterizes one of the c clusters composing the cluster substructure of the data set; a partition of X into c fuzzy clusters will be performed by minimizing the objective function [25-28].

$$J(P,L) = \sum_{i=1}^{c}\sum_{j=1}^{n}(A_i(x^j))^2 d^2(x^j, L^i) \tag{7}$$

where $P = \{A_1, ..., A_c\}$ is the fuzzy partition, $A_i(x^j) \in [0,1]$ represents the membership degree of feature point x^j to cluster A_i, $d(x^j, L^i)$ is the distance from a feature point x^j to the prototype of cluster A_i, defined by the Euclidean distance norm

$$d(x^j, L^i) = \left\| x^j - L^i \right\| = \left[\sum_{k=1}^{p} (x_k{}^j - L^i{}_k)^2 \right]^{1/2} \tag{8}$$

The optimal fuzzy set will be determined by using an iterative method where J is successively minimized with respect to A and L.

Supposing that L is given, the minimum of the function $J(\bullet, L)$ is obtained for:

$$A_i(x^j) = \frac{C(x^j)}{\sum_{k=1}^{c} \frac{d^2(x^j, L^i)}{d^2(x^j, L^k)}}, i = 1, \ldots, c \tag{9}$$

where C is a fuzzy set from X and

$$C(x^j) = \sum_{i=1}^{c} A_i(x^j) \quad . \tag{10}$$

It is easy to observe that $C(x^j) \leq 1$, $j = 1, 2, \ldots, n$.

For a given *P*, the minimum of the function *J*(*P*,•) is obtained for:

$$L^i = \frac{\sum_{j=1}^{n} \left[A_i(x^j)\right]^2 x^j}{\sum_{j=1}^{n} \left[A_i(x^j)\right]^2}, i = 1, \ldots, c \tag{11}$$

The above formula allows one to compute each of the p components of L^i (the center of the cluster i). Elements with a high degree of membership in cluster i (i.e., close to cluster i'c center) will contribute significantly to this weighted average, while elements with a low degree of membership (far from the center) will contribute almost nothing.

A cluster can have different shapes, depending on the choice of prototypes. The calculation of the membership values is dependent on the definition of the distance measure. According to the choice of prototypes and the definition of the distance measure, different fuzzy clustering algorithms are obtained. If the prototype of a cluster is a point – the cluster center – it will give spherical clusters, if the prototype is a line it will give tubular clusters [25-35], and so on.

The clusters are generally of unequal size. If we consider that a small cluster is in then neighbourhood of a greater one, than some points of the greater cluster would be "captured" by its smaller neighbour. This situation may be avoided if a data dependent metric is used. An adaptive may be induced by the radius or by the diameter of every fuzzy class A_i. The radius r_i of the fuzzy class A_i is defined by

$$r_i = \max d_i(x, L^i) = \max A_i(x) d(x, L^i) \tag{12}$$

Now, we may define the local adaptive distance, d_{ia}, where

$$d_{ia}(x, L^i) = \frac{A_i(x) d(x, L^i)}{r_i}, \text{ i} = 1, 2, \ldots, \text{c} \tag{13}$$

Cluster Validity

Cluster validation refers to procedures that evaluate the clustering results in a quantitative and objective function. Some kinds of validity indices are usually adopted to measure the adequacy of a structure recovered through cluster analysis [35].

Determining the correct number of clusters in a data set has been, by far, the most common application of cluster validity.

There are two approaches: The use of validity functionals, which is a post factum method, and the use of hierarchical algorithms, which produce not only the optimal number of classes (based on the needed granularity) but also a binary hierarchy that shows the existing relationships between the classes.

In the case of fuzzy clustering algorithms, some validity indices such as partition coefficient, partition entropy and Backer-Jain Index use only the information of fuzzy membership degrees to evaluate clustering results [35, 36]. Another category consists of validity indices (separation index, hypervolume and average partition density) that make use of not only the fuzzy membership degrees but also the structure of the data [35].

Considering the results obtained in our studies we appreciate that the four validity indices defined below are quite enough to characterize and compare the quality of the partition produced by the used fuzzy clustering methods.

Bezdek designed the partition coefficient (PC) to measure the amount of "overlap" between clusters. He defined the partition coefficient (PC) as follows [37]:

$$PC = \frac{1}{n}\sum_{i=1}^{c}\sum_{j=1}^{n}(A_i(x^j))^2 \tag{14}$$

where $A_i(x^j)$ is the membership of data point j in cluster i.

If there is a choice between several partitions, the maximum partition coefficient yields the partition with the "most unambiguous assignment" [35].

The next validity measure is based on Shannon's information theory and is very similar to the partition coefficient. The partition entropy (PE) about the clustering of data is estimated by the well-known relation:

$$PE = -\frac{1}{n}\sum_{i=1}^{c}\sum_{j=1}^{n}A_i(x^j)\ln A_i(x^j) \tag{15}$$

It is easy to observe that a good partition corresponds to small entropy and vice-versa.

Backer-Jain Index (IBJ) is based also only on membership degrees and can be computed with the following relation

$$IBJ = \frac{1}{n(c-1)}\sum_{k=1}^{c-1}\sum_{i=1}^{c}\sum_{j=1}^{n}\left|A_i(x^j) - A_k(x^j)\right|, \tag{16}$$

with IBJ = 0 for a uniform fuzzy partition and IBJ = 1 for a crisp partition.

The last index used in our studies, namely fuzzy hypervolume (FHV), is simply defined by

$$FHV = \sum_{i=1}^{c}\sqrt{\det(C_i)}, \tag{17}$$

where C_i is the fuzzy covariance matrix of cluster A_i (see below Equation 23).

FUZZY DIVISIVE HIERARCHICAL CLUSTERING

Using the FCM algorithm we may determine a binary fuzzy partition $\{A_1, A_2\}$ of the data set X. If this partition describes real clusters, we denote $P^1 = \{A_1, A_2\}$. Using the GFCM algorithm for two subclusters ($n = 2$) we may determine a binary fuzzy partition for each A_i of P^1. If this partition of A_i describes real clusters, these clusters will be attached to a new fuzzy partition, P^2. Otherwise, A_i will remain undivided. The cluster A_i will be marked and will be allocated to the partition P^2. The unmarked cluster members of P^2 will follow the same procedure. The divisive procedure will stop when all the clusters of the current partition P^l are marked, that is there are no more real clusters. This procedure is a divisive one and affords us a tool to perform a fuzzy hierarchy. A polarization measure of a fuzzy partition is used to appreciate the quality of the partition [26, 28]. We will consider that the binary fuzzy partition $P = \{C_1,C_2)$ of C describes real clusters if the polarization degree R(P) of P is greater than a prescribed threshold t and for every C_i exists an x from X that $C_i(x) > 0.5$. The polarization degree of P may be defined by

$$R(P) = \frac{\sum_{j=1}^{n} \max(C_1(x^j), C_2(x^j))}{\sum_{j=1}^{n} C(x^j)} \tag{18}$$

If the membership degree $C_1(x^j)$ and $C_2(x^j)$ are polarized towards the extreme values 0 and $C(x^j)$ then the R(P) value is high. If the membership degree $C_1(x^j)$ and $C_2(x^j)$ are near equal then R(P) is low. The polarization is associated with the possibility to have a cluster structure in data set.

In conclusion the binary divisive algorithm is an unsupervised procedure and may be used to obtain the cluster structure of the data set when the number of cluster is unknown.

THE GUSTAFSON-KESSEL ALGORITHM

The Gustafson-Kessel Algorithm (GK) generates a fuzzy partition that provides a degree of membership of each data point to a given cluster [38, 39]. To detect clusters of different geometrical shapes in a dataset, this method introduces an adaptive distance norm for each cluster. Each cluster A_i has its own norm-inducing matrix M_i, which affects the distance norm which takes the following expression.

$$d(x^j, L^i)_M = \|x^j - L^i\| = (x^j - L^i)^T M_i (x^j - L^i) \tag{19}$$

Use of the matrix M_i makes it possible for each cluster to adapt the distance norm to the geometrical structure of the data at each iteration. Based on the norm-inducing matrices, the objective of the GK method is obtained by minimizing the function J,

$$J(P,L,M)=\sum_{i=1}^{c}\sum_{j=1}^{n}(A_i(x^j))^2 d^2(x^j,L^i)_M \quad , \tag{20}$$

where $M = (M_1, M_2, ..., M_c)$ is a c-tuple of the norm-inducing matrices.

To obtain a feasible solution by minimizing Equation (20), the additional constraint is required for M_i.

$$\det(M_i) = \rho_i, \rho_i > 0, 1 \leq i \leq c \tag{21}$$

where P_i is a cluster volume for each cluster. Equation (21) guarantees that M_i is positive-definite, indicating that M_i can be varied to find the optimal shape of cluster with its volume fixed. Using the iterative method, minimization of Equation (20) with respect to M_i produces the following equation:

$$M_i = [\rho_i \det(C_i)]^{1/p} C_i^{-1}, \tag{22}$$

where

$$C_i=\frac{\sum_{j=1}^{n}\left[A_i(x^j)\right]^2(x_j-L_i)(x_j-L_i)^T}{\sum_{j=1}^{n}\left[A_i(x^j)\right]^2} \tag{23}$$

is the fuzzy covariance matrix of cluster A_i. The set of fuzzy covariance matrices is represented as a c-tuple of $C = (C_1, C_2, ..., C_c)$. There is no general agreement on what value to use for P_i; without any prior knowledge, a rule of thumb that many investigators use is $P_i = 1$ in practice [40].

ASSOCIATIVE SIMULTANEOUS FUZZY C-MEANS ALGORITHM

Let $X = \{x^1, ..., x^n\} \subset \mathbf{R}^d$ be the set of objects to be classified. A characteristic may be specified by its values corresponded to the n objects.

So, we may say that $Y = \{y^1, ..., y^d\} \subset \mathbf{R}^n$ is the set of characteristics. y_j^k is the value of the characteristic k with respect to the object j, so we may write $y_j^k = x_k^j$.

Let P be a fuzzy partition of the fuzzy set C of objects and Q be a fuzzy partition of the fuzzy set D of characteristics. The problem of the cross-classification (or simultaneous classification) is to determine the pair (P, Q) which optimizes a certain criterion function.

By starting with an initial partition P^0 of C and an initial partition Q^0 of D, we will obtain a new partition P^1.

The pair (P^1, Q^0) allows us to determine a new partition Q^1 of the characteristics. The algorithm consists in producing a sequence (P^k, Q^k) of pairs of partitions, starting from the initial pair (P^0, Q^0), in the following steps:

$$(P^k, Q^k) \rightarrow (P^{k+1}, Q^k)$$
$$(P^{k+1}, Q^k) \rightarrow (P^{k+1}, Q^{k+1})$$

The rationale of the hierarchical cross-classification method essentially supposes the splitting of the sets X and Y in two subclasses. The obtained classes are alternatively divided in two subclasses, and so on.

The two hierarchies may be represented by the same tree, having in each node a pair (C, D), where C is a fuzzy set of objects and D is a fuzzy set of characteristics.

As a first step we propose ourselves to simultaneously determine the fuzzy partitions (as a particular case, the binary fuzzy partitions) of the classes C and D, so that the two partitions should be highly correlated. With the GFCM algorithm, we will determine a fuzzy partition $P = \{A_1, ..., A_n\}$ of the class C, using the original characteristics.

In order to classify the characteristics, we will compute their values for the classes A_i, $i = 1, ..., c$. The value $\bar{y}_i^k$ of the characteristic k with respect to the class A_i is defined as:

$$\bar{y}_i^k = \sum_{j=1}^{n} A_i(x^j)x_k^j, i = 1,\ldots,c; k = 1,\ldots,d. \tag{24}$$

Thus, from the original dn-dimensional characteristics we computed d new c-dimensional characteristics which are conditioned by the classes A_i, $i = 1, ..., c$. We may admit that these new characteristics do not describe objects, but they characterize the classes A_i.

Let us consider now the set $\overline{Y} = \{\bar{y}^1,\ldots\bar{y}^d\}$ of the modified characteristics. We define the fuzzy set $\overline{D}$ on $\overline{Y}$ given by

$$\overline{D}(\bar{y}^k) = D(y^k), k = 1,\ldots,d. \tag{25}$$

The way the set $\overline{Y}$ has been obtained lets us to conclude that if we will obtain an optimal partition of the fuzzy set D, this partition will be highly correlated to the optimal partition of the class C. With the GFCM algorithm we will determine a fuzzy partition $Q = \{B_1, ..., B_c\}$ of the class D, by using the characteristics given by the relation (24). We may now characterize the objects in X with respect to the classes of properties B_i, $i = 1, ..., c$. The value $\bar{x}_i^j$ of the object j with respect to the class B_i is defined as:

$$\overline{x}_i^j = \sum_{k=1}^{d} B_i(\overline{y}^k) x_k^j, i = 1, \ldots, c; j = 1, \ldots, n. \tag{26}$$

Thus, from the original *nd*-dimensional objects we have computed *n* new c-dimensional objects, which correspond to the classes of characteristics B_i, $i = 1, ..., c$.

Let us consider now the set $\overline{X} = \{\overline{x}^1, \ldots \overline{x}^n\}$ of the modified characteristics. We define the fuzzy set $\overline{C}$ on $\overline{X}$ given by

$$\overline{C}(\overline{x}^j) = C(x^j), j = 1, \ldots, n. \tag{27}$$

With the GFCM algorithm we will determine a fuzzy partition $P' = \{A'_1, ..., A'_c\}$, of the class C by using the objects given by the relation (26). The process continues until two successive partitions of objects (or characteristics) are closed enough to each other.

Considering $P = \{A_1, ..., A_c\}$ is the fuzzy c-partition of X and $Q = \{B_1, ..., B_c\}$ is the fuzzy c-partition of Y produced after this step of our algorithm. Let us remark that we made no explicit association of a fuzzy set A_i on X with a fuzzy set B_j on Y, i.e., what is the fuzzy set B_j that best describes the essential characteristics corresponding to the fuzzy set A_i. It only supposes that A_i is to be associated with B_i, and this is not always true.

Let us denote by S_c the set of all permutations on {1, ..., c}. We wish to build that permutation s $\in S_c$ which best associates the fuzzy set A_i with the fuzzy set $B_{s(i)}$, for every i = 1, ..., c. Our aim is to build some function J: $S_c \rightarrow R$ so that the optimal permutation s is that which maximizes this function. Let us consider the matrix $Z \in R^{c,c}$ given by

$$z_{kl} = \sum_{j=1}^{n} \sum_{i=1}^{d} A_k(x^j) B_l(y^i) x_i^j, k, l = 1, \ldots, c. \tag{28}$$

Let us remark the similarity between the way we compute the matrix Z in (28) and the way we computed the new objects and characteristics in relations (26) and (27).

The experience enables us to consider the function J as given by

$$\widetilde{J} = \prod_{i=1}^{c} z_{i,\sigma(i)}. \tag{29}$$

Thus, supposing that the permutation s maximizes the function J defined above, we will be able to associate the fuzzy set A_i with the fuzzy set $B_{s(i)}$, $i = 1, ..., c$. As we will see in the comparative study below, this association is more natural than the association of A_i with B_i, i = 1, ..., c. Based on these considerations we are able to introduce the following algorithm, that we have called the *Associative Simultaneous Fuzzy c-Means Algorithm* (ASF) [41, 42]:

- *S1*: Set $l = 0$. With the GFCM algorithm we determine a fuzzy c-partition $P^{(l)}$ of the class C by using the initial objects.

- *S2*: With the GFCM algorithm we determine a fuzzy c-partition $Q^{(l)}$ of the class D by using the characteristics defined in (26).
- *S3*: With the GFCM algorithm we determine a fuzzy c-partition $P^{(l+1)}$ of the class C by using the objects defined in (26).
- *S4*: If the partitions $P^{(l)}$ and $P^{(l+1)}$ are close enough, that is if

$$\|P^{(l+1)} - P^{(l)}\| < \varepsilon,$$

where ε is a preset value, then go to S5, else increase l by 1 and go to S2.

- *S5*: Compute the permutation s that maximizes the function J given in relation (29).
- *S6:* Assign the fuzzy sets B_i, so that $B_{s(i)}$ becomes B_i, i = 1, ..., c.

Let us remark now that, after the steps S5 and S6, we are able to associate the fuzzy set A_i with the fuzzy set B_i, i = 1, ..., c.

Fuzzy Modeling of Environmental Data

Different fuzzy clustering algorithms mentioned above were successfully applied to the study of several German and Romanian natural mineral waters using data obtained from samples collected from different sampling sites covering, for example, a large percentage of the Romanian natural mineral waters. The characteristic clustering technique produces fuzzy partitions of the mineral water properties involved and thus is a useful tool for studying (dis)similarities between different ions (i.e., speciation).

The cross-classification algorithm provided not only a fuzzy partition of the mineral waters analysed, but also a fuzzy partition of the characteristics considered. In this way it is possible to identify which ions and other physico-chemical features are responsible for the similarities or differences observed between different groups of mineral waters [43].

Various clustering procedures, - hierarchical and non-hierarchical crisp and fuzzy clustering, fuzzy cross-clustering and PCA (varimax rotation) - combined with receptor modeling were applied to a data set consisting of wet deposition loads of major ions from five sampling sites in Central Austria collected in a period of 10 years (1984-1993).

The data classification with respect to the sampling sites as objects has shown that a distinct separation between the northern Alpine rim and in the inner Alpine region is achieved. This indicates the role of geographical disposition of the sampling sites [44].

In a similar way fuzzy cluster analysis and fuzzy recognition were successfully used to model a large set of data from the Min River (China) [45].

The fuzzy analysis was employed as graphical tool to compare ecosystems with respect to their pollution [46], to identify individual aerosol particles containing Cr, Pb, and Zn above the North Sea [47] and at a rural site in Central Europe during winter [48], characterisation of particulate matter from the Kara Sea using electron probe X-ray micro analysis [49], correlation between element concentrations in basal till and the uppermost layers of mineral soil at forest sites in Finland [50] or heavy metals in urban soil [51], qualitative identification of trace elements in clinker [52], for the prediction of petrophysical rock parameters [53] and to process on-line data streaming from five monitoring stations in Chile for forecasting SO_2 levels [54] or data processing in on-line laser mass spectrometry of inorganic, organic or biological airborne particles [55] and, finally, also for evaluation of PCDD/F emissions during solid waste combustion [56].

CHARACTERIZATION AND CLUSTERING OF SOIL DATA

The soil data collection was performed in the northern part of Romanian Carpathians Mountains: the western part of Bistrita Mountains (b; 1,..., 36), the north-western part of Ignis-Oas Mountains (i; 37,...,86) and Maramures (m; 87,...,125) area, according to the standardized methods for sampling, sample preparation and chemical analysis of soils.

Thirteen different soil metal ions were checked (lead, copper, manganese, zinc, nickel, cobalt, chromium, cadmium, calcium, magnesium, potassium, iron and aluminum). The standardized flame absorption spectrophotometric method for the following chemical elements Fe, Al, Ca, Mg, K, Cu, Pb, Mn, Zn and spectrophotometry with atomization in the graphite owen for Cd and Cr were used. The statistical results (central location and spread of data) obtained from the initial data set (125 samples and 13 ion concentrations) are presented Table 1. The hard partition corresponding to the fuzzy successive partition of the 125 soil samples produced by using fuzzy divisive hierarchical cross-clustering algorithm (autoscaled data) are presented in Figure 1.

Table 1. Table of statistics concerning the concentration (mg/kg) of samples

Metal	Mean	Median	Minimum	Maximum	S. D.	Skewness	Kurtosis
Pb	68.15	41.01	12.38	830.2	104.31	4.89	28.61
Cu	10.07	8.88	1.22	29.1	5.41	1.19	1.38
Mn	472	299	13.23	1506	386	0.74	-0.56
Zn	69.03	63.78	1.08	252.1	43.68	1.10	2.54
Ni	24.38	17.08	1.32	118.5	24.04	2.43	5.90
Co	10.10	8.99	0.30	39.9	8.43	1.47	2.19
Cr	14.28	10.97	1.89	62.3	11.32	2.57	7.47
Cd	0.36	0.10	0.10	2.4	0.42	2.22	5.76
Ca	929	262	34.71	6040	1396	2.17	4.26
Mg	5094	2703	100	29447	6546	2.58	6.20
K	1286	906	259	9860	1469	3.85	15.81
Fe	39836	30282	100	533359	49092	8.33	83.47
Al	30022	24926	1461	87807	19694	1.14	1.09

The hard partition is obtained by defuzzification of fuzzy partition. The samples, in our case, are assigned to the cluster with the highest membership degree. In Figure 2 are depicted the results obtained by applying horizontal fuzzy clustering with a predefined number of three clusters according to the location of samples.

Comparing the partitions in Figure 1 and 2 and the membership degrees (MDs) to the final fuzzy partition obtained by both fuzzy procedures one can observe that the results are unsatisfactory i.e., the partitions obtained are not in a good agreement with the nature of samples. However, the samples from Ignis-Oas area better separated by fuzzy hierarchical cross-clustering (cluster A_{11}) and the samples from Maramures and Bistrita area form two clusters and seem to be the most heterogeneous group.

By applying the fuzzy Gustafson-Kessel algorithms different and interesting results were obtained. The results are shown in Figure 3 and 4 and illustrate a better separation of samples especially in the case of fuzzy hierarchical clustering, in a very good agreement to the location of sampling scenario.

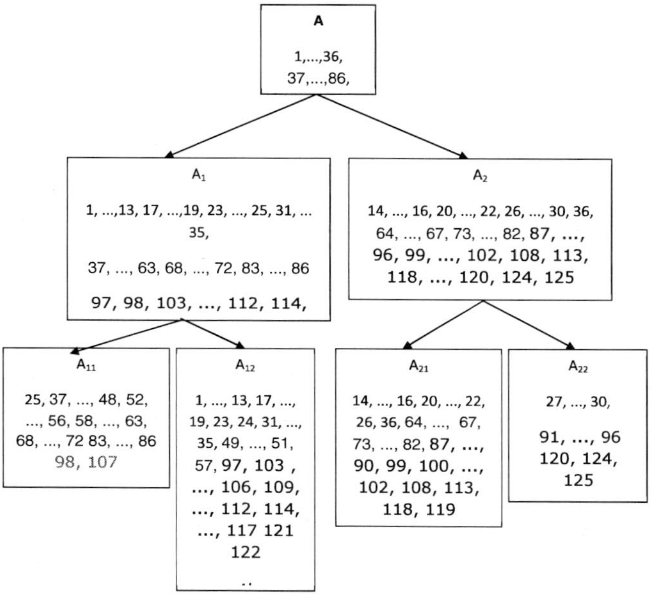

Figure 1. The hard partition tree corresponding to the fuzzy hierarchical cross-clustering of the 125 soil samples (Hierarchical GFCM).

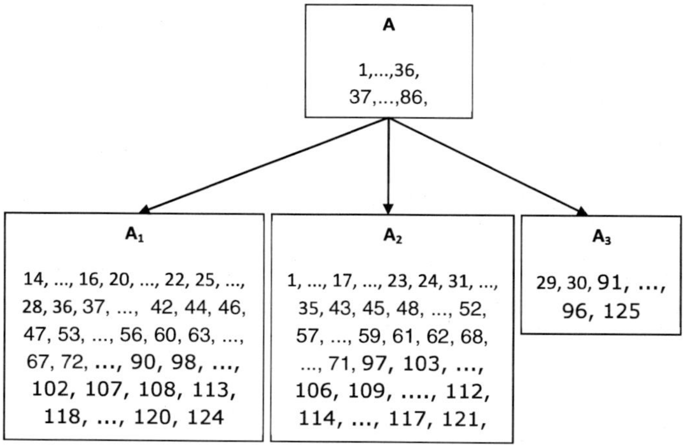

Figure 2. The hard partition tree corresponding to the fuzzy horizontal clustering of the 125 soil samples (FCM).

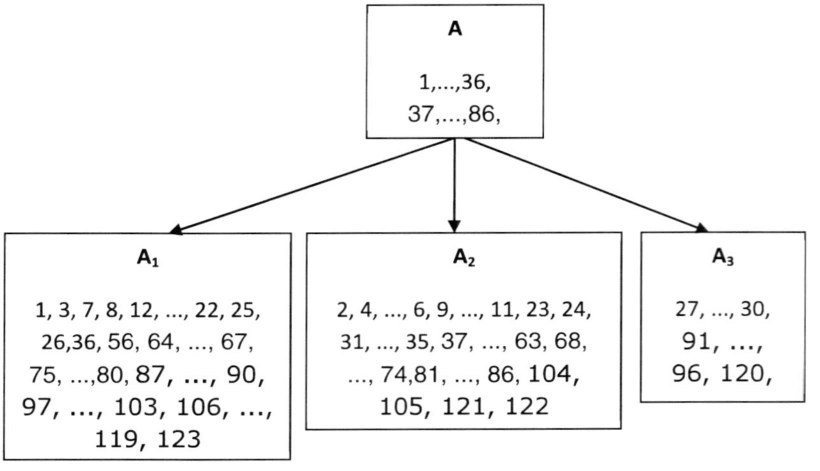

Figure 3. The hard partition tree corresponding to the fuzzy horizontal clustering of the 125 soil samples (GK).

Table 2. The cluster validity indices values

Cluster Validity Index	Horizontal (three defined clusters)		Hierarchical	
	GFCM	GK	GFCM	GK
PC	0.530	0.561	0.371	0.372
PE	0.768	0.730	1.15	1.29
IBJ	0.570	0.595	0.444	0.588
FHV	$8.8x10^{25}$	$4.6x10^{24}$	$2.27x10^{27}$	$1.47x10^{24}$

This statement is well supported by a two dimensional and a three dimensional representations of the membership degrees and by the values of cluster validity indices depicted in Table 2; higher values for partition coefficient and Backer-Jain Index, and lower values for partition entropy and fuzzy hypervolume in the case of the two both GK procedures (horizontal and hierarchical) by comparing to GFCM counterparts.

In addition, the coordinates of the prototypes (fuzzy means of each ion concentration and each partition) permit the identification of the most characteristic concentrations of chemical elements for each fuzzy partition, namely the sampling location.

By comparison of results presented in Table 3 it is easy to observe, for example, that the concentration of lead and cadmium is highest for the samples originated from Ignis-Oas area, (the closest to the metallurgical plant) and lowest concentrations characterize the soil samples originated from Maramures and Bistrita areas (more or less close to the point source).

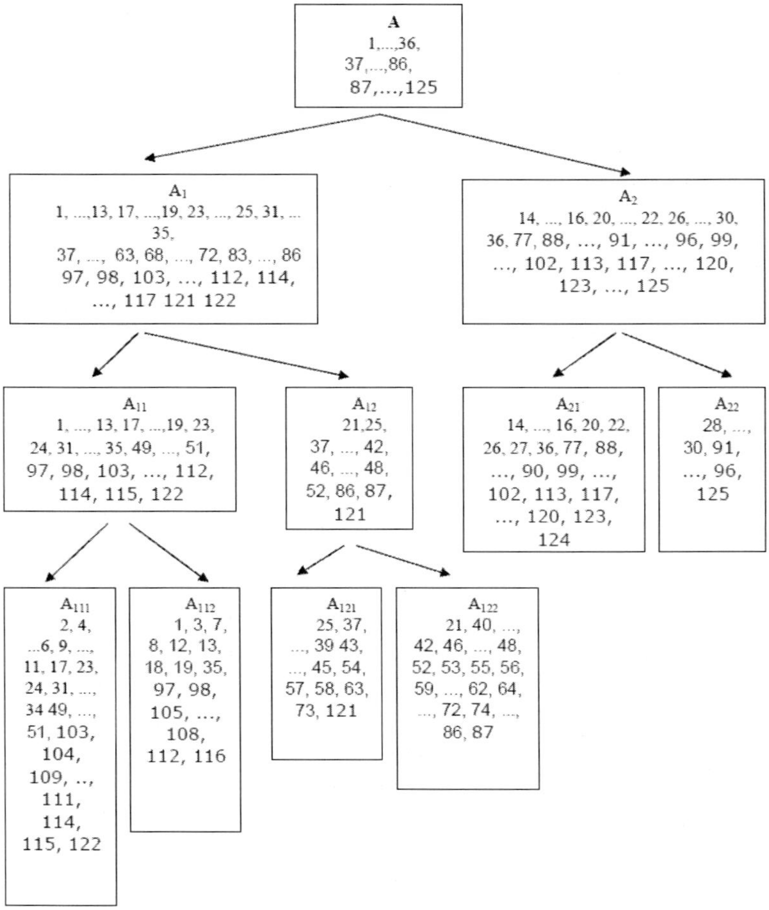

Figure 4. The hard partition corresponding to the fuzzy hierarchical clustering of the 125 soil samples (GK).

FUZZY CLUSTERING OF THE GRASS AND SOIL SAMPLES

In this example we will analyze a set of 35 grass and soil samples discussed in [57] and depicted in Table 4. In order to apply the fuzzy algorithms presented above to classify the grass and soil samples, we used the concentration of 37 chemical elements (Na, Mg, Al, K, Sc, Ca, Cr, Ti, V, Mn, Ni, Fe, Co, Zn, Se, As, Br, Sr, Rb, Mo, Ag, Sb, I, Ba, Cs, La, Ce, Eu, Sm, Tb, Dy, Yb, Hf, Ta, W, Th and U) determined by conventional and epithermal neutron activation analysis at IBR-2 pulsed fast reactor FLNP JINR (Dubna, Russia) [58].

The hard partition corresponding to the fuzzy successive partition of the 35 grass and soil samples produced by using the autoscaled data are presented in Table 5.

Comparing the partitions in Table 5 and the membership degrees (MDs) to the final fuzzy partition obtained using the hierarchical and horizontal fuzzy clustering algorithms (Table 6), we can observe that the results are in a good agreement with the nature and the protocol of samples in Table 4.

Table 3. Fuzzy mean of metal ions concentration corresponding to each fuzzy partition

Fuzzy partition	Pb	Cu	Mn	Zn	Ni	Co	Cr	Cd	Ca	Mg	K	Fe	Al
A_{111}	83.05	5.73	124.06	28.21	8.17	2.74	8.12	0.41	472	2026	1054	22862	14510
A_{112}	46.13	10.07	269.13	41.26	16.02	4.61	12.64	0.19	316	2835	1146	120977	19046
A_{121}	245.6	9.44	361.73	86.92	15.99	7.43	10.06	0.95	672	2500	742	30340	36433
A_{122}	44.87	7.97	606.08	72.04	16.95	10.78	9.41	0.35	822	2566	697	35044	47481
A_{21}	40.24	14.98	590.57	85.03	33.96	12.96	18.71	0.19	770	6830	1267	45406	26112
A_{22}	28.69	16.20	900.59	134.35	98.45	32.37	46.58	0.16	3031	25854	5967	61290	39473

Table 4. Legend to the tables and figures - sample number and description (distance from the point source)

Nr. crit.	Location[1]	Nr. crit.	Location[1]	Nr. crit.	Location[1]	Nr. crit.	Location[1]	Nr. crit.	Location[1]
1	200mP	8	500mS	15	1300mE	22	200mP	29	300mE
2	200mP	9	800mS	16	grass blank[2]	23	200mP	30	400mE
3	200mP	10	1300mS	17	0mP	24	300mS	31	500mE
4	200mP	11	300mE	18	0mP	25	400mS	32	800mE
5	200mP	12	400mE	19	200mP	26	500mS	33	1300mE
6	300mS	13	500mE	20	200mP	27	800mS	34	soil blank[2]
7	400mS	14	800mE	21	200mP	28	1300mS	35	industrial mean

[1]P - plant; S - Southern direction; E - Eastern direction; industrial mean- mean of all industrial samples.
[2]blank values provided from grass and soil samples collected in Ustka clean region, Poland.

Table 5. The hard partition corresponding to the fuzzy hierarchical and horizontal partition of the 35 grass and soil samples

Final hierarchical fuzzy partition	Grass and soil samples	Fuzzy horizontal clustering with four predefined classes	Grass and soil samples
A_{11}	20 21 26 27 31 32 34	A_1	17 18 19 29 30 33
A_{1211}	1 2 5 7	A_2	20 21 26 27 28 31 32
A_{1212}	3 4 6 13 14 15	A_3	1 2 3 4 5 6 7 8 9 10 11 12 13 14 15 16
A_{22}	8 9 10 11 12 16	A_4	22 23 24 25 35
A_{21}	17 18 19 28 29 30 33		
A_{22}	22 23 24 25 35		

Thus, for example, it appears clear that the majority of the grass samples form a well defined group. Also, it is interesting to observe that grass blank is very similar with grass samples having a membership degree higher than 0.5 (0.612) to the well-defined class of grass samples obtained by fuzzy horizontal clustering with four predefined classes.

This finding supports the idea that the concentrations of the considered chemical elements are very similar in both grass samples and "grass blank". Much more, the membership degrees in Table 6 illustrates a large dissimilarity between the grass samples and soil samples and also between the soil samples.

In addition we have to observe that "soil blank" sample belong to the grass samples class (MD = 0.416) in the case of fuzzy horizontal clustering, and with a lower membership degree (0.351) to soil class in the case of fuzzy hierarchical clustering.

In order to develop a classification of characteristics (element concentrations) we applied the FHiCC procedure to the 37 chemical elements concentrations obtained for grass and soil samples. The characteristics clustering with the 37 elements concentrations and the 35 grass and soil samples, with autoscaling of data, gave the final partition presented in Table 7. The first chemical elements separated from the others are Al and Ca, followed by K and Fe, and then by Na, Mg, Zn and Ba.

The cluster containing the other elements (Sc, Cr, Ti, V, Mn, Ni, Co, Se, As, Br, Sr, Rb, Mo, Ag, Sb, I, Cs, La, Ce, Eu, Sm, Tb, Dy, Yb, Hf, Ta, W, Th and U) is not subjected to any more splitting (their membership degrees, MD, to this cluster are all near 1).

The horizontal characteristics clustering also with autoscaling of data, (see also Table 7) illustrates the same aspect i.e., the high similarity of the last chemical elements and a large dissimilarity among the first eight chemical elements (Al, Ca, K, fe, Na, Mg, Zn, and Ba). It is interesting to remark that all the divisions in Table 7 are clear-cut, the membership degrees to the different final partition are in the large majority of cases near 1 or 0.

Table 6. The results of the hierarchical and horizontal (four predefined classes) fuzzy clustering of the grass and soil samples

Sample	Final hierarchical fuzzy partition						Fuzzy horizontal clustering with four predefined classes			
	A_{11}	A_{1211}	A_{1212}	A_{122}	A_{21}	A_{22}	A_1	A_2	A_3	A_4
1	0.058	**0.636**	0.128	0.149	0.019	0.008	0.025	0.050	**0.913**	0.011
2	0.043	**0.643**	0.167	0.118	0.019	0.008	0.019	0.037	**0.934**	0.008
3	0.031	0.421	**0.425**	0.090	0.021	0.010	0.014	0.027	**0.951**	0.006
4	0.016	0.327	**0.550**	0.082	0.016	0.007	0.007	0.014	**0.975**	0.003
5	0.073	**0.353**	0.265	0.265	0.029	0.014	0.034	0.065	**0.884**	0.016
6	0.028	0.209	**0.596**	0.132	0.023	0.011	0.013	0.024	**0.955**	0.006
7	0.078	**0.381**	0.230	0.296	0.010	0.004	0.028	0.064	**0.895**	0.011
8	0.071	0.229	0.236	**0.426**	0.024	0.011	0.031	0.063	**0.889**	0.014
9	0.084	0.055	0.076	**0.746**	0.025	0.012	0.035	0.075	**0.872**	0.016
10	0.054	0.073	0.125	**0.712**	0.023	0.010	0.023	0.049	**0.915**	0.010
11	0.064	0.120	0.230	**0.537**	0.032	0.015	0.030	0.058	**0.896**	0.014
12	0.291	0.101	0.105	**0.381**	0.082	0.038	0.125	0.252	**0.563**	0.057
13	0.024	0.129	**0.436**	0.384	0.020	0.009	0.010	0.022	**0.961**	0.005
14	0.045	0.144	**0.377**	0.388	0.029	0.014	0.021	0.041	**0.926**	0.010
15	0.032	0.139	**0.529**	0.261	0.025	0.012	0.015	0.029	**0.947**	0.007

Table 6. (Continued)

Sample	Final hierarchical fuzzy partition						Fuzzy horizontal clustering with four predefined classes			
	A_{11}	A_{1211}	A_{1212}	A_{122}	A_{21}	A_{22}	A_1	A_2	A_3	A_4
16	0.256	0.109	0.115	**0.413**	0.071	0.033	0.111	0.222	**0.612**	0.052
17	0.244	0.032	0.028	0.058	**0.426**	0.210	**0.400**	0.286	0.138	0.174
18	0.274	0.049	0.044	0.088	**0.343**	0.200	**0.343**	0.280	0.191	0.184
19	0.217	0.022	0.020	0.042	**0.506**	0.191	**0.444**	0.285	0.105	0.164
20	**0.606**	0.019	0.017	0.039	0.247	0.068	0.114	**0.763**	0.089	0.032
21	**0.494**	0.039	0.037	0.087	0.247	0.094	0.196	**0.545**	0.181	0.075
22	0.119	0.019	0.017	0.035	0.202	**0.606**	0.195	0.113	0.065	**0.625**
23	0.123	0.019	0.017	0.035	0.222	**0.581**	0.219	0.130	0.074	**0.576**
24	0.066	0.008	0.007	0.015	0.123	**0.778**	0.117	0.054	0.024	**0.804**
25	0.159	0.024	0.022	0.044	0.339	**0.410**	0.317	0.210	0.115	**0.356**
26	**0.446**	0.035	0.030	0.066	0.326	0.095	0.273	**0.498**	0.147	0.080
27	**0.528**	0.035	0.031	0.070	0.250	0.083	0.183	**0.597**	0.157	0.061
28	0.292	0.047	0.043	0.089	**0.332**	0.194	0.312	**0.317**	0.195	0.175
29	0.199	0.031	0.029	0.060	**0.354**	0.325	**0.324**	0.246	0.142	0.285
30	0.188	0.014	0.012	0.025	**0.631**	0.128	**0.542**	0.275	0.069	0.113
31	**0.578**	0.048	0.043	0.100	0.170	0.058	0.154	**0.578**	0.213	0.053
32	**0.348**	0.040	0.038	0.080	0.335	0.157	0.286	**0.400**	0.178	0.133
33	0.170	0.022	0.021	0.043	**0.379**	0.363	0.340	0.236	0.114	0.308
34	**0.351**	0.109	0.106	0.215	0.146	0.070	0.187	0.308	**0.416**	0.087
35	0.217	0.048	0.046	0.092	0.262	**0.332**	0.259	0.217	0.188	**0.333**

Table 7. The results of the hierarchical and horizontal (three predefined classes) fuzzy clustering of characteristics (element concentrations)

E	Hierarchical clustering of characteristics (element concentrations)									Horizontal clustering		
	A_{11}	A_{12}	A_{21}	A_{2211}	A_{2212}	A_{22211}	A_{222112}	A_{22212}	A_{2222}	A_1	A_2	A_3
Na	0.032	0.013	0.186	**0.713**	0.000	0.007	0.007	0.015	0.025	0.111	0.018	**0.871**
Mg	0.059	0.022	0.310	0.000	**0.542**	0.009	0.008	0.017	0.030	*0.196*	0.027	**0.776**
Al	**0.404**	0.117	*0.403*	0.022	0.025	0.004	0.003	0.007	0.013	**0.828**	0.044	0.128
K	0.001	**0.746**	0.140	0.029	0.031	0.007	0.006	0.013	0.025	0.002	**0.996**	0.002
Sc	0.000	0.000	0.000	0.000	0.000	0.000	0.001	0.000	**0.996**	0.000	0.000	**0.999**
Ca	**0.749**	0.042	0.132	0.020	0.023	0.004	0.004	0.008	0.017	**0.560**	*0.203*	*0.237*
Cr	0.000	0.000	0.000	0.000	0.000	0.001	0.003	0.001	**0.993**	0.000	0.000	**0.999**
Ti	0.001	0.000	0.005	0.023	0.014	0.000	0.000	**0.606**	0.348	0.003	0.001	**0.996**
V	0.000	0.000	0.000	0.000	0.000	0.001	0.004	0.007	**0.992**	0.000	0.000	**0.999**
Mn	0.000	0.000	0.000	0.003	0.002	0.091	*0.406*	0.049	**0.446**	0.000	0.000	**0.999**
Ni	0.000	0.000	0.000	0.000	0.000	0.000	0.001	0.000	**0.998**	0.000	0.000	**0.999**
Fe	*0.317*	0.079	**0.558**	0.013	0.016	0.002	0.002	0.004	0.008	**0.805**	0.032	*0.162*
Co	0.000	0.000	0.000	0.000	0.000	0.000	0.001	0.000	**0.997**	0.000	0.000	**0.999**
Zn	0.000	0.000	0.002	0.011	0.007	**0.666**	0.000	0.044	*0.269*	0.001	0.000	**0.999**
Se	0.000	0.000	0.000	0.000	0.000	0.000	0.001	0.000	**0.997**	0.000	0.000	**0.999**
As	0.000	0.000	0.000	0.000	0.000	0.001	0.006	0.001	**0.990**	0.000	0.000	**0.999**

E	Hierarchical clustering of characteristics (element concentrations)									Horizontal clustering		
	A_{11}	A_{12}	A_{21}	A_{2211}	A_{2212}	A_{22211}	A_{222112}	A_{22212}	A_{2222}	A_1	A_2	A_3
Br	0.000	0.000	0.000	0.000	0.000	0.000	0.001	0.000	**0.997**	0.000	0.000	**0.999**
Sr	0.000	0.000	0.000	0.000	0.000	0.019	0.095	0.017	**0.866**	0.000	0.000	**0.999**
Rb	0.000	0.000	0.000	0.000	0.000	0.002	0.007	0.002	**0.989**	0.000	0.000	**0.999**
Mo	0.000	0.000	0.000	0.000	0.000	0.000	0.001	0.000	**0.997**	0.000	0.000	**0.999**
Ag	0.000	0.000	0.000	0.000	0.000	0.000	0.001	0.000	**0.997**	0.000	0.000	**0.999**
Sb	0.000	0.000	0.000	0.000	0.000	0.000	0.001	0.000	**0.997**	0.000	0.000	**0.999**
I	0.000	0.000	0.000	0.000	0.000	0.000	0.001	0.000	**0.997**	0.000	0.000	**0.999**
Ba	0.000	0.000	0.000	0.003	0.002	0.043	**0.536**	0.052	*0.363*	0.000	0.000	**0.999**
Cs	0.000	0.000	0.000	0.000	0.000	0.000	0.001	0.000	**0.997**	0.000	0.000	**0.999**
La	0.000	0.000	0.000	0.000	0.000	0.000	0.000	0.001	**0.998**	0.000	0.000	**0.999**
Ce	0.000	0.000	0.000	0.000	0.000	0.000	0.001	0.000	**0.998**	0.000	0.000	**0.999**
Eu	0.000	0.000	0.000	0.000	0.000	0.000	0.001	0.000	**0.996**	0.000	0.000	**0.999**
Sm	0.000	0.000	0.000	0.000	0.000	0.000	0.001	0.000	**0.997**	0.000	0.000	**0.999**
Tb	0.000	0.000	0.000	0.000	0.000	0.000	0.001	0.000	**0.997**	0.000	0.000	**0.999**
Dy	0.000	0.000	0.000	0.000	0.000	0.000	0.001	0.000	**0.997**	0.000	0.000	**0.999**
Yb	0.000	0.000	0.000	0.000	0.000	0.000	0.001	0.000	**0.996**	0.000	0.000	**0.999**
Hf	0.000	0.000	0.000	0.000	0.000	0.000	0.001	0.000	**0.997**	0.000	0.000	**0.999**
Ta	0.000	0.000	0.000	0.000	0.000	0.000	0.001	0.000	**0.996**	0.000	0.000	**0.999**
W	0.000	0.000	0.000	0.000	0.000	0.000	0.001	0.000	**0.997**	0.000	0.000	**0.999**
Th	0.000	0.000	0.000	0.000	0.000	0.000	0.001	0.000	**0.997**	0.000	0.000	**0.999**
U	0.000	0.000	0.000	0.000	0.000	0.000	0.001	0.000	**0.997**	0.000	0.000	**0.999**

Table 8. The cross-clustering of the grass and soil samples (a) and chemical elements determined (b) produced with autoscaled data

Fuzzy class	Grass and soil sample (a) and elements associated (b)
A_{11}	(a) 20 21 26 27 31 32 (b) Mg, Sc, Cr, V, Mn, Ni, Fe, Sr, Sb, Cs, La, Ce, Sm, Tb, Dy, Yb, Ta, Th
A_{12}	(a) 1 2 3 4 5 6 7 8 9 10 11 12 13 14 15 16 34 (b) Na, Al, Ba
A_{21}	(a) 17 18 19 28 29 30 33 (b) Ca, Ti, Co, Zn, Se, As, Rb, Mo, I, Eu, Hf, W, U
A_{22}	22 23 24 25 35 (b) K, Br, Ag

The clustering hierarchy produced by applying cross-clustering algorithm for objects (35 grass and soil samples (a)) and for characteristics (37 elements concentrations (b)) using again autoscaled data is presented in Table 8. The partitioning of the grass and soil samples in different classes is very similar with those obtained above by the usual fuzzy clustering algorithm. What is different is the partitioning of the chemical elements and their association with different grass and soil samples. The majority of chemical elements are associated to the class A_{11} which includes only a part of soil samples namely 20, 21, 26, 27, 31 and 32. The fuzzy class A_{12} contains all the grass samples including also the grass and soil blank samples (16 and 34) and the chemical elements associated are Na, Al and barium. This is a very interesting finding because it appears clearly that grass is mainly much lower contaminated by the phosphatic fertilizer plant. The third class A_{21} including the other soil samples (17, 18, 19, 28, 29, 30 and 33) have similar concentrations for the following chemical elements: Ca, Ti, Co, Zn, Se, As, Rb, Mo, I, Eu, Hf, W and uranium. The soil samples 22, 23, 24, 25 and 35 from class A_{22} appear to have similar concentrations for K, Br and silver. Again, a good agreement with the nature and the protocol of the samples can be observed.

CONCLUSION

Fuzzy sets and fuzzy logic represent a useful and powerful tool in chemistry and related fields. Cluster analysis is a large field, both within fuzzy sets and beyond it. Many algorithms have been developed to obtain hard clusters from a given data set. Among those, the c-means algorithms are probably the most widely used. Hard c-means execute a sharp classification, in which each object is either assigned to a class or not. The membership to a class of objects therefore amounts to either 1 or 0. The application of Fuzzy sets in a classification function causes this class membership to become a relative one and consequently an object can belong to several classes at the same time but with different degrees. The c-means algorithms are prototype-based procedures, which minimize the total of the distances between the prototypes and the objects by the construction of a target function. Both methods, sharp and fuzzy clustering, determine class centers and minimize, e.g., the sum of squared distances between these centers and the objects, which are characterized by their features. Fuzzy Generalized c-Means and Fuzzy Gustafson-Kessel algorithms are easy and well improved tools, which have been applied more or less in fields of chemistry and other fields. The applications section of this chapter demonstrates advantages of fuzzy clustering with respect to clustering and characterization of different soil samples. Although several applications of fuzzy clustering and classification have already been reported in the field of chemistry and other fields, many potential chemometric relevant applications are still waiting to be exploited.

REFERENCES

[1] Massart, D. L. and Kaufman, L., *The Interpretation of Analytical Chemical Data by the Use of Cluster Analysis,* Wiley, New York 1983.

[2] Auf der Heyde, T. P. E., *J. Chem. Ed.*, 1990, 67, 461-469.

[3] Shenkin, P. S. and McDonald, D. Q., *J. Comput. Chem.*, 1994, 15, 899-916.

[4] Gordon, A. D., *Classification, Champan and Hall*, London 1981.

[5] Manly, B. F. J., *1986, Multivariate Statistical Methods,* Champan and Hall, London 1986.

[6] Massart, D. L., Vandeginste B. G. M., Deming, S. N., Michotte, Y. and Kaufman, L., *Chemometrics: a Textbook*, Elsevier, Amsterdam 1988.

[7] Brereton, R. G., *Chemometrics: Applications of Mathematics and Statistics to the Laboratory,* Ellis Horwood, Chichester 1990.

[8] Meloun, M., Mlitky, J. and Forina, M., *Chemometrics for Analytical Chemistry*, vol. I: PC- aided statistical data analysis, Ellis Horwood, Chichester 1992.

[9] Einax, J., Zwanziger, H. and Geiß, S., *Chemometrics in Environmental Analysis*, John Wiley Sons Ltd, Chichester 1997.
[10] Otto, M., *Chemometrics*, Wiley-VCH, Weinheim 1998.
[11] Zadeh, L. A., *Inf. Control*, 1965, 8, 338-353.
[12] Zadeh, L. A. *Fuzzy Sets and Their Applications to Pattern Classification and Clustering Analysis in Classification and Clustering*, J. van Ryzin (Ed.), Academic Press, New York, 1997, p. 251-299.
[13] Kandel, A., *Fuzzy Techniques in Pattern Recognition*, Wiley, New York 1982.
[14] Bezdek, J. C., *Pattern Recognition with Fuzzy Objective Function Algorithms*, Plenum Press, New York 1987.
[15] Bezdek, J. C., Ehrlich, R. and Full, W., *Comput. Geosci*. 1984, 10, 191-203.
[16] Lowen, R., *Fuzzy Set Theory*, Kluwer Academic Publishers 1995.
[17] Zimmermann, H. J., *Fuzzy Set Theory-and Its Applications*, Kluwer Academic Publishers 1996.
[18] Tanaka, K., *An Introduction to Fuzzy Logic for Practical Applications*, Springer-Verlag, New York 1997.
[19] Rouvray, D. H. (Ed.), *Fuzzy Logic in Chemistry*, Academic, San Diego 1997.
[20] Ehrentreich, F., *Fuzzy Methods in Chemistry in Encyclopedia of Computational Chemistry,* P. Schleyer, (Ed.), Wiley, Chichester, 1998, p.1090.
[21] Dunn, J., *J. Cybernetics*, 1973, 3, 32-57.
[22] Ruspini, E., *Inf. Control*, 1969, 15, 22-32.
[23] Babuska, R., *Fuzzy Modelling for Control*, Kluwer Academic Publishers, The Netherlands, 1998.
[24] Jong, J. S., Sun, C. T. and Mizutani, E., *Neuro-Fuzzy and Soft Computing*, Pretince-Hall, SUA, 1997.
[25] Sârbu, C., Dumitrescu, D. and Pop, H. F., *Rev. Chim*. (Bucharest), 1993, 44, 450-459.
[26] Dumitrescu, D., Sârbu, C. and Pop, H. F., *Anal. Lett*., 1994, 27, 1031-1054.
[27] Pop, H. F., Dumitrescu, D. and Sârbu, C., *Anal. Chim. Acta* 1995, 310, 269-279.
[28] Pop, H. F., Sârbu, C., Horowitz, O. and Dumitrescu, D., *J. Chem. Inf. Comput. Sci*., 1996, 36, 465-482.
[29] Bezdek, J. C., Coray C., Gunderson, R. and Wantson, J. *SIAM J. Appl. Math*., 1981, 40, 358-369.
[30] Pop, H. F. and Sârbu, C., *Anal. Chem*., 1996, 68, 771-780.
[31] Sârbu, C. and Pop, H. F., *Rev. Chim*. (Bucharest), 1997, 48, 732-737.
[32] Pop, H. F. and Sârbu, C., 1997, *Rev. Chim*. (Bucharest), 1997, 48, 888-891.
[33] Sârbu, C., *J. AOAC Int*., 2000, 83, 1463-1467.
[34] Cundari, T. R., Sârbu, C. and Pop, H. F. *J. Chem. Inf. Comput. Sci*., 2002, 42, 310-321.
[35] Höppner, F., Klawonn, F., Kruse, R. and Runkler, T., *Fuzzy Cluster Analysis*, John Wiley Sons, Chichester, 1999.
[36] Backer, E. and Jain, A. K., *IEEE Trans. Patt. Anal. Machine Intell. PAMI-3*, 1981, 1, 66.
[37] Bezdek, J. C., *J. Cybernetics*, 1974, 3, 58-73.
[38] Gustafson, E. and Kessel, W., *Proc. IEEE Conf. Decision Control*, 1979, pp. 761-766.
[39] Horn, D. and Axel, I., *Bioinformatics*, 2003, 19, 1110-1115.

[40] Bezdek, J. C., Keller, J., Krisnapuram, R. and Pal, N. R., *Fuzzy Models and Algorithms for Pattern Recognition and Image Processing*. Kluwer Academy Publishers, Boston, 1999.
[41] Dumitrescu, D., Pop, H. F. and Sârbu, C., *J. Chem. Inf. Comput. Sci.,* 1995, 35, 851-857.
[42] Sârbu, C., Horowitz, O. and Pop, H. F., *J. Chem. Inf. Comput. Sci.*, 1996, 36, 1098-1108.
[43] Sârbu, C. and Zwanziger, H. W., *Anal. Lett.*, 2001, 34, 1541-1553.
[44] Simeonov, V., Puxbaum, H., Tsakovski, S., Sârbu, C. and Kalina, M., *Environmetrics*, 1999, 10, 137-152.
[45] Chen, G. N., *Anal. Chim. Acta*, 1993, 271, 115-124.
[46] Pudenz, S., Brüggemann, R., Luther, B., Kaune, A. and Kreimes, K., *Chemosphere*, 2000, 40, 1373-1382.
[47] Van Malderen, H., Hoornaert, S. and van Grieken, R., *Environ. Sci. Technol.* 1996, 30, 489-498.
[48] Held, A., Hinz, K. P., Trimborn, A., Spengler, B. and Klemm, O., *Aerosol Sci.*, 2002, 33, 581-594.
[49] Jambers, W., Smekens, A., van Grieken, R., Shevchenko, V. and Gordeev, V., Colloids and Surfaces A: Physicochem. *Eng. Aspects*, 1997, 120, 61-75.
[50] Lahdenperä, A. M., Tamminen, P. and Tarvainen, T., *Appl. Geochem.*, 2001, 16, 123-136.
[51] Markus, J. A. and McBratney, A. B., *Aust. J. Soil Res.*, 1996, 34, 453-465.
[52] Tamás, F. D., Abonyi, J., Borszéki, J. and Halmos, P. *Cem. Conc. Res.*, 2002 32, 1325-1330.
[53] Finol, J., Guo, Y. K. and Jing, X. D., *J. Petrol. Sci. Eng.*, 2001, 29, 97-113.
[54] Pérez-Correra, J. R., Letelier, M. V., Cipriano, A., Jorquera, H., Encalada, O. and Solar, I., In: *Air Pollution VI,* Brebbia, C. A., Ratto, C. F. and Power, H. (Eds.), WIT Press, 1998, p. 257-266.
[55] Hinz, K. P., Greweling, M., Drews, F. and Spengler, B., *J. Am. Soc. Mass Spectrom.,* 1999, 10, 648-660.
[56] Samaras, P., Kungolos, A., Karakasidis, T., Georgiou, D. and Perakis, K., *J. Environ. Sci. Health, A* 2001, 36, 153-161.
[57] Szczepaniak, K., Sârbu, C., Astel, A., Witkowska, E., Biziuk, M., Culicov, O., Frontasyeva, M. and Bode, P., *Central European Journal of Chemistry*, 2006, 4, 29-55.
[58] Frontasyeva, M. and Pavlov, S., *Analytical Investigations at IBR-2 Reactor in Dubna, JINR Preprint E14-2000-177*, Dubna 2000.

In: Current Applications of Chemometrics
Editor: Mohammadreza Khanmohammadi
ISBN: 978-1-63463-117-4

Chapter 4

ACCURACY IN SELF MODELING CURVE RESOLUTION METHODS

Golnar Ahmadi and Hamid Abdollahi*
Faculty of Chemistry, Institute for Advanced Studies
in Basic Sciences, Zanjan, Iran

ABSTRACT

Self Modeling Curve Resolution (SMCR) or Multivariate Curve Resolution (MCR) is a powerful tool in qualitative and quantitative analysis of miscellaneous mixtures. The ability of analyte determination in presence of unknown interferences is one of the most exploited aspects of some of SMCR methods. However the rotational ambiguity is the intrinsic problem of all these methods which leads to a band of the feasible solutions instead of one specific answer. This problem can affect the accuracy of both qualitative and quantitative results. Lots of struggles are made to find ways at least to decrease or at last to eliminate this phenomenon in order to have more accurate results in analysis of a system.

In this chapter, a brief description about the MCR methods and conditions to obtain unique solutions is offered. Quantitative analysis with the help of MCR methods in two-way data sets are explained in details. The effects of the rotational ambiguity on the accuracy of quantitative results are discussed in details. A simulated example helps to better understanding the situation.

INTRODUCTION

Nowadays, in most cases, mixtures are multi-component natural samples in which analyte signals are hidden by strong interferents and matrix effects [1-8].

* Corresponding author: Hamid Abdollahi. Faculty of Chemistry, Institute for Advanced Studies in Basic Sciences, P.O. Box 45195-159, Zanjan, Iran. E-mail: abd@iasbs.ac.ir.

To deal with these problems chemometric methods and multivariate data analysis techniques in different branches of analytical chemistry have been developed during the last two decades [9, 10].

Mathematical resolution of analyte signals by chemometric methods is much cheaper, faster and environmentally friendly, than any kind of chemical or physical separation analytical method.

Among chemometric methods, model free multivariate curve resolution (MCR), known as soft-modeling methods [11] in combination with hyphenated systems, offer the possibility of resolution, analysis and quantitation of multiple components in unknown mixtures without their previous chemical or physical separations. The principles of it are astonishingly simple: the analysis is based on the decomposition of a matrix of data into the product of two smaller matrices which are interpretable in chemical terms.

Generally, one resulting matrix is related to the concentrations of the species involved in the process and the other matrix contains their response vectors [12].

MCR was first introduced by Lawton and Silvestre with the denomination of Self Modeling Curve Resolution, (SMCR) [13]. In that work a mixture analysis resolution problem was described in mathematical terms for the case of a simple two-component spectral mixture. By introducing several precious concepts such as "subspace spanned by true solutions" and "feasible solutions", this pioneering work provided a landscape for the most of the struggles done in that field afterwards.

These methods can generally be grouped as non-iterative and iterative algorithms.

Most non-iterative resolution methods rely on combining information of small sections of a data set (subspaces) constructed from global and local rank information to obtain the pure component profiles. Window Factor Analysis (WFA) [14], Sub-window Factor Analysis (SFA) [15], or Heuristic Evolving Latent Projections (HELP) [16] are among the first and most significant approaches within this category. Most recent algorithms are usually evolutions of the parent approaches, such as Orthogonal Projection Resolution (OPR) from WFA [17] or Parallel Vector Analysis (PVA) from SFA [18].

These methods are based on the correct definition of the concentration windows.

Iterative resolution approaches are currently considered the most popular MCR methods due to the flexibility to cope with many kinds of data structures and chemical problems and to the ability to accommodate external information in the resolution process. Iterative Target Transformation Factor Analysis, ITTFA, [19, 20] and Multivariate Curve Resolution-Alternating Least Squares, MCR-ALS [21-24] and Resolving Factor Analysis, RFA [25] are some of the approaches of this category. All of these methodologies share a common step of optimization (of $\mathbf{C}$ and/or $\mathbf{S}^T$ matrices) that starts from initial estimates of $\mathbf{C}$ or $\mathbf{S}^T$ that evolve to yield profiles with chemically meaningful shapes, tailored according to chemical or mathematical information included in the optimization process under the form of constraints.

KEY CONCEPTS

MCR techniques are based on the bilinear decomposition of a measured mixed signal into their pure contributions according to the following equation:

$$\mathbf{D}=\mathbf{C}\mathbf{S}^{T} + \mathbf{E}_{MCR} \tag{1}$$

D (I,J) is the measured data matrix with I different samples and J different variables. **C** (I,N) is a matrix which contains the contribution of N species in I samples of the data set and $\mathbf{S}^{T}$ (N,J) is the matrix of pure responses of N components at J measured variables. Residuals or non-modeled part of **D** are collected in $\mathbf{E}_{MCR}$ which has the dimension the same as **D**.

In a bilinear model the contributions of the components in the two orders of the measurements are additive.

To clarify this decomposition problem, let us consider a chemical example. **D** can be a chromatographic system with DAD detection. This data set is the collection of the signals obtained from an n-component mixture by a detector during the elution process as a function of time. Each row contains the full spectra in a specific time and each column is the elution profiles at each channel of the response detector. Base on the additive of the absorbance in Beer-Lambert's law, the total raw signal is the sum of the signals related to each one of the compounds (see Figure X_1a). Each one of these pure contributions ($\mathbf{D}_i$) can be expressed by a pure spectral profile ($\mathbf{s}_i$), which is weighted according to the concentration of the related compound along the elution direction. The column of weights or concentration values ($\mathbf{c}_i$) is actually the elution profile or chromatographic peak of the compound. Thus, the signal contribution of each eluting compound is described by a dyad of profiles, $\mathbf{c}_i$ and $\mathbf{s}_i$, which are the chromatographic peak and the related pure spectrum, respectively (see Figure X_1b). This additive model of individual contributions is equivalent to the representation of the bilinearity concept which was introduced above (figure X_1c).

Another bilinear decomposition of the D matrix can be performed based on mathematical factors by applying PCA [26]:

$$\mathbf{D}=\mathbf{U}\mathbf{V}^{T}+\mathbf{E}_{PCA} \tag{2}$$

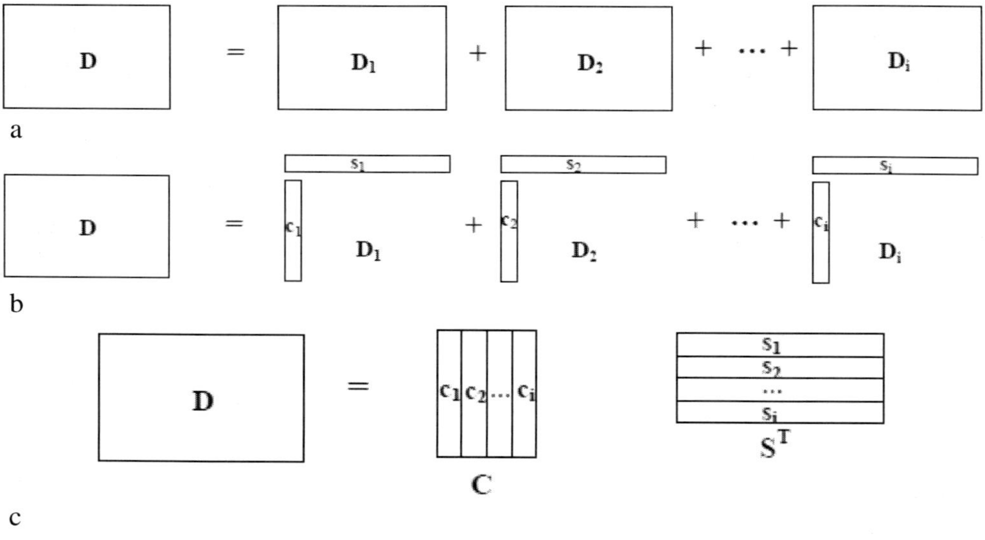

Figure 1. The measurement model of a chromatographic system. (a) Described as an additive model of pure signal contributions, (b) described as a model of additive dyads of pure concentration profile and spectrum, (c) described as a bilinear model of concentration profiles and spectra.

U and $\mathbf{V}^T$ are respectively the scores and loadings of matrix **D** and $\mathbf{E}_{PCA}$ is the matrix of residuals. The solutions obtained with this mode of decomposition are not physically meaningful, but **U** and $\mathbf{V}^T$, can be then used as estimations of the basis set of the column and row spaces of data matrix **D**, respectively.

The operations which decompose a matrix into bilinear model of factors are generally famed as "Factor Analysis" methodologies [27].

AMBIGUITIES

The main drawback of all MCR methods is the presence of ambiguities in the solutions obtained by these methods. This is known and well documented since the very beginning of their development in chemometrics [13]. There are two classes of ambiguities associated with curve resolution methods: Intensity and Rotational ambiguities.

Rotational Ambiguity

Going back to MCR model, for any non-singular matrix **T**, which is called "transformation matrix", the identity matrix $\mathbf{I}=\mathbf{T}^{-1}\mathbf{T}$ can be inserted in equation 1 and it can be rewritten as:

$$\mathbf{D}=\mathbf{C}\mathbf{S}^T=\mathbf{C}\,(\mathbf{T}\mathbf{T}^{-1})\,\mathbf{S}^T \tag{3}$$

This gives:

$$\mathbf{D}=\mathbf{C}_{new}\mathbf{S}^T{}_{new} + \mathbf{E} \tag{4}$$

where $\mathbf{C}_{new}=\mathbf{CT}$ and $\mathbf{S}_{new}=\mathbf{T}^{-1}\mathbf{S}^T$

In this equation, by linear combination of C and S^T (initial solutions) using a transformation matrix T, a new set of solutions $\mathbf{C}_{new}$ and $\mathbf{S}^T{}_{new}$ will be obtained. It is clear that any nonsingular invertible matrix T can do the transformation in Equation (3). The sentences above are the mathematical formulation of the rotational ambiguity [28]. In the other words, an infinite number of solutions ($\mathbf{C}$, $\mathbf{S}^T$) can be created; all of them are different in shape but resulting in the same matrix **D**. It is usually possible to reduce considerably this infinite number of possible solutions by means of constraints derived from the physical nature and previous knowledge of the problem under study. By implementation of different natural constraints such as non-negativity [13], local rank, selectivity [29], unimodality [30], various hard constraints based on physicochemical models [31] (i.e., equilibrium or kinetic) or etc., the number of the **T** matrices can significantly be reduced. If the incorporated constraints make not unique solutions for the profiles of all components in the system, a range of solutions for concentration and/or spectral profiles of each component are obtained, each of them fit the data equally well, although only one of them is the true one. The obtained ranges are called feasible solutions [13, 28]. There have been numerous attempts to reduce the rotational ambiguity with the ultimate goal to arrive at a unique solution [29, 32].

Intensity Ambiguity

Even if there is a unique matrix **T**, there can still be another form of ambiguity in the **C** and **S** matrices, known as intensity ambiguity, illustrated by the following equation:

$$D = \sum_{n=1}^{N} (k_N \mathbf{C_N})(\frac{1}{k_N} \mathbf{S_N^T}) \qquad (5)$$

For a particular species N the concentration profile $\mathbf{c}_N$ can be arbitrarily increased in an unlimited way by multiplying it by an arbitrary scalar number k if at the same time its spectrum $\mathbf{s}^T{}_N$ is decreased by the same amount by dividing it by the same number k. However, multiplication of any concentration profile with a scalar and the corresponding spectrum with the inverse of the scalar has no net effect on the shape of the profiles. By applying a normalization or closure constraints this problem can be solved [28].

Resolution Theorems

Getting more initial information about a system and applying them in a suitable way will result in less ambiguity and under certain circumstances even unique answers will attain. Among all the constraints, the most important ones are the selectivity and the local rank constraints that considerably limit or even eliminate the rotational ambiguity. Tauler et al. studied rotational and intensity ambiguities as well as the effect of the selectivity and the local rank constraints on successful applications of MCR methods [29]. They suggested that the presence and detection of the selectivity and appropriate use of constraints are the cornerstones of the individual MCR analyses. In 1995, Manne proposed the general conditions for the correct resolution of concentration and spectral profiles by three theorems namely "Resolution Theorems" [32]. These theorems can be used for predicting the uniqueness of any resolved profile obtained from every MCR method.

To achieve this goal, it is essential to have local rank information. There are several methods to detect the local rank and selectivity based mostly on evolving factor analysis (EFA) [33] and also heuristic evolving latent projection type of approaches [34-36].

The theorems has been proven by Window Factor Analysis (WFA) and Subwindow Factor Analysis (SWFA) methods which are non-iterative self-modeling methods for extracting the concentration profiles of individual components from evolutionary processes.

Theorem 1:

If all interfering compounds that appear inside the concentration window of a given analyte also appear outside this window, it is possible to calculate the concentration profile of the analyte.

It is attempted to describe this theorem by figure X_2a.

As it is shown in the figure, the solid window shows a concentration region that contains the hypothetical analyte A (labeled by green color) and two interferences B and C (labeled by blue and red colors).

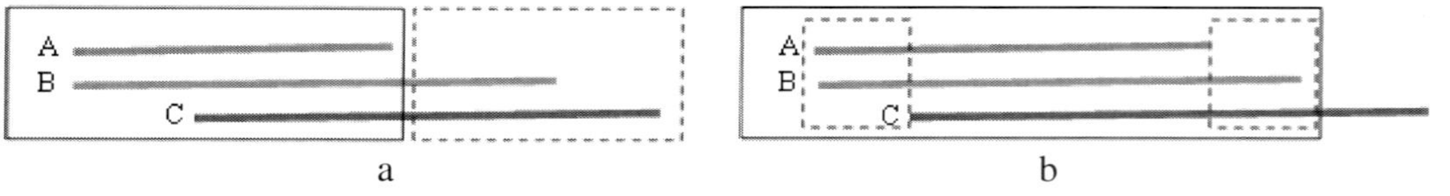

Figure 2. the concentration window of each component is displayed by a continuous line in each case; in (a) the absence window of the analyte is shown by a dashed line rectangle. In (b) the absence subwindows for each interferent, within the analyte concentration window, are displayed by dashed line rectangles.

Further the second window, dashed line, enfolds a region where all the interferences of the analyte (the blue and red components) are present. Clearly, the green component is qualified for all conditions that are required by Theorem 1.

Therefore, it is possible to calculate the concentration profile of the green component uniquely assuming this theorem.

Theorem 2:

If for every interferent the concentration window of the analyte has a subwindow where the interferent is absent, then it is possible to calculate the spectrum of the analyte. This theorem gives an instruction for unique calculation of the spectral profiles without first resolving the concentration profiles.

Consider the component B (blue component) as a hypothetical analyte. The concentration window of the analyte has two subwindows where in one of them the green component (A) and in another one the red component (C) is absent. This is represented graphically in figure X_2b. Therefore, it is possible to calculate the spectrum of the blue component based on Theorem 2.

Theorem 3:

For a resolution based only upon rank information in the concentration direction the conditions of Theorems 1 and 2 are not only sufficient but also necessary. In accordance to this theorem the complete resolution of a specific data set occurs when all components satisfy the all conditions of Theorems 1 and 2.

An important consideration in utilizing the concept of resolution theorems in analyses is the correct determination of the rank of the data and its local ranks. It is also important to implement zero concentration regions constraint during iterative soft-modeling methods [37].

Characterization and Visualization of the Rotational Ambiguity

As it was pointed out (sec. X_3), rotational ambiguity is the intrinsic problem of all soft-modeling methods regardless of the algorithm used. While the conditions of the resolution theorems are not fulfilled in a data set, the sets of the feasible solutions exist instead of unique solutions. In these cases the iterative methods converge to any of the feasible solutions, and many users will take it as a 'fact' without realizing that it is only one of a possibly wide range of solutions.

So, Detection of rotational ambiguity and appropriate warning is of great importance.

The main interests in calculation and analysis of the rotational ambiguity are evaluation of the accuracy of the results of the model-free methods and investigation of experimental noise effects on the results. In addition the effect of different constraints to decrease the ambiguity can be discussed.

The first approach of SMCR introduced by Lawton and Sylvestre represents a good visualization of the inner and the outer bounds of the estimated profiles in the abstract eigenvector space [13]. After that many efforts have been made to calculate the feasible band boundaries that surround all the feasible solutions [28, 38-40].

Visualization is the process of transforming information into a visual display, allowing users to observe the unseen features for better understanding the information. In scientific visualization, computer graphics are applied to scientific data for getting insight, examining hypothesis, and general elucidation [41]. The resulting graphs or visual displays enable the scientists to identify the hidden structures of data, which are needed for further data analysis. In the following subsections with respect to all valuable pioneering works done for calculation and analysis of the rotational ambiguity, some improved algorithms for the complete computation and visualization of the rotational ambiguity will be discussed.

Grid Search Minimization Algorithm and Error Maps

I. Two-Component Systems [42]

Considering D matrix with signals related to a simple two-component mixture problem, its bilinear decomposition based on Equation (1) will provide C matrix with two columns and S^T matrix with two rows. In the other hand by performing a principal component analysis (PCA) [26] on the data matrix D, another bilinear decomposition similar to what is given in Equation (2) is obtained. Based on resolving factor analysis (RFA) [25], by ignoring small differences in the error matrices E of Equations (1) and (2), columns of **C** and **S** can be written as a particular linear combination of **U** and **V** columns, respectively:

$$\mathbf{D}=\mathbf{U}(\mathbf{T}^{-1}\mathbf{T})\mathbf{V}^T = (\mathbf{U}\mathbf{T}^{-1})(\mathbf{T}\mathbf{V}^T) = \mathbf{C}\mathbf{S}^T \tag{6}$$

So that:

$$\mathbf{S}^T=\mathbf{T}\mathbf{V}^T \tag{6a}$$

and

$$\mathbf{C}=\mathbf{U}\mathbf{T}^{-1} \tag{6b}$$

The goal in RFA is to determine the **T** matrix for which the correct concentration and spectra are obtained. Note that the identity matrix can also be inserted in Equation (6) such that $\mathbf{S}=\mathbf{T}^{-1}\mathbf{V}^T$ and $\mathbf{C}=\mathbf{U}\mathbf{T}$.

In two-component systems the transformation matrix T is a 2×2 matrix. The introduced grid search minimization approach takes advantage of the fact that there is a set of transformation matrix T instead of one which fully defines set of feasible solutions of the system.

By normalizing the diagonal elements of 2×2 T matrix to one, it is not only defined with two off-diagonal elements instead of four, but also the intensity ambiguity is eliminated.

$$T = \begin{bmatrix} t_{11} & t_{12} \\ t_{21} & t_{22} \end{bmatrix} \longrightarrow T_n = \begin{bmatrix} 1 & tn_{12} \\ tn_{21} & 1 \end{bmatrix} \tag{7}$$

Note that, there are different possible options for normalization; the one chosen here is to normalize the diagonal elements to one. The inverse of T_n is then:

$$T_n^{-1} = \frac{1}{1 - t_{12}t_{21}} \begin{bmatrix} 1 & -tn_{12} \\ -tn_{21} & 1 \end{bmatrix} \tag{8}$$

Considering Equations 6a and b, the first row of T_n multiplied with V^T, defines the first row of S^T, so tn_{12} defines the first spectrum. Similarly, tn_{21} defines the second spectrum. tn_{12}, the element of T_n that defines the first spectrum, delineates the second concentration profile and also tn_{21} which defines the second spectrum, identifies the first concentration profile.

Practically this procedure starts by selecting a random selection of a pair of T (tn_{12}, tn_{21}). By arranging the values in a normalized 2×2 matrix, C and S^T are calculated based on Equations 6a and b. non-negativity and any other constraints are imposed and the new profiles are used to reproducing the data set.

The main goal of the grid search minimization procedure is to find all the **T** matrices for which all the constraints are fulfilled and the residuals between the original data and the data set reproduced by the calculated **C** and **S** is the minimum value within the error limit of the data set:

$$R = D - \widehat{D} = D - U\,T^{-1}\,TV^T \tag{9}$$

$$ssq = \sum_{i=1}^{I}\sum_{j=1}^{J} R_{ij}^2 = f(t_{12}, t_{21}) \tag{10}$$

The sum of squares (ssq) distribution as a function of the two relevant elements of **T** may be visualized as a three-dimensional contour or surface plot. Rotational ambiguity is indicated as the minimal flat area within the t_{12} and t_{21} landscape where ssq is constant. These contour or mesh plots will be referred to as error maps. These error maps are schematically shown in figures X_3a and b.

There are notes on these maps that can be referred to:

As there is no inherent order in the rows of $\mathbf{S}^T$ and the corresponding columns of **C**, there is a general 'symmetry' in the figure (it is not shown here).

So by expanding the range of the grid search, two minima, for example two rectangular regions with identical minimal ssq can be observed.

If there is a unique solution, the minimal areas are reduced to sharp minima which will be observed as a point in the two dimensional plot.

In some instances, one of the elements, tn_{12} or tn_{21}, is uniquely defined whereas the other is not; then the rectangular area is reduced to a linear range.

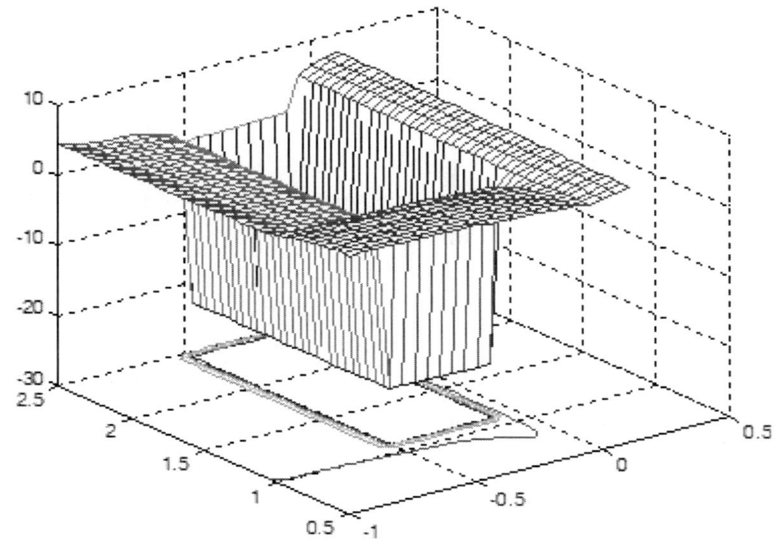

a

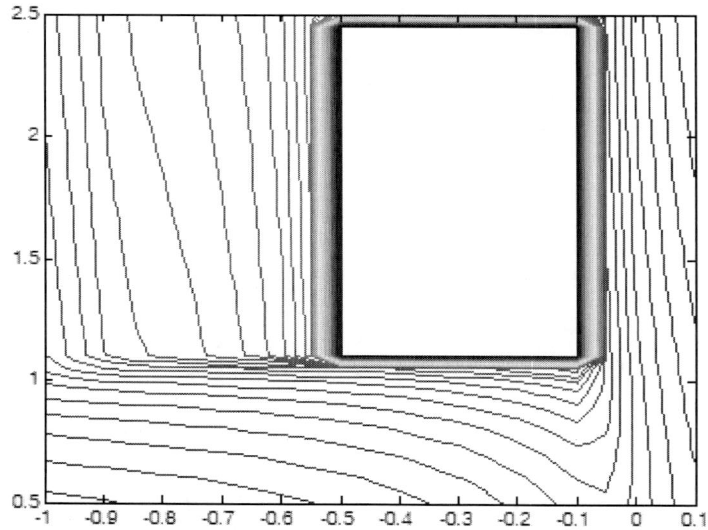

b

Figure 3. Representation of: ssq vs (t_{12} and t_{21}) (a) and two-dimensional contour plot (t_{12} vs t_{21}) (b).

II. Three-Component Systems

The idea of the grid search minimization procedure can be generalized to three component systems [43].

In three components systems a normalized T matrix with the first column equal to one is defined for which only 6 remained elements is going to be calculated instead of 9:

$$\mathbf{T} = \begin{bmatrix} 1 & t_{12} & t_{13} \\ 1 & t_{22} & t_{23} \\ 1 & t_{32} & t_{33} \end{bmatrix} \tag{11}$$

This kind of normalization provide the possibility of visualizing the rows of T and hence the related spectra in a 2-D plane, in addition to eliminating intensity ambiguity.

At the first step, **C** and **A** are estimated by the normalized transformation matrix according to equations (4a) and (4b). In the next step all the physical constraints are imposed on the **C** and **A** profiles which results in new matrices $\hat{C}$ and $\hat{A}$.

Finally, the sum of the square of the residual is computed by equation (6) as the difference between the reconstructed data (from $\hat{C}$and $\hat{A}$) and the original data set (**D**). This value will be the minimum for the accurate matrices **T** (equation 12)

$$R = D - \widehat{D} = D - \hat{C}\,\hat{A} \tag{12}$$

The procedure of finding the minimum values of the sum of squares (ssq) is performed by scanning of two parameters of each row of T matrix while the other four elements are calculated independently by using a fitting algorithm such as simplex. The feasible range of t_{k2} and t_{k3} is the range in which the values obtained for ssq are the minimal. Outside the feasible range, the fitting results in large ssq.

$$ssq = \sum_{i=1}^{I}\sum_{j=1}^{J} R_{ij}^{2} = f(t_{k2}, t_{k3}) \quad \text{k=1, 2 and 3} \tag{13}$$

The ssq can be plotted in a 3-D graph as a function of the t_{k2} and t_{k3}. Three clusters of points as the Area of Feasible solutions (AFS) of three components with irregular and convex shapes exhibit the boundaries between feasible and infeasible solutions. A three and two-dimensional plot are shown schematically for this procedure.

The procedure above can be performed by an alternative algorithm by calculating only the border of the AFSs. As points inside and outside the AFS do not convey any additional useful information the latter algorithm is quicker since a substantial amount of irrelevant information is calculated.

By inspiring from the algorithm above a new fast algorithm is suggested that is based on the inflation of polygons. Starting with an initial triangle located in a topologically connected subset of the AFS, an automatic extrusion algorithm is used to form a sequence of growing polygons that approximate the AFS from the interior. The polygon inflation algorithm can be generalized to systems with more than three components [44].

MCR-Band Method

For more complex multi-component mixtures, different optimization strategies have been proposed [38, 40, 45, 46]. Among them, a new strategy based on the optimization of an objective function has been introduced [28, 38].

Gemperline and Tauler introduced a new objective function with the intention of calculation the boundaries of feasible bands, and evaluation of the rotation ambiguity associated with MCR solutions.

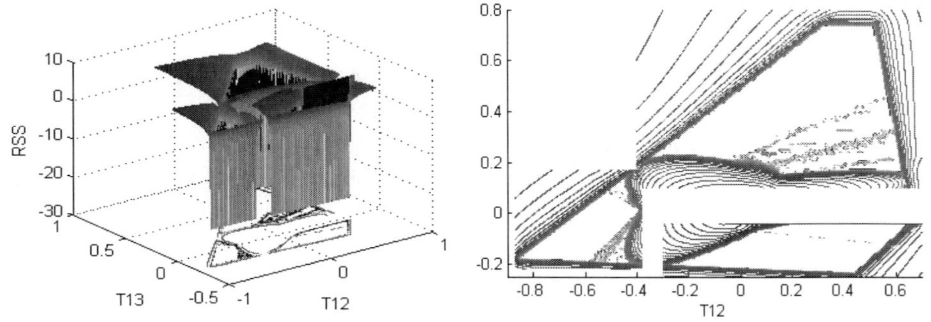

Figure 4. Representation of a) mesh and b)contour plot of a three component system.

The function to be optimized based on definition at Reference [28] is defined in equation (14):

$$f_k(\mathbf{T}) = \frac{\left\| c_k s_k^T \right\|}{\left\| \mathbf{CS}^T \right\|} \qquad \text{k=1,2,...,N} \tag{14}$$

This function which depends on the transformation matrix is defined as the Frobenius norm of the signal contribution of a particular species (k) with respect to the Frobenius norm of the whole signal of all the considered species. The optimization of this objective function allows the calculation of the maximum and minimum band boundaries of the feasible solutions. It should be mentioned this method can be applied to any number of components.

Geometrical Calculation of Feasible Solution

Although most of the references for visualization and characterization of the rotational ambiguity are based on algebraic definitions [28, 38, 42-44], computational geometry provides a more understandable concept of this intrinsic phenomenon. This concept contains approaches to find inner and outer polygons (abstract spaces) which form the basis for analytically finding the feasible regions [40]. These approaches provide new possibilities to compare different methodologies in literature of multivariate curve resolution. This outlook to the feasible solutions was instituted with Lawton-sylvestre's work in a simple two-component system at 1971 [13] and was developed to three component systems by Borgen and his co-worker which is now famed as Borgen plot [47]. However, because of the complexity of the demonstrated concepts the method was disregarded by chemometricians for two decades. The interesting sight of the concerning of the abstract space geometry is to discover new information about structure of data set, for example local rank. Applying knowledge that revealed from data structure may increase the accuracy of resolution of data set by soft methods. Consideration of the abstract space geometry aids to provide approaches for imposing constraints and to detect presence of the unique solution for a data set.

QUANTITATIVE ANALYSIS AND CHEMOMETRICS

Chemometric methods coupled with modern analytical instruments have been widely used for the rapid quantitative analysis of target analytes in complicated systems. Analyte determination in the presence of unexpected sample constituents, i.e., those not taken into account in the calibration phase, is one of the important subjects in the field of chemometrics. In most cases, the mixtures are multi-component natural samples in which analyte signals are hidden by strong interferents and matrix effects. To deal with these problems chemometric methods and multivariate data analysis techniques in different branches of analytical chemistry have been developed during the last two decades [48-54].

One of the main challenges in analytical chemistry is to design and developing of the affordable procedures which produce more correct and reliable results.

Data Structures

In 1990s, Booksh and Kowalski used tensor algebra to classify instrumental data as zeroth-order, first-order, second-order, and higher-order according to the type of data, which analytical instruments provide [55]. "Order" is usually employed to denote the number of data dimensions for a single sample. Scalars are zeroth-order tensors, vectors are first-order tensors, a matrix for which the elements obey certain relationships is a second-order tensor, etc. "Ways" is reserved for the number of dimensions of a number of joined data arrays, measured for a group of samples. Zero-order sample data lead to one-way data sets. Correspondingly, First-order sample data (vectors) lead to two-way data sets, second-order sample data (matrices) to three-way data sets, third-order sample data (three-dimensional arrays) to four-way data sets, etc. A representation of data of increasing complexity, with order and number of ways is shown in figure X_5.

This framework is general and permits to classify not only the data, but also the instruments that deliver them and the calibration methods for their analysis.

When the aim of the analytical process is to determine the concentration of some of the species present in the sample, a calibration must be tuned. Zero-order data are usually fitted to a straight line by least squares. This is known as univariate or zero-order calibration and requires full selectivity for the analyte of interest.

When the order of the data increases, mathematical data processing becomes more complex. However, some special properties such as first- and second-order advantage will be provided.

The ability of simultaneous analysis of more than one component with overlapping signals is the first advantage of multivariate data sets.

In first-order calibration methodologies there is the possibility of extracting useful analytical information from intrinsically unselective data. First-order calibration is able to flag unknown interferences (those not taken into account in calibraton samples) as outliers, because it cannot be modeled with a given calibration data set, warning that analyte prediction is not recommended. This property is known as the first-order advantage. However, in these calibrations a high number of samples are required and they must be of the same constitution of the unknown sample.

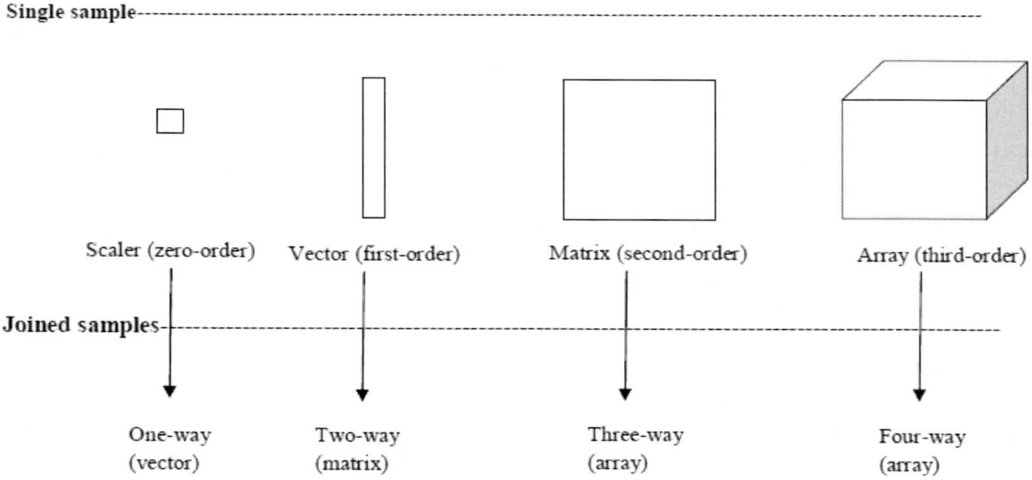

Figure 5. Orders and ways of different kind of the data sets for samples.

Principal component regression (PCR) [56] and Partial Least Square (PLS) [57] are two well-established algorithms in first-order multivariate calibrations.

Second- and higher-order calibration methodologies allow for analyte quantification even in the presence of unexpected sample constituents; those are not included in the calibration set. This is universally recognized as the second-order advantage. Second-order calibrations can be done using a few standard samples. GRAM [58], RAFA [59], PARAFAC [60], N-PLS [61], MCR-ALS [24], etc. are some of different algorithm used in second-order calibrations practices.

Extended Multivariate Curve Resolution

Second-order data sets are natural type of data to be analyzed with soft-modeling methods. When a single data set is analyzed with a soft-modeling method the qualitative information linked to the present components may be extracted. By applying more restrictions it is tried to obtain the results with less or even no ambiguities at all. However, there are limits in the quality of the profiles resolved that do not depend on the resolution method applied, but rather on the overlap among compounds profiles in both directions.

Simultaneous analysis of several data sets with MCR methods known as "Extended MCR" brings multiple advantages [29, 62]. The possibility to obtain quantitative information of the analyte(s) of interest is one of most exploited achievement of this form of the analysis which is due to the inherent property of second-order advantage in second-order data sets.

A single data matrix can be row-wise, column-wise, or row- and column-wise augmented to form a multi-set structure when other matrices with the same number of rows, columns, or both are appended in the appropriate direction. To have a new meaningful data structure, all individual data matrices (slices) in this type of data set should share some information with the other appended matrices; otherwise it does not make sense to produce such a new data arrangement. All types of matrix augmentation are explained below:

Suppose that k data matrices (D_1, D_2, D_3, ..., D_k) are obtained for a system by spectroscopic monitoring of a kinetic process analyzed under various initial conditions (e.g., different initial concentrations of the chemical constituents).

In column-wise augmentation the different data matrices are supposed to share their column vector space. This can be written simply as:

$$\begin{bmatrix} D_1 \\ D_2 \\ \vdots \\ D_K \end{bmatrix} = \begin{bmatrix} C_1 \\ C_2 \\ \vdots \\ C_K \end{bmatrix} S^T + \begin{bmatrix} E_1 \\ E_2 \\ \vdots \\ E_K \end{bmatrix} = C_{aug} S^T + E_{aug} \tag{15}$$

This data arrangement makes no assumption about how the concentration profiles of the species (C_1, C_2, C_3, ..., C_k) change in various matrices analyzed, but they have or share the same pure spectra (S^T). This extended bilinear model is valid for data sets formed by matrices with equal or unequal number of rows. Because of the freedom in shape of the concentration profiles, matrices with different chemical nature in row spaces can also be appended.

The notation for row-wise augmentation is:

$$\begin{bmatrix} D_1 & D_2 & \cdots & D_k \end{bmatrix} = C\begin{bmatrix} S_1^T & S_2^T & \cdots & S_k^T \end{bmatrix} + \begin{bmatrix} E_1 & E_2 & \cdots & E_k \end{bmatrix} = CS_{aug}^T + E_{aug} \tag{16}$$

In this data arrangement, the different matrices are supposed to share their row vector (concentration) space. This is the case for example when a chemical reaction is monitored with different spectral instruments. In this case the species have the same pure concentration profiles in all sub-matrices, but their spectral responses will be completely different (in shape) depending on the considered spectroscopy. In this case, it is not necessary that the different appended data matrices have the same number of columns; only the same number of rows (samples) is needed to allow for the row-wise matrix augmentation strategy.

The mathematical notation of column- and row-wise augmentation is:

$$\begin{bmatrix} D_{11} & D_{12} & \cdots & D_{1L} \\ D_{21} & D_{22} & \cdots & D_{2L} \\ \vdots & & & \\ D_{K1} & D_{K2} & \cdots & D_{KL} \end{bmatrix} = \begin{bmatrix} C_1 \\ C_2 \\ \vdots \\ C_K \end{bmatrix} \begin{bmatrix} S_1^T & S_2^T & \cdots & S_L^T \end{bmatrix} + \begin{bmatrix} E_{11} & E_{12} & \cdots & E_{1L} \\ E_{21} & E_{22} & \cdots & E_{2L} \\ \vdots & & & \\ E_{K1} & E_{K2} & \cdots & E_{KL} \end{bmatrix} = C_{aug} S_{aug}^T + E_{aug} \tag{17}$$

This means that some of the matrices are supposed to share the same column (spectra) vector space and some others their row vector (concentration) space. Going back to matrix notation above (17), it can be understood that sub-matrices with identical indices k share the common row space and submatrices with identical indices L share the common column space. This bilinear model is usually applied in the analysis of different complex samples such as biological systems with complex matrix effects [63, 64].

Multi-set data analysis with MCR methods provides some valuable aspects:

- In case of one dimension in common, even when there is no selectivity, the simultaneous analysis of different matrices will still be better because it forces the profiles of the species in the common dimension to be the same for the different data matrices. This works as a strong restriction which decreases freedom of the rotations of the profiles in the relevant dimension.
- Multi-set data analysis is one of the possible ways to complete resolution of some of the component's profiles. Let us consider a chromatographic data set, containing two components (A and B), one of them embedded under the major one. By appending a run of a standard of the major compound (for example A) to the original data set, elution profile of B and consequently the spectral profile of A can be resolved uniquely. This happens because the conditions of the resolution theorems are fulfilled. As it is shown on the figure X_6a, component A which can be considered as interference for B is present outside the concentration window of B too (the presence of A outside the concentration window of B is shown with dashed line). By setting zero for the elements of elution profile of component B in the C matrix (figure X_6b), this region is introduced to the algorithm used and unique concentration profile of B can be obtained based on first resolution theorem. Correspondingly, second resolution theorem will be valid for A and unique spectral profile of A will be obtained. Setting zero for the elements of some components in a multi-set data analysis is known as 'corresponding among species' which fixes the sequence and the presence or absence of components in a particular $\mathbf{C}_k$ and/or $\mathbf{S}_l^T$ matrices when Equations (15)–(17) are solved. This presence/absence information is coded in binary format and introduced into the MCR algorithm. As a consequence, when a particular component does not exist in a particular $\mathbf{C}_k$ and/or $\mathbf{S}_l^T$, the elements in the related profile are set to zero.

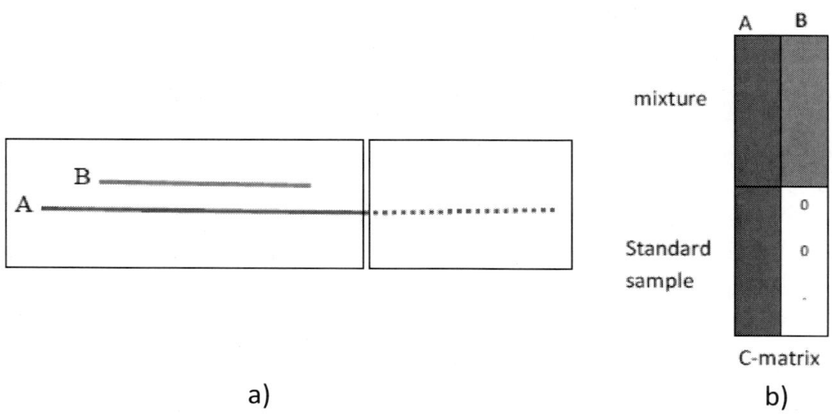

Figure 6. Representation of augmentation of calibration set of A to the mixture in a) elution window form and b) matrix form.

- Breaking rank deficiencies by matrix augmentation is another problem that could be solved in multi-set data analysis [65].
- Although the improvement in the qualitative information in a multi-set data analysis, the possibility to obtain quantitative information is the most important achievements of the issue. The main reason is the so-called second-order advantage, which allows for the quantification of the analytes in the presence of unknown interferences.

Quantitative Estimations Using Two-Way Soft-Modeling Methods

Suppose a data set obtained from a sample containing an analyte of interest and several unknown interferences. Preparing pure standard data of the analyte enable us to estimate the concentration of the desired components. The unknown and the calibration set can be augmented in one direction and be analyzed by a soft-model such as MCR-ALS. Once augmented data set resolving is achieved, the information linked to the analyte(s) of interest can be separated from the information associated with the interferents. The quantitative information is hidden in the augmented mode of the data set. For example, if the spectral profiles are common among the sub-matrices, quantitative information can be extracted from the augmented concentration profiles. Taking only the resolved elution profiles of the analyte in the different sub-matrices and estimating a quantitative parameter (peak area (A) or peak height (h)), we may construct a classical calibration line with the pure standards and derive the concentration of the analyte in the unknown sample, as it is shown in figure X_7.

In the extreme case, quantitative estimations can be performed with one standard sample. Relative quantitative information of one particular compound can be derived easily by comparing the area or height of the resolved profiles of standard and unknown sample:

$$C_u = \frac{A_u}{A_k} C_k \tag{18}$$

where C_u and C_k are the concentrations of the analyte in the unknown and standard samples, and A_u and A_k are the areas below the resolved profiles in the unknown and standard sample respectively. A_u and A_k can be replaced by h_u and h_k when using peak height in calculations.

Soft-Modeling Calibration of First-Order Data Sets

Methods such as Partial Least Square (PLS) [57], Multivariate Linear Regression (MLR) [66], Classical Least Square (CLS) [67] and etc. are regularly used as standard first-order calibration methodologies.

Recently two-way soft-modeling methods especially MCR-ALS are applied for quantitative analysis of first-order data although second-order data sets are natural type of the data presented to these techniques [68-71].

In order to explain the methodology, the iterative MCR-ALS method is chosen as widely used soft-modeling method in data analysis.

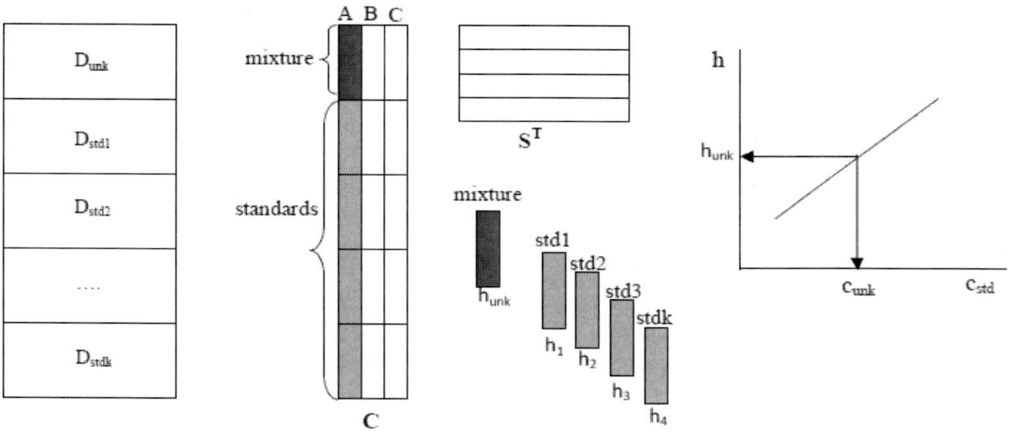

Figure 7. Calibration procedure for component A with soft-modeling methods; column-wise augmentation of unknown sample and calibration set. B and C are considered interferences not considered in calibration set. hs are the peak heights.

This optimization algorithm to perform the mathematical decomposition of Equation 1 has been described in details elsewhere [21-24]. The correlation constraint previously developed and used in different works [68-71] is applied during the MCR-ALS analysis, to establish calibration models for the quantitative determination of analytes in first-order data sets. In order to perform these analysis a number of samples with different concentrations of the analytes (usually random values) are used as calibration samples and another set of samples are used as the unknown set for which the amounts of the analytes are unknown and also may contain interferences. Similar to other first-order calibration methodologies, calibration set should contain all the interferences of the unknown set. The signals obtained from all samples are arranged in a single data matrix with rows equal to the total number of samples (calibrations and unknowns) and columns equal to recorded instrumental signals (wavelengths, channels…). Concentration values of the analyte in calibration samples obtained at each ALS iteration are then correlated with their known true concentration values (reference values) by a linear regression. Using the obtained linear equation, all concentration values of the particular analyte in calibration and unknown samples can be updated for the next ALS optimization step until the convergence is achieved. This constraint may be applied for any number of calibrated analytes. The known values of the analyte in calibration samples can be either implemented directly as an equality constraint during each iteration or by sequential regression procedures named as correlation constraint. In both cases concentration values of the analyte in unknown set are predicted at the end of the iterations but in the latter case the effect of the errors in reference values are also considered.

Accuracy of Chemical Quantitative Analysis Using Soft-Modeling Methods

In performing an analyte quantitation, the accuracy of the obtained result is one of the important issues [72]. The main source of inaccuracy in the quantitative results obtained by soft-modeling methods is rotational ambiguity problem.

The presence of rotation ambiguities and of non-unique solutions in MCR results decreases their reliability and makes their assessment difficult. There is, therefore, an urgent need for a deeper investigation of this topic and for the quality assessment of MCR results. In most cases when soft-modeling methods are used for a chemical analysis, rotational ambiguity is reduced or eliminated by imposing as many constraints as possible for the system. Resolution theorems [32] indicate clearly the conditions needed to recover uniquely the pure concentration and signal profiles of a compound in a data set.

These conditions depend mainly on the degree of selectivity in the column mode or row mode of the measurements. So, in the same system, some profiles can be recovered uniquely, and some others will necessarily be affected by a certain ambiguity. When ambiguity exists, a compound is represented by a band of feasible solutions instead of a unique profile. Here, the main question is to determine the accuracy of the obtained quantitative results in the presence of the non-eliminated part of rotational ambiguity.

To answer this question let us simulate an experimental situation without any random error and Consider a multivariate second-order chromatographic data set corresponding to a mixture of two co-eluting compounds (see figure X_8).

Also we suppose a calibration set of second-order signals is built with one of the constituents with different concentrations. The created data matrices were joined together with one matrix on top of the other, keeping in common the column space. Further investigations are performed on the augmented data matrix.

Since the strategy of column-wise augmentation of the data matrices forces common species to have the same spectrum in all data matrices, the quantitative information of the systems can be found in the C-matrix of the augmented data set. In order to evaluate the effect of the rotational ambiguity on the quantitative estimation, we need to calculate the feasible band of the present components. By conducting grid search minimization procedure [42] (sec. X_5a_I) in the U-subspace of the augmented data set, the ranges of the feasible solutions for the C-matrix or elution profiles are obtained. In the calculations, the following constraints are imposed on the model: non-negativity of spectra and concentration profiles and correspondence among species (see section X_6_b). In this example, this constraint supplies the information on the absence of interference in all the calibration samples and is allowing it to only appear in the test sample.figures X_9a and b shows the mesh and contour plots of the log of the sum of squares of the residuals as a function of t_{12} and t_{21}, respectively, where t_{12} and t_{21} are the normalized coordinates of two components in the U-subspace. The whole range of t_{12} values in the minimum region of figure X_9b produces the whole set of feasible solutions of the analyte elution profile as the first species, also the spectra of the interference as the second one. As can be seen, minimal region became a line in this example which means that coordinate t_{21} is uniquely obtained. This coordinate produces the elution profile of the interference as well as the spectrum of the analyte.

The obtained results are in accordance with the first and second resolution theorems. Although the applied constraints, in particular correspondence among species, provide a unique solution for the elution profile of the interferent, it can be seen (from Figure X_9) that the situation is not restricting enough for the analyte elution profile to be uniquely determined.

In order to validate the quantitative results of the analyte, all the feasible solutions for that component have to be recovered. To do this, each t_{12} value belonging to the minimum region is translated to the elution profile as illustrated in figure X_10a.

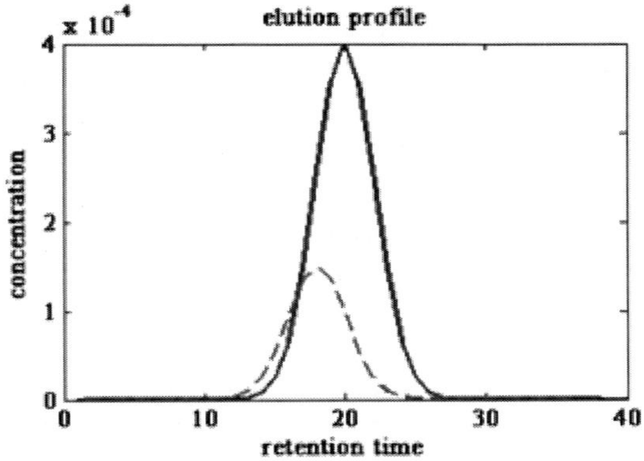

a

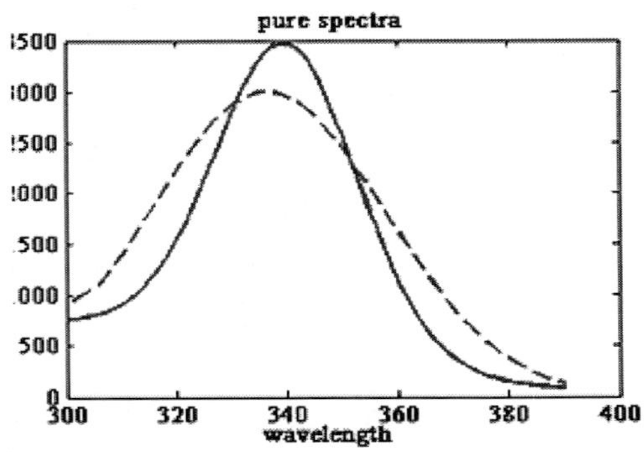

b

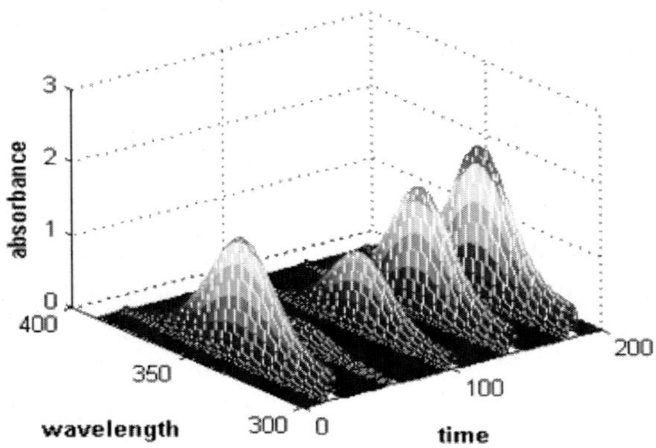

c

Blue: analyte, green: interference.

Figure 8. Simulated a) elution profiles b) spectral profile of two components, c) augmented created data set.

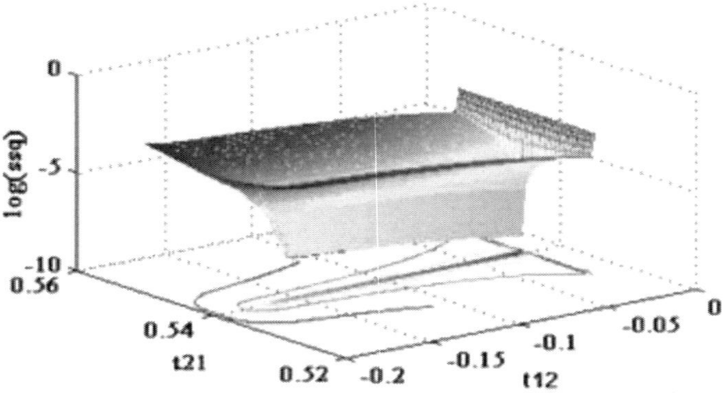

a

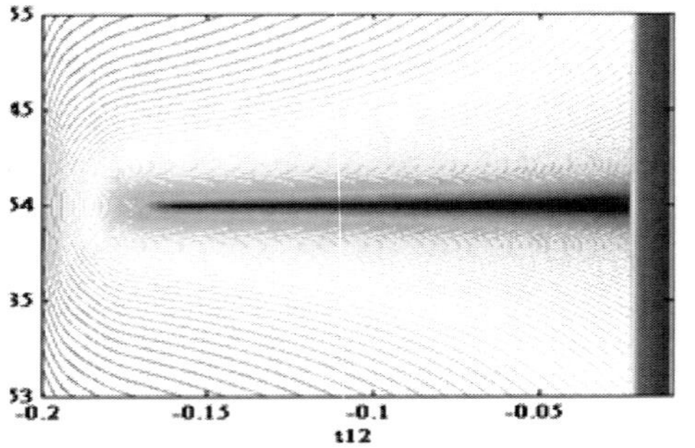

b

Figure 9. a) mesh b) contour plot obtained by grid search minimization procedure in U-space of the data set.

This means that when the solutions are not unique due to the presence of the rotational ambiguity, a linear combination of the soft-modeling solutions fulfilling the constraints can be obtained, and unfortunately the true solution cannot be recognized, unless more information is provided to the system. Note that all profiles are normalized to a maximum of one and for better visualization a part of the feasible solutions related to the analyte profile in the unknown sample is magnified. For better understanding the system, the t_{21} value of the interference can be translated. Unique elution profile of this component (figure X_10b) confirms the unique spectra of the analyte (figure X_10c) (based on the cross-relationship of the elements of T matrix). Also, the range of the interference spectral profile is demonstrated in figure X_10d. The results are completely in accordance with resolution theorems.

As was explained in sec. X_6c, taking the resolved elution profiles of the analyte in the different sub-matrices (unknown and calibration set), quantitative estimations can be performed. As there is a feasible range for the analyte of interest, these calculations may be repeated for each of the profiles of the band respectively.

Distinct calibration line can be calculated corresponding to each of the profiles related to pure standard samples and the concentration of the analyte in the unknown sample corresponding to that profile can be derived.

So, different concentration value will be obtained for the analyte of interest.

Based on the mentioned procedure, the concentrations related to each of the profiles can be estimated separately and sorted in a range from their minimum to maximum amount.

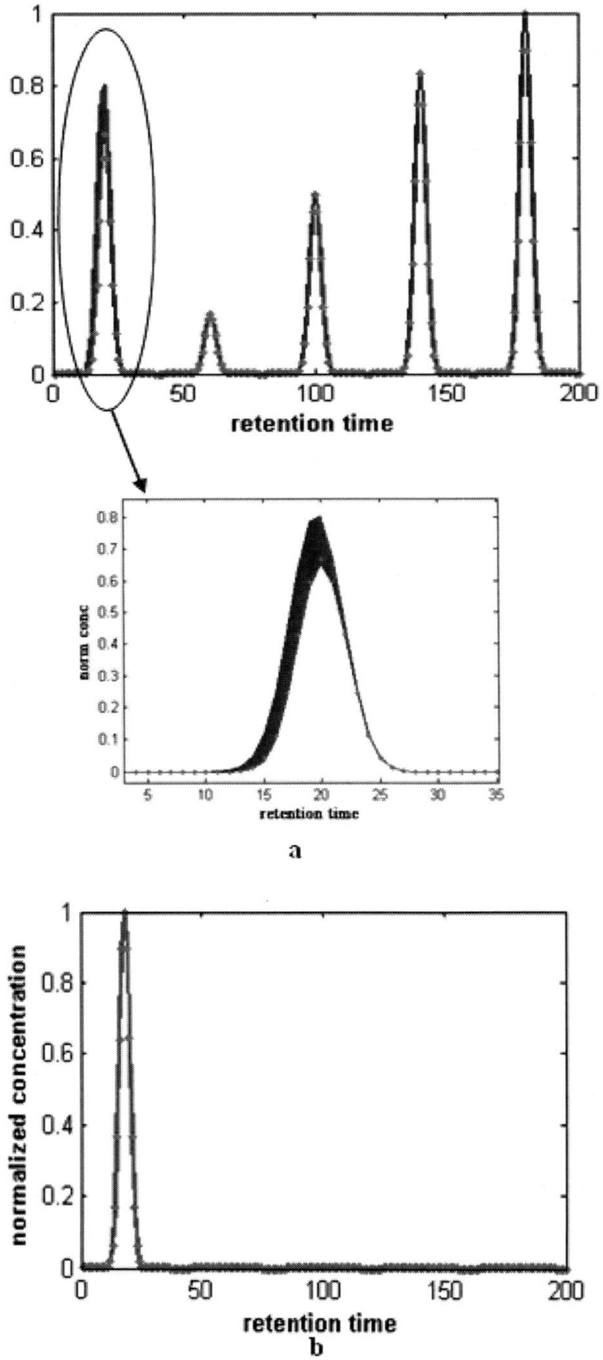

Figure 10. (Continued).

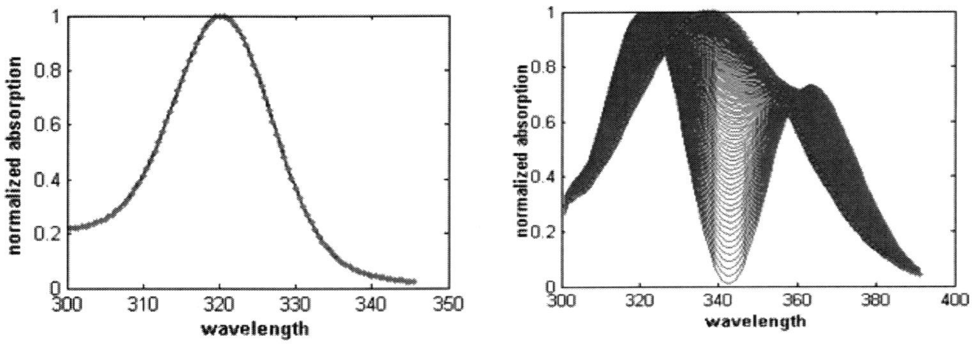

c d

Figure 10. a) Feasible solutions of the analyte's elution profile b) unique elution profile of the interference c) unique spectrum of the analyte d) feasible range of the analyt's spectra. Analyte's feasible range is magnified and the true solution is shown with red dots.

Table 1. Percentage of quantitation error for upper and lower level of concentration in feasible region

Quantitative Results	Upper Level Concentration	Lower Level Concentration
	5.26×10^{-4}	3.87×10^{-4}
% Error	31.50	-3.25

% quantitation error = $[(C_{obtained} - C_{true}) / C_{true}\times100]$.
C_{true} is the theoretical value of the analyte in the simulated data, which is considered 4×10^{-4} (a.u).

As we know the real amount of the analyte in this simulated case, the relative error prediction corresponding to upper and lower amount of predicted concentrations of the analyte can be obtained. The results are reported in Table X_1.

Since the data set are noise free, the magnitude of the concentration range depends on the extent of the rotational ambiguity which still has remained in the system after applying all possible constraints.

The obtained concentration range can be accepted as a criterion for quantitative evaluation of the rotational ambiguity. In addition, the calculated range expresses the reliability of the results for the system when it is resolved by a soft-modeling method.

Generally, iterative soft-modeling methods will converge towards one particular solution within the region of feasible solutions.

Accurate concentrations will be obtained if and only if the applied method converges to the true solution of the system; on the other hand, the unavoidable problem of rotational ambiguity will lead to an inaccurate result. The calculated range for the relative error proves that the quantitative result generated by soft-modeling method may be affected considerably due to a systematic error of rotational ambiguity.

Obviously, in real systems, when quantitative analysis is an important aspect of the investigation, deviation of the final results from the true quantity may be troublesome for the analyst.

The strategy used for calibration here, offers a way to evaluate the accuracy of the quantitative results obtained by a soft-modeling method.

Thus, when a soft-modeling method is used as a quantitative tool in analyzing a system, checking the range of the feasible solution and evaluating the error prediction range is highly recommended. As a suggestion, calculating the concentration range based on profiles of the feasible region can be considered as a criterion for evaluating the extent of the rotational ambiguity. In the case where this range is large, introducing more restrictions to the system should be tried in order to decrease the feasible region.

Uncertainty in the Results

In the previous section we talked about the reliability of the quantitative results in presence of the rotational ambiguity. Even in absence of the rotational ambiguity experimental errors and noise are propagated from experimental data to parameter estimations when a soft-modeling method is used in data analysis. Providing methodologies to evaluation of error estimates and of propagation of experimental errors is a necessary requirement to enable the use of resolution methods in standard analytical procedures [73].

Because of the non-linear nature of the most curve resolution methods, finding analytical expressions to assess the error associated with resolution results is an extremely complex problem [74].

Different statistical procedures have been proposed to provide error parameter estimations and propagation of experimental errors for those cases where formulae for analytical evaluation of errors are not available owing to the strong non-linear behavior of the proposed model. These procedures are known under the general name of resampling methods, such as noise-added method [75, 76], Monte Carlo simulations [77] and jackknife [78].

Although ambiguity and noise are two distinct sources of uncertainty in resolution, their effect on the resolution results cannot be considered independently. The boundaries of feasible solutions can be affected, being unclear in presence of the noise. Interaction of the experimental error and rotational ambiguity on the area of the feasible solutions and feasible range of components is one of the interesting topics for investigation in chemometric fields.

Conclusion

In this chapter the conditions of obtaining unique solutions in analysis of different systems was discussed. In analysis of a chemical mixture with a soft-modeling method, it is important to find or provide situations (considering resolution theorems and constraining the system) to get unique solutions. However, the effect of the rotational ambiguity on the accuracy of qualitative and quantitative results is unavoidable.

Specifically, reliability of quantitative results strongly depends on non-eliminated rotational ambiguity of the system under study. In real systems where the random error is unavoidable, the deviation of the quantitative calculations from the true value can be quite large. Evaluating the extent of the rotational ambiguity is one of the interesting subjects in MCR field. This can help when the accuracy of the quantitative results is highly regarded.

As a suggestion, calculating the concentration range (as was explained in the chapter) can be considered as a criterion for evaluating the extent of the rotational ambiguity.

In the case where this range is large, introducing more restrictions to the system can be tried in order to decrease the feasible region.

REFERENCES

[1] Gabrielsson, J., Trygg, J. *Crit. Rev. Anal. Chem.* 2006, 36, 243-255.
[2] Peré-Trepat, E., Lacorte, S., Tauler, R. *Anal. Chim. Acta* 2007, 595, 227-237.
[3] Bortolato, S. A., Arancibia, J. A., Escandar, G. M. *Anal. Chem.* 2008, 80, 8276-8286.
[4] Lozano, V. A., Ibanez, G. A., Olivieri, A. C. *Anal. Chem.* 2010, 82, 4510-4519.
[5] Piccirilli, G. N., Escandar, G. M. *Analyst* 2010, 135, 1299-1308.
[6] Diewok, J., de Juan, A., Maeder, M., Tauler, R., Lendl, B. *Anal. Chem.* 2003, 75, 641-647.
[7] Saurina, J., Tauler, R. *Analyst* 2000, 2000, 2038-2043.
[8] Maggio, E. M., Damiani, P. C., Olivieri, A. C. *Talanta* 2011, 83, 1173-1180.
[9] Gomez, V., Callao, M. P. *Anal. Chim. Acta* 2008, 627, 169-183.
[10] Lavine, B. K. Chemometrics, *Anal. Chem.* 2000, 72, 91R-97R.
[11] De Juan, A., Tauler, R. *Anal. Chim. Acta* 2003, 500, 195-210.
[12] De Juan, A., Tauler, R. J. *Chromatogr. A.* 2007, 1158, 184-195.
[13] Lawton, W. H., Sylvestre, E. A., *Technometrics* 1969, 13, 617-633.
[14] Malinowski, E. R. *J. Chemom.* 1992, 6, 29-40.
[15] Manne, R., Shen, H., Liang, Y. *Chem. Intell. Lab. Syst.* 1999, 45, 171-176.
[16] Kvalheim, O. M., Liang, Y. Z. *Anal. Chem.* 1992, 64, 936-946.
[17] Xu, C. J., Liang, Y. Z., Jiang, J. H. *Anal. Let.* 2000, 33, 2105-2128.
[18] Jiang, J. H., SasiÊ, S., Yu, R., Ozaki, Y. *J. Chemom.* 2003, 17,186-197.
[19] Gemperline, P. J., *J. Chem. Inf. Comput. Sci.* 1984, 24, 206-212.
[20] Vandeginste, B. G. M., Derks, W., Kateman, G. *Anal. Chim. Acta* 1985, 173, 253-264.
[21] Tauler, R., Casassas, E. *Anal. Chim. Acta* 1989, 223, 257-268.
[22] Tauler, R. *Chemom. Intell. Lab. Syst.* 1995, 30, 133-146.
[23] Karjalainen, E. J. *Chemom. Intell. Lab. Syst.* 1989, 7, 31-38.
[24] Jaumot, J., Gargallo, R., de Juan, A., Tauler, R. *Chemom. Intell. Lab. Syst.* 2005, 76, 101-110.
[25] Mason, C., Maeder, M., Whitson, A. *Anal. Chem.* 2001, 73, 587-1594.
[26] Jolliffe, I. T. *Principal Component Analysis*, 2nd ed., Springer: New York; 2002; pp. 150-165.
[27] Malinowski, E. R. *Factor analysis in Chemistry*, third ed., Wiley: New York, 2002.
[28] Tauler, R. *J. Chemom.* 2001,15, 627-646.
[29] Tauler, R., Smilde, A., Kowalski, B. R. *J. Chemom.* 1995, 9, 31-58.
[30] De Juan, A., Heyden, Y. V., Tauler, R., Massart, D. L. *Anal. Chim. Acta* 1997, 364, 307-318.
[31] De Juan, A., Maeder, M., Martínez, M., Tauler, R. *Anal. Chim. Acta* 2001, 442, 337-350.
[32] Manne, R. *Chemom. Intell. Lab. Syst.* 1995, 27, 89-94.
[33] Maeder, M., Zuberbuehler, A. D. *Anal. Chim. Acta* 1986, 181, 287-291.
[34] Keller, H., Massart, D., *Anal. Chim. Acta* 1991, 246, 379-390.

[35] Whitson, A. C., Maeder, M., *J. Chemom.* 2001, 15, 475-484.
[36] Zeng, Z.-D., Xu, C. J., Liang, Y. Z., Li, B. Y. *Chem. Intell. Lab. Syst.* 2003, 69, 89-101.
[37] Akbari, M., Abdollahi, H. *J. Chemom.* 2013, 27, 278-286.
[38] Gemperline, P. *J. Anal. Chem.* 1999, 71, 5398-5404.
[39] Leger, M. N., Wentzell, P. D. *Chem. Intell. Lab. Syst.* 2002, 62, 171-188.
[40] Rajko, R., Istvan, K., *Chemom.* 2005, 19, 448-463.
[41] Pickover, C. A., Tewksbury, S. K. *Frontiers of scientific visualization.* Wiley: New York, 1994, 8-20.
[42] Vosough, M., Mason, C., Tauler, R., Jalali-Heravi, M., Maeder, M., *J. Chemom.* 2006, 20, 302-310.
[43] Golshan, A., Abdollahi, H., Maeder, M., *Anal. Chem.* 2011, 83, 836-841.
[44] Sawall, M., Kubis, C., Selent, D., Börner, A., Neymeyr, K., *J. Chemom.* 2013, 27,106-116.
[45] Henry, R. C., Kim, B. M., *Chem. Intell. Lab. Syst.* 1990, 8, 205-216.
[46] Kim, B. M., Henry, R. C., *Chem. Intell. Lab. Syst.* 1999, 49, 67-77.
[47] Borgen, O. S., Kowalski, B. R., *Anal. Chim. Acta* 1985, 174, 1-26.
[48] Parastar, H., Radovic, J. R., Heravi, M. J., Diez, S., Bayona, J. M., Tauler, R. *Anal. Chem.* 2011, 83, 9289-9297.
[49] De Godoy, L. F. A., Hantao, L. W., Pedrozo, M. P., Poppi, R. J., Augusto, F. *Anal. Chim. Acta* 2011, 699, 120-125.
[50] Gargallo, R., Tauler, R., Cuesta-Sánchez, R., Massart, D. L., *Trends Anal. Chem.* 1996, 15, 279-286.
[51] Mas, S., de Juan, A., Tauler, R., Olivieri, A. C., Escandar, G. M., *Talanta* 2010, 80, 1052-1067.
[52] Schenon, A. V., Culzoni, M. J., Marsili, N. L., Goicoechea, H. C., *Food Chem.* 2013, 138, 1928-1935.
[53] Abdollahi, H., Yaftian, M. R., Zeinali, S., *Anal. Chim. Acta* 2005, 531, 153-160.
[54] Saurina, J., Hernandez-Cassou, S., Tauler, R., Izquierdo-Ridorsa, A., *Anal. Chem.* 1999, 71, 126-134.
[55] Booksh, K. S., Kowalski, B. R. Theory of analytical chemistry.; *Anal. Chem.* 1994, 66, 782A-791A.
[56] Beebe, K. R., Kowalski, B. R., *Anal. Chem.* 1987, 59, 1007A-1017A.
[57] Lindberg, W., Person, J. A., Wold, S., *Anal. Chem.* 1983, 55, 643-648.
[58] Sanchez, E., Kowalski, B. R., *Anal. Chem.* 1986, 58, 496-499.
[59] Ho, C. N., Christian, G. D., Davidson, E. R., *Anal. Chem.* 1978, 50, 1108-1113.
[60] Bro, R. PARAFAC. Tutorial and applications, *Chem. Intell. Lab. Syst.* 1997, 37, 149-171.
[61] Bro, R. Multiway calibration, Multilinear PLS, *J. Chemom.* 1996, 10, 47-61.
[62] Maeder, M., de Juan, A. Multiset Data Analysis: Extended Multivariate Curve Resolution. In: S. D. Brown, R. Tauler, B. Walczak, (Eds.), *Comprehensive Chemometrics,* Elsevier: Oxford, UK, 2009; Vol. 2, pp. 473-501.
[63] Jaumot, J., Eritja, R., Tauler, R., Gargallo, R., *Nucleic Acids Res.* 2006, 34, 206-216.
[64] Navea, S., de Juan, A., Tauler, R., *Anal. Chem.* 2002, 64, 6031-6039.
[65] Ruckebusch, C., de Juan, A., Duponchel, L., Huvenne, J. P., *Chem. Intell. Lab. Syst.* 2006, 80, 206-214.
[66] Faber, N. M., Buydens, L. M. C., Kateman, G. J., *J. Chemom.* 1993, 7, 495-526.

[67] Mark, H. *Principles and Practice of Spectroscopic Calibration*. Wiley: New York, 1991.

[68] Antunes, M. C., Simão, J. E. J., Duarte, A. C., Tauler, R., *Analyst* 2002, 127, 809-817.

[69] Azzouz, T., Tauler, R., *Talanta* 2008, 74, 1201-1210.

[70] Goicoechea, H., Olivieri, A. C., Tauler, R., *Analyst* 2010, 135, 636-642.

[71] Lyndgaard, L. B., Berg, F. V., de Juan, A., *Chem. Intell. Lab. Syst.* 2013, 125, 58-66.

[72] Ahmadi, G., Abdollahi, H., *Chem. Intell. Lab. Syst.* 2013, 120, 59-70.

[73] Jamout, J., Gargallo, R., Tauler, R., *J. Chemom.* 2004, 18, 327-340.

[74] Tauler, R., de Juan, A. In: *Practical Guide to Chemometrics*; Gemperline, P., 2nd ed., CRC Taylor Francis: New York, 2006; pp. 446-448.

[75] Derks, E. P. P. A., Sanchez, P. M. S., Buydens, L. M. C., *Chem. Intell. Lab. Syst.* 1995, 28, 49-60.

[76] Faber, N. M., *J. Chemom.* 2001, 15, 169-188.

[77] Massart, D. L., Buydens, L. M. C., Vandegiste, B. G. M. *Handbook of Chemometrics and Qualimetrics,* 1st ed., Elsevier: Amsterdam, Netherlands, 1997.

[78] Shao, J., Tu, D. *The Jackknife and Bootstrap*. Springer: New York, 1995.

In: Current Applications of Chemometrics
Editor: Mohammadreza Khanmohammadi
ISBN: 978-1-63463-117-4

Chapter 5

CHEMOMETRIC BACKGROUND CORRECTION IN LIQUID CHROMATOGRAPHY

Julia Kuligowski*[1], *Guillermo Quintás*[2] *and Miguel de la Guardia*[3]

[1]Neonatal Research Group, Health Research Institute Hospital La Fe, Valencia, Spain
[2]Leitat Technological Center, Bio In Vitro Division, Valencia, Spain
[3]Department of Analytical Chemistry, University of Valencia, Burjassot, Spain

ABSTRACT

The analysis of compounds in complex samples is challenging and often requires the use of chromatographic techniques in combination with appropriate detectors. The obtained data quality can be influenced by background changes due to instrumental drifts or the selected chromatographic conditions. Therefore chemometric tools for the compensation of such effects have been developed. This chapter is focused on chemometric background correction from data obtained from on-line hyphenated LC systems with spectrophotometric detectors, because in those systems changes in the background are frequently observed. A selection of univariate and multivariate approaches that have been described in the literature including Cubic Smoothing Splines (CSS), background correction based on the use of a Reference Spectra Matrix (RSM), Principal Component Analysis (PCA), Simple-to-use Interactive Self-modeling Mixture Analysis (SIMPLISMA), Multivariate Curve Resolution – Alternating Least Squares (MCR-ALS), Eluent Background spectrum Subtraction (EBS) and Science Based Calibration (SBC) among others, are briefly introduced including examples of applications. Finally, in the conclusion a brief guide for selecting the most appropriate tool for each situation is provided.

INTRODUCTION

Spectroscopic methods are versatile tools in research as well as in dedicated systems due to the molecular specific information provided. Under certain circumstances, direct analysis

* Corresponding Author: miguel.delaguardia@uv.es

of complex samples is feasible employing spectroscopic detectors often in combination with chemometric tools [1].

Yet, dealing with complex samples, the separation of the target analytes prior to detection is unavoidable for achieving accurate and precise results. Consequently, numerous applications can be found that still require a separation step prior to spectroscopic detection. The most common chromatographic techniques employed in modern analytical laboratories are gas chromatography (GC) and liquid chromatography (LC).

Both techniques require the use of an appropriate detection system. Liquid phase separation systems, i.e., LC and also, capillary electrophoresis (CE) may additionally require the elimination of unwanted background signal variation due to the solvent absorption. During the past years, a number of chemometric tools for improving the outcome of chromatographic results have been developed and applied.

Especially in the field of comprehensive two-dimensional chromatography, a variety of chemometric methods are employed to the resolution of highly complex sample mixtures as well as for the analysis of the information rich data obtained during their analysis [2, 3, 4, 5, 6].

Spectroscopic detection should be compatible with LC systems providing appropriate data acquisition frequency, selectivity and sensitivity levels. Besides, the use of multichannel detectors increases the accessible qualitative and quantitative information. The most commonly employed spectroscopic detection systems in LC are ultraviolet (UV), diode array detection (DAD) and molecular fluorescence detection (FLD). Alternative detection methods that can be used are infrared (IR) and Raman spectroscopy as well as nuclear magnetic resonance (NMR).

Two fundamentally different approaches for hyphenating liquid phase separation systems and spectroscopic detectors can be identified: i) off-line coupling, involving the physical elimination of the mobile phase before spectral acquisition, and ii) on-line coupling, where detection is carried out without effluent elimination, e.g., by using flow-cells [7].

The need for chemometric background correction depends on both, the detection and the chromatographic conditions employed for sample analysis. Under isocratic conditions the mobile phase composition is maintained constant during the chromatographic run, and minor changes in the background signal deriving from e.g., instrumental drifts can be easily corrected by direct subtraction of a spectrum of the mobile phase measured at the beginning of the chromatographic run or directly before the elution of the analytes of interest. In contrast, under gradient conditions the changing mobile phase composition can have a strong effect if the mobile phase components or additives contribute to the signal.

For example, in LC-IR, typical mobile phases employed in reversed and normal phase LC strongly absorb infrared radiation and hence, changes in the absorption caused by the mobile phase can be up to several orders of magnitude more intense than the signal derived from the analyte [8, 9].

This chapter is focused on chemometric background correction from data obtained from on-line hyphenated LC systems with spectrophotometric detectors such as UV, DAD and IR, because those detection techniques are considered to be prone to interferences from unstable or changing background signals.

However, many of the described principles can be extrapolated to other separation and detection systems where similar effects can be observed.

Chemometric Background Correction in Liquid Chromatography

Hyphenated LC instruments provide data that can be written as $D = CS^T + E$, where D $(z \times c)$ is the data matrix with the overall spectral information of all absorbing species, C $(z \times k)$ is the vector of the concentration profiles of the k absorbing species, S^T $(k \times c)$ is a matrix containing the pure spectra of the k components, E $(z \times c)$ is the error matrix and, c and z are the number of detected variables and spectra, respectively.

Under isocratic conditions, the eluent composition and spectrum is constant. Accordingly, direct subtraction of the solvent spectrum in the raw data matrix D allows reducing the complexity of the system, remaining the spectroscopic active eluting analytes and matrix compounds as the only components in the background corrected data matrix. In practice this approach has some important limitations: minor changes instrumental conditions (e.g., temperature, changes in the mobile phase composition delivered by the LC pump) might lead to changes in intensity and shape of the eluent absorbance spectrum. Therefore, the direct subtraction of the reference spectrum might lead to inaccurate background compensation obtaining noisy chromatograms and distorted analyte spectra. On the other hand, under gradient conditions, chemometric background correction is not straightforward, because eluent composition during the run changes and the changes in the background absorbance intensity can be combined with shifts in the eluent band shapes [10].

Univariate Methods for Chemometric Background Correction

A number of univariate approaches aiming at on the fly chemometric background correction have been proposed. Using cubic smoothing splines (CSS) based approach [11] spectra measured in the region before and after chromatographic peak clusters were used as knots to model the variation of the eluent absorption intensity with time.

The approach can be divided into two steps: i) peak detection and ii) calculation of the background signal. The first step can be carried out either manually, or employing peak detection algorithms described in literature or in commercial chromatographic software packages. In the second step using the selected knots, cubic smoothing splines are fitted to model eluent absorption. Then, background correction is accomplished by subtracting the interpolated eluent absorption from the raw data. Cubic smoothing splines were calculated using the MATLAB® csaps function, where a smoothing parameter, that determines how closely the spline follows the given data, has to be defined. The csap function is very sensitive to the choice of p, which influences the accuracy of the achieved correction and has to be optimized. In this approach, accuracy of the chemometric correction depends in a great extent on the selection of the knots which and it can be troublesome when the peaks are not baseline resolved, as it frequently occurs in the analysis of complex samples.

To circumvent this limitation, an alternative strategy based on the use of a reference-spectra matrix (RSM) was proposed [12].

As shown in Figure 1, this is a straightforward strategy that exploits the spectral information obtained during the equilibration of the LC system or a blank gradient run to perform background correction during the LC run.

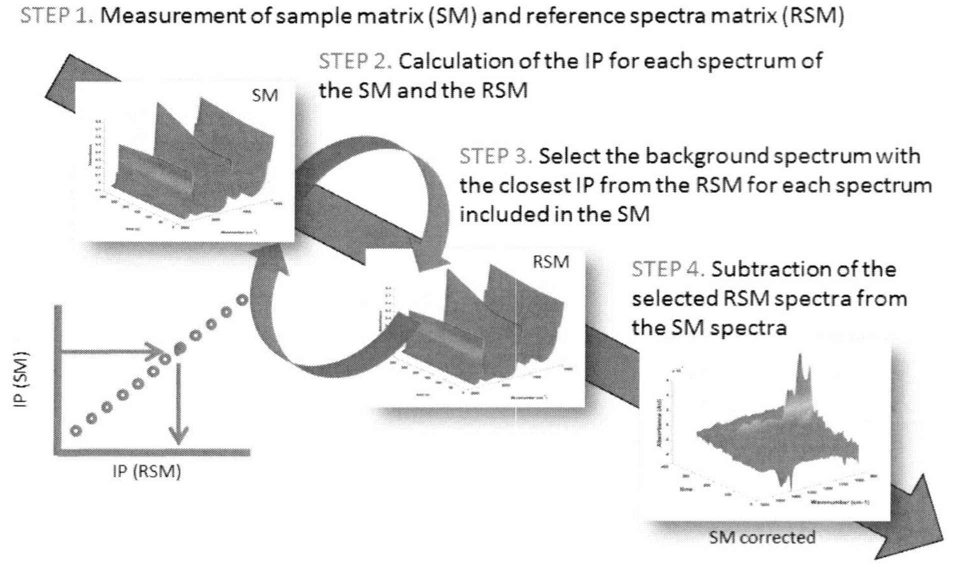

Figure 1. Step-by-step description of the workflow employing RSM based background correction.

The RSM consists of a set of spectra of different mobile phase compositions within the same composition range as that employed during the sample analysis. Then, for each spectrum measured during the sample analysis, a reference spectrum of matching composition is identified in the RSM and used as reference for background correction. Based on this simple strategy, a number of identification parameter (IP) characteristics of the mobile phase composition can be used. The usefulness of the IP depends on the ability of the statistic to discriminate subtle differences in set of spectra. Besides, the identification and correction process must be fast in order to be used on the fly. Different IPs have been evaluated including the absorbance ratio (AR) [12] and the relative absorbance (RW) [13] at two defined wavenumbers $r1$ and $r2$ calculated for each spectrum s in the SM and RSM as shown in equations 1 and 2, respectively:

$$AR_s = \frac{y_{r1}^s}{y_{r2}^s} \quad (1)$$

$$RW_s = y_{r1}^s - y_{r2}^s \quad (2)$$

Similarity or distance methods based on point-to-point matching can also be used as IPs. The correlation coefficient on the mean centered absorbance has been proposed as spectral similarity index [14]. The similarity index COR between two spectra s_A and s_B can be calculated as described in equation 3:

$$COR = \frac{z_A^T z_B}{\|z_A\| \cdot \|z_B\|}, \text{ where } z_n = s_n - \bar{s_n} \quad (3)$$

The usefulness of the different proposed IPs has been assessed under different chromatographic conditions using different reversed and normal mobile phase systems and under both, isocratic and gradient conditions (see Table 1).

Table 1. Examples of applications employing univariate BGC methods

BGC Approach	Application	Chromatography			Detection	Reference
		Column	System	Organic modifier [%] v/v		
CSS	Polymers (PEG)	C4 (150x2.1 mm, 4 μm)	Methanol:H_2O	30-90	FTIR, flow-cell 10 μm	[11]
CSS	Polymers (PEG)	C4 (150x2.1 mm, 4 μm)	2-Propanol:H_2O	10-25	FTIR, flow-cell 10 μm	[11]
CSS	Polymers (PEG)	C4 (150x2.1 mm, 4 μm)	Ethanol:H_2O	10-40	FTIR, flow-cell 10 μm	[11]
AR-RSM	Pesticides	C18 (150×2.1 mm, 3 μm)	Acetonitrile:H_2O	40-99	FTIR, flow-cell 25 μm	[12]
AR-RSM	Polymers (PEG)	C18 (250x2 mm, 5 μm)	Methanol:H_2O	5-100	FTIR, flow-cell 10 μm	[15]
AR-RSM	Carbohydrates	NH_2 (250x2 mm, 5 μm)	Acetonitrile:H_2O	75-55	FTIR, flow-cell 10 μm	[16]
RW-RSM	Nitrophenols	C18 (150x0.3 mm, 3 μm)	Acetonitrile:H_2O	50-65	FTIR, micromachined nL flow-cell 25 μm	[13]
p2p-RSM	Nitrophenols	C18 (150x2 mm, 5 μm)	Acetonitrile:H_2O	35-85	FTIR, flow-cell 16.5 μm	[14]
Polyfit-RSM	Carbohydrates	NH_2 (250x2 mm, 5 μm)	Acetonitrile:H_2O	75-55	FTIR, flow-cell 10 μm	[17]
Polyfit-RSM	Carbohydrates	NH_2 (250x2 mm, 5 μm)	Acetonitrile:H_2O	75-55	FTIR, flow-cell 10 μm	[17]

Note: BGC stands for background correction, RSM stands for reference spectra matrix, conc stands for concentration, AR stands for absorbance ratio, RW stands for relative absorbance, p2p stands for point to point matching, Polyfit stands for polynomial fit, PEG stands for polyethylene glycol and FTIR stands for Fourier transform infrared.

For example, the application of the AR as IP was evaluated in a method for pesticides quantification in reverse phase gradient LC [12], for the determination of the critical eluent composition for poly-ethylene analysis [15] as well as for the quantitative determination of sugars in beverages [16].The absorbance ratio and a point to point matching algorithm was employed for the quantification of nitrophenols employing a micro-LC and a micro-machined nL flow cell [13] as well using standard LC-IR instrumentation [14].

Figure 2 illustrates the background correction process of a nitrophenol standard mixture containing 4-nitrophenol (4-NP), 3-methyl-4-nitrophenol (3m4-NP), 2,4-dinitrophenol (2,4-dNP) and 2-nitrophenol (2-NP). Figure 2a shows the change in the relative absorbance signal at the wavenumbers selected for calculating the RW and Figure 2b shows the same data set after background correction. After elimination of the interfering background contribution, the elution of analytes can be observed in the chromatogram at retention times between 8 and 11 min. A close-up view of the elution time window is shown in Figure 2c, where characteristic nitrophenol bands can be observed. After background correction, a chromatogram can be extracted and used for quantification as shown in Figure 2d.

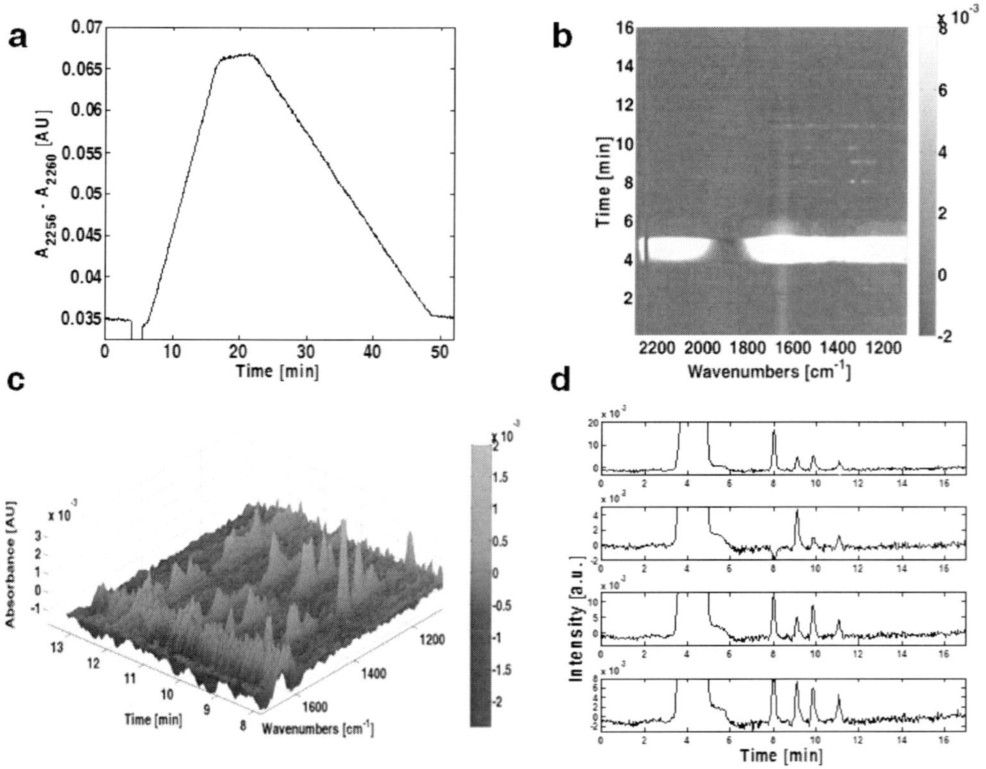

Figure 2. LC-IR data obtained during gradient elution after the injection of a nitrophenol standard solution. Change in RW observed during the gradient (a), contour plot of the background corrected spectra (b), 3D plot of the background corrected spectra in the retention time window 8-13 min (c) and extracted chromatograms (d).

The background correction methods have an important limitation due to the influence on the background correction accuracy of the size of the RSM. In order to limit the effect of this drawback, an alternative approach based on the use of a reference spectral surface calculated from a series of polynomial fittings of the background absorption at each wavenumber in the RSM was proposed [17]. As the absorbance of the eluent at each wavenumber is dependent on its composition, this approach correlates the eluent composition during the gradient with a defined spectral feature related to the mobile phase composition. Two alternative parameters were evaluated: i) the absorbance at a defined single wavenumber (SW-Polyfit-RSM) and ii) the absorbance ratio at two wavenumbers (AR-Polyfit-RSM). The selected parameters have to be characteristic for the absolute or relative concentration of at least one of the components of the mobile phase. Data included in the RSM are used for the calculation of polynomial fits modelling background absorption at each wavenumber. The optimum degree of the calculated polynomial fits might vary within a user-defined range and is selected automatically employing the degrees of freedom adjusted R-square statistic. Then, for each spectrum include in the SM, a set of reference values is measured by using either the SW or AR parameter, and the background contribution is calculated employing the previously calculated polynomial regression model for this wavenumber. In the final step, a simple subtraction of the calculated background signal from the SM is carried out. This method was tested on acetonitrile: water reversed phase gradient LC data sets obtaining background correction accuracies comparable to that provided by AR-RSM.

MULTIVARIATE METHODS FOR CHEMOMETRIC BACKGROUND CORRECTION

As shown above, hyphenated LC instruments provide two-dimensional bilinear data and because of that methods based on factor analysis have been thoroughly employed to identify or correct the different sources of variation in LC data.

Principal component analysis (PCA), simple-to-use interactive self-modeling analysis (SIMPLISMA) and evolving factor analysis (EFA) have been used to extract both, spectral information and elution profiles of analytes.

Besides, they have also been used for the determination of the number of eluting compounds in LC runs with linearly independent spectra and/or eluting profiles (i.e., the chemical rank of the chromatographic data set), and for the chemometric resolution of overlapping analytes. For example, using SIMPLISMA for background correction, a critical step is the determination of the target analytes, as the over or under estimation of that value would obviously affect background correction accuracy. Here, the use of the log-(eigenvalues) obtained from a singular value decomposition (SVD) analysis of the RSM matrix has been successfully used.

On the other hand, parallel factor analysis (PARAFAC, PARAFAC2) and multivariate curve resolution-alternating least-squares (MCR-ALS) have been also employed to resolve the elution profiles and pure instrumental responses. A remarkable feature of all PARAFAC and MCR-ALS strategies is their second order advantage, which allows analyte quantification in the presence of unknown interferences. Table 2 summarizes the most innovative works published on the field of background correction in gradient LC set-ups during the last years, indicating the analytes under study and the characteristics of the chemometric data treatment applied.

Table 2. Examples of applications employing multivariate BGC methods

BGC approach	Application	Chromatography			Detection	Reference
		Column	System	Organic modifier [%] v/v		
PLS-RSM	Pesticides Caffeine	C18 (150×2.1 mm, 3 μm)	Acetonitrile:H_2O	40-100	FTIR, flow-cell 25 μm	[24]
EBS	Polymers	C18 (150×2.1 mm, 5 μm)	Acetonitrile:THF		DAD (flow cell)	[20]
SIMPLISMA	Carbohydrates / Nitrophenols	NH_2 (250x2 mm, 5 μm) / C18 (150x2 mm, 5 μm)	Acetonitrile:H_2O	75-55 / 35-85	FTIR, flow-cell 10 and 16.5 μm	[22]
PCA	Carbohydrates / Nitrophenols	NH_2 (250x2 mm, 5 μm) / C18 (150x2 mm, 5 μm)	Acetonitrile:H_2O	75-55 / 35-85	FTIR, flow-cell 10 and 16.5 μm	[22]
SBC	Nitrophenols	C18 (150x2 mm, 5 μm)	Acetonitrile:H_2O	35-85	FTIR, flow-cell 16.5 μm	[23]
SBC	Pharmaceutical substances	C18 (150×4.6 mm, 5 μm)	KH_2PO_4 buffer:Methanol:Acetonitrile	85:15 - 55:25:20	DAD (flow cell)	[25]

Note: BGC stands for background correction, RSM stands for reference spectra matrix, conc stands for concentration, PLS stands for partial least squares, EBS stands for eluent background spectrum subtraction, SIMPLISMA stands for simple-to-use interactive self-modeling analysis, PCA stands for principal component analysis, SBC stands for science based calibration, FTIR stands for Fourier transform infrared and DAD stands for diode array detector.

Edelmann et al. [18] showed the usefulness of MCR–ALS for the quantification of co-eluting analytes in wine samples analyzed under isocratic conditions in an on-line LC-Fourier transform infrared spectrometry (LC-FTIR) system. Using isocratic conditions, background correction was initially performed by subtracting the average baseline spectrum acquired before co-elution of the target analytes and also using first-derivative spectra to remove baseline drifts. Ruckebusch et al. [19] proposed the use of MCR-ALS for the analysis of gel permeation chromatography-FTIR (GPC-FTIR) data obtained during the analysis of butadiene rubber and styrene butadiene rubber blends. Results obtained provided in-depth knowledge of polymers along the molecular weight distribution.

In 2004 Boelens et al. [20] developed an innovative and efficient method for the eluent background spectrum subtraction (EBS) in LC coupled to spectroscopic techniques (e.g., Raman, UV-Vis). In this method, data measured during the LC run was split into two blocks. The first block consists of the matrix where no analytes are present (i.e., the background spectral subspace or B-subspace), and the other part is formed by spectra in which eluent and analytes are present (i.e., the elution spectra). Initially, variation in the eluent spectra at baseline level is modeled in the background spectral subspace by PCA. Here, the number of principal components is selected according to the empirical IND algorithm. Then, eluent background correction of the elution spectra the PCA is carried out using the PCA loading vectors. This approach assumes, as abovementioned, that the elution spectra are a linear combination of analyte and background spectra, latter being described by the PCA loading vectors. For background correction, these are fitted under the elution spectra by an asymmetric least-squares (asLS) method. The asLS algorithm assumes that the analyte spectrum consists of positive values only and negative residuals of the least-squares fit are penalized while performing the regression in an iterative way. For application of the EBS method, the elution time window of the analytes has to be known in advance. This prerequisite, might limit its applicability. If the identification of the elution window of the analytes is carried out using an additional detection system (e.g., DAD), the presence of non-UV absorbing analytes might lead to a wrong estimation of the number of principal components required to describe the background absorption of the analytes subspace. Besides, a second limitation is that it cannot be carried out on-the-fly because the data set has to be split and analyzed prior to background correction.

In 2006, István et al. [21] employed PARAFAC and PARAFAC2 chemometric background correction on simulated on-line LC-IR runs. Although PARAFAC2 performed better than PARAFAC, it did not give correct decompositions and a new method named objective subtraction of solvent spectrum with iterative use of PARAFAC and PARAFAC2 (OSSS–IU–PARAFAC and OSSS–IU–PARAFAC2, respectively) was developed to improve results. The OSSS–IU–PARAFAC2 was found to be useful even in the presence of extensive spectral overlap between the target analytes and the background spectra, and also for analytes at low concentrations. Whilst the new method improved the accuracy of the decomposition, the restrictions imposed by this approach (e.g., constant eluent composition or constant elution profiles of any given component among different LC runs) limited its practical applicability as the trilinear assumption excludes its application of gradient elution. Besides, for a proper use of the algorithm it requires the use of more than one chromatographic run with different compositions.

The use of MCR-ALS is troublesome when the background absorption is very intense compared to the signal of the analytes of interest as it occurs in hyphenated LC-IR systems.

While chemometric background correction is difficult under both, isocratic and gradient elutions, changes in shape and spectral intensity during gradient runs might increase the chemical rank of the data matrices complicating MCR-ALS data analysis. In this situation, the previous removal of a significant part of the solvent background contributions has shown to be effective to facilitate MCR-ALS analysis and so, to increase the accuracy of the resolution. Two strategies approaches based on SIMPLISMA and PCA were evaluated for chemometric background correction (see Figure 3).

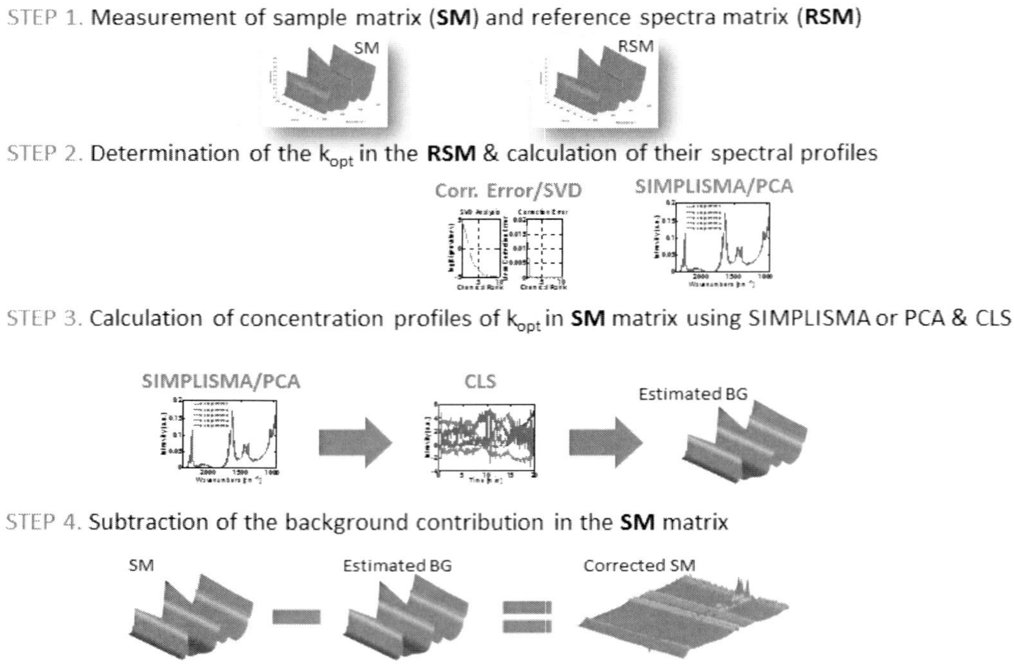

Figure 3. Step-by-step description of the workflow employing PCA/SIMPLISMA based background correction.

After an initial chemometric background correction step, MCR-ALS provided improved signal-to-noise ratios, removed remaining mobile phase and background signal contributions and allowed the resolution of overlapped chromatographic peaks [22]. Moreover, MCR-ALS can be applied exploiting the use of augmented data matrices and/ or equality constraints for concentrations or spectra that might improve results.

Another alternative involving Science Based Calibration (SBC) [23] has been proposed. Here, a reference spectra matrix and a reference spectrum of the analyte of interest are used to extract quantitative chromatographic profiles. The SBC approach is based on the fact that a spectrum can be viewed as the sum of the analyte signal in dependence on its concentration and absorbance that is not derived from the analyte (i.e., instrumental noise, interfering matrix compounds etc.). In the same way, for a set of spectra, the signal can be seen as the sum of the analyte signal described by the mean analyte spectrum and its standard deviation and the spectral noise defined as the mean noise value and its covariance matrix. When the analyte concentration is kept constant, differences between the spectra are only due to i) variations in the concentrations of other sample components and ii) instrumental noise.

Table 3. Method comparison

BGC approach	RSM required	Reference spectra required	Instrument stability	Gradient	Recovery of corrected spectra	On-the-fly BGC	# User-defined variables
CSS	No	No	Not critical	Yes	Yes	No	Low
RSM-AR RW, RSM-p2p	Yes	No	Highly critical	Yes	Yes	Yes	Low
RSM-Polyfit	Yes	No	Highly critical	Yes	Yes	Yes	Low
PCA SIMPLISMA-RSM	Yes	No	Critical	Yes	Yes	No	Medium
EBS	No	No	Critical	Yes	Yes	No	Medium
MCR-ALS	No	No	Not critical	No	Yes	No	High
PARAFAC/PARAFAC2	No	No	Critical	No	Yes	No	High
SBC	No	Yes	Not critical	Yes	No	No	Medium

Consequently, after mean centering, under these conditions the spectra represent only noise. Using this so called "noise matrix", and a reference spectrum from the analyte, a regression model can be calculated which can be employed to predict the concentration of the analyte in new sample spectra.

The performance of the method depends mainly on the representativeness of the employed noise matrix and the quality of the available reference spectrum. Therefore, SBC is well fitted to important characteristics of on-line LC-IR systems.

On the one hand, the target analytes are often known and their spectra can be obtained and so, used as the 'analyte signal' and, on the other hand, if the IR spectrum of the mobile phase is acquired during a blank gradient LC run or during the re-equilibration of the system after the run, the resulting spectral matrix can be used as the 'noise' matrix as it includes both interfering spectra corresponding to the changes in the background absorption due to the LC gradient and instrumental noise.

The usefulness of this approach for background correction has been assessed by means of the injection of a standard solution containing four nithophenols into an on-line LC-IR system.

The advantage of this multivariate approach is that background correction and resolution of analytes of interest and other matrix compounds can be achieved in one step. However, it is not possible to recover background corrected analyte spectra.

METHOD COMPARISON

In Table 3 an overview over the characteristics and constraints of each of the discussed background correction methods can be found. Fundamentally, the described approaches are restrained by the information necessary for computing background correction as well as the instrumental conditions of applicability.

For example, RSM-based methods require the availability of a suitable RSM, whereas for the SBC approach, a reference spectrum of the analytes is necessary.

On the other hand side, for some approaches instrumental stability is of great importance for accurate background correction, whereas other methods can only be employed efficiently under isocratic elution conditions, hence limiting drastically the number of possible applications.

Furthermore, it should be clear for a proper selection of the correction method, which information has to be recovered from the background corrected data. Most of the described methods allow the calculation of background corrected spectra or spectral profiles. However, for example the SBC approach do not provides this information. It might also be relevant to know if on-the-fly background correction is necessary for implementing an analytical tool. There have been developed only very few methods allowing to correct spectra real-time in on-line gradient systems.

Finally, it can be relevant to consider the number of user defined variables that are implied in the correction process. Generally, univariate methods are simpler and require less user-interaction than multivariate methods. In some cases, the optimization of the background correction step can be time consuming and challenging for each new application.

Conclusion

In summary, a panel of univariate and multivariate background correction methods for spectroscopic detection in LC has been proposed. The selection of the most appropriate method depends on the experimental conditions and the available information and it may vary depending on the analytical problem. As a rule of thumb, mathematically simple methods are preferable over more complex approaches because they usually have fewer restrictions and require a lower number of user-defined variables. Besides, no complex data pre-treatments are needed and the need for previous knowledge on matrix compounds is limited. Furthermore, the instrumental conditions as well as the spectral characteristics of analytes, matrix and mobile phase have to be taken into account. Some approaches are more sensitive to detector drifts or retention time changes. Another important point to consider is the need for on-the-fly background correction as well as the need for recovering background corrected analyte spectra.

References

[1] J. Moros, S. Garrigues, M. de la Guardia, *TrAC Trends in Anal. Chem.* 29 (2010) 578-591.

[2] Z.D. Zeng, H.M. Hugel, P.J. Marriott, *Anal. Bioanal Chem* 401 (2011) 2373-2386.

[3] 4. Z.D. Zeng, J. Li, H.M. Hugel, G.W. Xu, P.J. Marriott, TrAC – Trends Anal. Chem 53 (2014) 150-166.

[4] 7. K.M. Pierce, J.C. Hoggard, *Analytical Methods* 6 (2014) 645-653.

[5] C. Ruckebusch, L. Blanchet, *Anal. Chim. Acta* 765 (2013) 28-36.

[6] L.W. Hantao, H.G. Aleme, M.P. Pedroso, G.P. Sabin, R.J. Poppi, F. Augusto, *Anal. Chim. Acta.* 731 (2012) 11-23.

[7] J. Kuligowski, G. Quintás, M. de la Guardia, B. Lendl, *Anal. Chim. Acta.* 679 (2010) 31-42.

[8] J. Kuligowski, G. Quintás, S. Garrigues, B. Lendl, M. de la Guardia, *TrAC Trends in Anal. Chem.* 29 (2010) 544-552.

[9] J. Kuligowski, B. Lendl, G. Quintás in: Liquid Chromatography: Fundamentals and Instrumentation, Fanali, P.R. Haddad, C.F. Poole, P. Schoenmakers, D. Lloyd (Eds.), Elsevier 2013.

[10] J. Kuligowski, On-line Coupling of Liquid Chromatography and Infrared Spectroscopy, Lambert Academic Publishing 2012.

[11] J. Kuligowski, D. Carrión, G. Quintás, S. Garrigues, M. de la Guardia, *J. Chromatogr A* 1217 (2010) 6733-6741.

[12] G. Quintás, B. Lendl, S. Garrigues, M. de la Guardia, *J. Chromatogr. A* 1190 (2008) 102-109.

[13] G. Quintás, J. Kuligowski, B. Lendl. *Anal. Chem* 81 (2009) 3746-3753.

[14] J. Kuligowski, G. Quintás, S. Garrigues, M. *de la Guardia Talanta* 80 (2010) 1771-1776.

[15] J. Kuligowski, G. Quintás, S. Garrigues, M. de la Guardia, *Anal. Chim. Acta.* 624 (2008) 278-285.

[16] J. Kuligowski, G. Quintás, S. Garrigues, *M. de la Guardia, Talanta* 77 (2008) 779-785.
[17] J. Kuligowski G. Quintás, S. Garrigues, M. de la Guardia *J. Chromatogr A* 1216 (2009) 3122-3130.
[18] Edelmann, J. Diewok, J. Rodriguez Baena, B. Lendl, *Anal. Bioanal. Chem.* 376 (2003) 92-97.
[19] Ruckebusch, F. Vilmin, N. Coste, J.-P. Huvenne, *Applied Spectroscopy* 62 (2008) 791-797.
[20] H.F.M. Boelens, R.J. Dijkstra, P.H.C. Eilers, F. Fitzpatrick, J.A. Westerhuis, *J. Chromatogr. A* 1057 (2004) 21-30.
[21] K. Istvan, R. Rajko, G. Keresztury, *J. Chromatogr., A* 1104 (2006) 154.
[22] J. Kuligowski, G. Quintás, R. Tauler, B. Lendl, *M. de la Guardia Analytical Chemistry* 83 (2011) 4855-4862.
[23] J. Kuligowski, G. Quintás, S. Garrigues, *M. de la Guardia, J. Sep. Sci.* 32 (2009) 1-7.
[24] G. Quintás, J. Kuligowski, B. Lendl, *Applied Spectroscopy* 63 (2009) 1363-1369.
[25] J. Kuligowski, M. Martinez Galera, M.D. Gil Garcia, M.J. Culzoni, H.C. Goicoechea, S. Garrigues, G. Quintas, *M. de la Guardia, Talanta* 83 (2011) 1158-1165.

In: Current Applications of Chemometrics
Editor: Mohammadreza Khanmohammadi

ISBN: 978-1-63463-117-4

Chapter 6

How to Compute the Area of Feasible Solutions A Practical Case Study and Users' Guide to *FAC-PACK*

***M. Sawall*[1,*] *and K. Neymeyr*[1,2,†]**
[1]Institut für Mathematik, Universität Rostock, Rostock, Germany
[2]Leibniz-Institut für Katalyse, Rostock, Germany

Abstract

The area of feasible solutions (AFS) is a low dimensional representation of the set of all possible nonnegative factorizations of a given spectral data matrix. The AFS can be used to investigate the so-called rotational ambiguity of MCR methods in chemometrics, and the AFS is an excellent starting point for feeding in additional information on the chemical reaction system in order to reduce the ambiguity up to a unique factorization.

The *FAC-PACK* program package is a freely available and easy-to-use Matlab toolbox (with a core written in C and FORTRAN) for the numerical AFS computation for two- and three-component systems. The numerical algorithm of *FAC-PACK* is based on the inflation of polygons and allows a stable and fast computation of the AFS. In this contribution we explain all functionalities of the *FAC-PACK* software and demonstrate its application to a consecutive reaction system with three components.

1. Introduction

Multivariate curve resolution techniques in chemometrics suffer from the non-uniqueness of the nonnegative matrix factorization $D = CA$ of a given spectral data matrix $D \in \mathbb{R}^{k\times n}$. Therein $C \in \mathbb{R}^{n\times s}$ is the matrix of pure components concentration profiles and $A \in \mathbb{R}^{s\times n}$ is the matrix of the pure component spectra. Further, s denotes the number of independent chemical components. This non-uniqueness of the factorization is often

*E-mail address: mathias.sawall@uni-rostock.de
†E-mail address: klaus.neymeyr@uni-rostock.de

called the rotational-ambiguity. The set of all possible factorizations of D into nonnegative matrices C and A can be presented by the so-called Area of Feasible Solutions (AFS).

The first graphical representation of an AFS goes back to Lawton and Sylvestre in 1971 [10]. They have drawn for a two-component system the AFS as a pie-shaped area in the plane; this area contains all pairs of the two expansion coefficients with respect to the basis of singular vectors, which result in nonnegative factors C and A. In 1985 Borgen and Kowalski [4] extended AFS computations to three-component systems by the so-called Borgen plots, see also [14, 22, 1, 2].

In recent years, methods for the numerical approximation of the AFS have increasingly gained importance. Golshan, Abdollahi and Maeder presented in 2011 a new idea for the numerical approximation of the boundary of the AFS for three-component systems by chains of equilateral triangles [5]. This idea has even been extended to four-component systems [6].

An alternative numerical approach for the numerical approximation of the AFS is the *polygon inflation method* which has been introduced in [18, 19, 20]. The simple idea behind this algorithm is to approximate the boundary of each segment of the AFS by a sequence of adaptively refined polygons. A numerical software for AFS computations with the polygon inflation method is called *FAC-PACK* and can be downloaded from

http://www.math.uni-rostock.de/facpack/

The current software revision 1.1 of *FAC-PACK* appeared in February 2014.

The aim of this contribution is to give an introduction into the usage of *FAC-PACK*. In Section 2 a sample model problem is introduced, the initial data matrix is generated in MATLAB and some functionalities of the software are demonstrated. In Section 3 this demonstration is followed by the Users' guide to *FAC-PACK*, which appears here for the first time in printed form.

2. A Practical Case Study

2.1. Generation of Test Data

Let us consider the consecutive system of reactions $X \to Y \to Z$ with kinetic constants $k_1 = 0.75$ and $k_2 = 0.25$. The time interval for the concentration factor C is assumed to be $t \in [0, 20]$ with an equidistant discretization by $k = 21$ grid points. The spectral factor A is constructed on the wavelength interval $\lambda \in [0, 100]$ with an equidistant discretization which uses $n = 101$ points. The three Gaussian functions (plus a constant)

$$a_1(\lambda) = 0.95 \exp(-\frac{(\lambda - 20)^2}{500}) + 0.3,$$
$$a_2(\lambda) = 0.9 \exp(-\frac{(\lambda - 50)^2}{500}) + 0.25,$$
$$a_3(\lambda) = 0.7 \exp(-\frac{(\lambda - 80)^2}{500}) + 0.2$$

define the rows of A by their evaluation along the discrete wavelength axis. According to the bilinear Lambert-Beer model the spectral data matrix $D = CA$ for this reaction is a

21×101 matrix with the rank 3. This model problem is a part of the *FAC-PACK* software; see the data set example2.mat in Section 3.3.3. The pure component concentration profiles and spectra together with the mixture spectra for the sample problem are shown in Figure 1.

The MATLAB code to generate data matrix D reads as follows:

```
t = linspace(0,20,21)';
x = linspace(0,100,101)';
k = [0.75 0.25];
C(:,1)  = exp(-k(1)*t);
C(:,2)  = k(1)/(k(2)-k(1))*(exp(-k(1)*t)-exp(-k(2)*t));
C(:,3)  = 1-C(:,1)-C(:,2);
A(1,:)  = 0.95*exp(-(x-20).^2./500)+0.3;
A(2,:)  = 0.9*exp(-(x-50).^2./500)+0.25;
A(3,:)  = 0.7*exp(-(x-80).^2./500)+0.2;
D = C*A;
save('example2', 'x', 't', 'D');
```

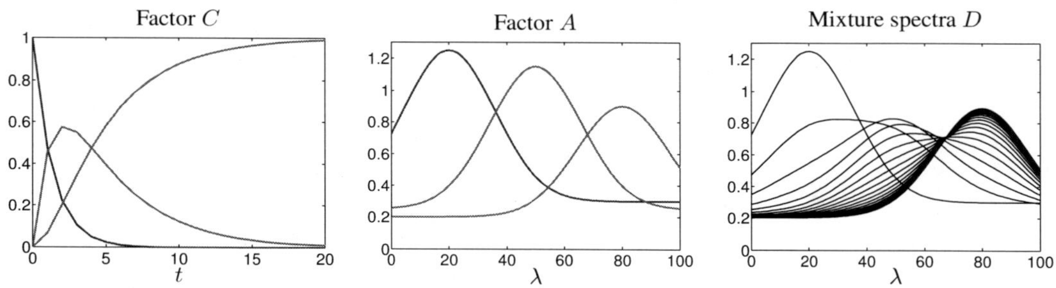

Figure 1. Pure component concentration profiles and spectra together with the mixture spectra for the sample problem presented in Section 2.1.

2.2. AFS Computation with *FAC-PACK*

Here we consider the *factorization problem* which is an inverse problem compared to the construction of D in Section 2.1. For given D we are looking for matrix factorizations $D = CA$ with nonnegative factors C and A. The pair of matrices (C, A) as constructed in Section 2.1 is one possible factorization of D within the set of all nonnegative matrix factorizations of D. The set of all nonnegative factorizations is described by the area of feasible solutions (AFS), which is a certain low dimensional representation of these feasible matrix pairs. For geometric techniques to compute the AFS see [4, 14] and the references therein. For numerical methods to compute the AFS see [1, 5, 2, 6, 18, 19].

In *FAC-PACK* the computation of the AFS is based on the polygon inflation algorithm, which is explained in [18, 19]. The users' guide to *FAC-PACK* is contained in Section 3; a quick introduction is given in Section 3.1 and a detailed introduction is given in Section 3.2. The program starts with the import of the spectral data, see Section 3.3.2. In the following we consider the model problem from Section 2.1. For this sample data there is no need for

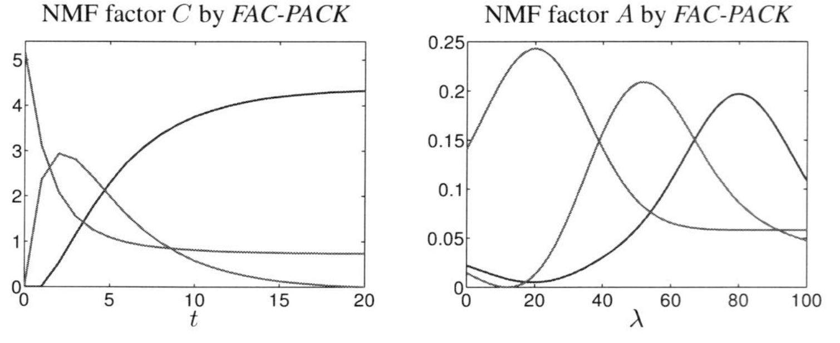

Figure 2. Initial nonnegative matrix factorization (NMF) for the model problem as computed by *FAC-PACK*.

a baseline correction. Otherwise the baseline correction module as introduced in Section 3.1.1 can be used.

The first step of the polygon inflation algorithm is the computation of a singular value decomposition (SVD). The SVD is computed automatically after the data matrix is loaded. For the given sample problem the four largest singular values are

$$\sigma_1 = 23.920, \quad \sigma_2 = 7.444, \quad \sigma_3 = 2.991, \quad \sigma_4 = 0.000.$$

Since $\sigma_4 = 0$, the rank of D equals 3 and D represents a three-component system. After selecting 3 as the number of components in the AFS computation module, an initial nonnegative matrix factorization (NMF) can be computed, see Section 3.5.3. This initial NMF often has no physical or chemical meaning and is sometimes called an abstract factorization. However, this initial NMF provides the starting point from which the polygon inflation algorithm inflates the polygons. The initial NMF for the model problem is presented in Figure 2.

The set of all nonnegative matrix factorizations of D is represented by the area of feasible solutions (AFS). For a three-component system the AFS is a subset of the two-dimensional plane, see Section 3.5.4 and [18] for the mathematical background of this low-dimensional representation. Section 3.5.4 additionally contains the setting options for the various control parameters of the polygon inflation algorithm and also the options on the algorithmic variants of the polygon inflation algorithm. The resulting area of feasible solutions (AFS) for the spectral factor A is denoted by $\mathcal{M}_A$ and is shown in Figure 3. This AFS consists of three separated segments (isolated subsets).

Each of the three segments of the spectral AFS $\mathcal{M}_A$ is approximated by an adaptively generated polygon. The AFS $\mathcal{M}_C$ for the concentration factor C can be computed in a similar way; then the complete algorithm is applied to the transposed data matrix D^T. In some cases the AFS consists of only one segment with a hole or of more than three segments. Then the AFS is computed by a modification of the method which is called the inverse polygon inflation algorithm [19]. In this modified algorithm first the superset FIRPOL of the AFS is computed and in a second step points not belonging to the AFS are eliminated by inflation a second polygon in the interior of FIRPOL.

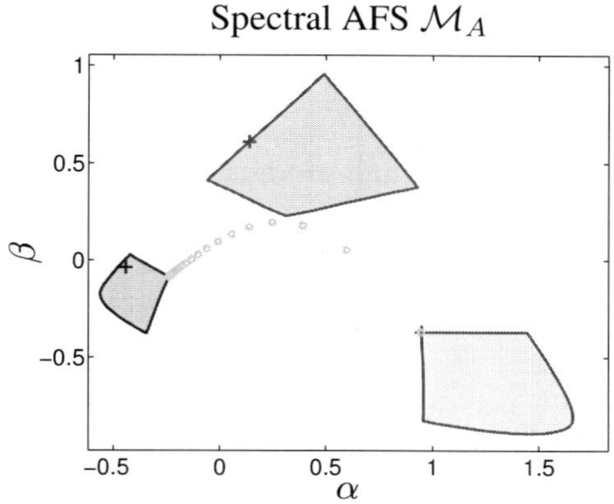

Figure 3. The spectral AFS $\mathcal{M}_A$ for the model problem from Section 2.1. The three markers (+) indicate the components of the initial NMF. These three points are the starting points for the polygon inflation method. Furthermore, the series of points marked by the symbol ∘ represent the single spectra, which are rows of the spectral data matrix D.

2.3. Reduction of the Rotational Ambiguity by the Complementarity Theorem

The *bilinear* Lambert-Beer model $D = CA$ results in *two* factors and so *two* areas of feasible solutions can be constructed. The spectral AFS $\mathcal{M}_A$ has been considered in Section 2.2. The concentrational AFS $\mathcal{M}_C$ can be constructed in the same way if the algorithm is applied to the transposed data matrix D^T. The simultaneous representation of $\mathcal{M}_A$ and $\mathcal{M}_C$ can be very advantageous. In this way restrictions can be analyzed of the partial knowledge of one factor on the other factor. This mutual dependence of the factors is a consequence of the factorization

$$D = U\Sigma V^T = \underbrace{U\Sigma T^{-1}}_{C}\underbrace{TV^T}_{A}.$$

where partial knowledge of A implies some restrictions on the matrix elements of the $s \times s$ matrix T, this implies some restrictions on T^{-1} and so the factor C is to some extent predetermined. Therein s is the rank of D, which is the number of independent components. A systematic analysis of these restrictions appeared in [17] and has led to the complementarity theorem. The application of the complementarity theorem to the AFS is analyzed in [20], see also [3].

For the important case of an $s = 3$-component system a given point in one AFS (either a given spectrum or concentration profile) is associated with a straight line in the other AFS. The form of this straight line is specified by the complementarity theorem. Only points which are located on this line and which are elements of the AFS are consistent with the pre-given information. In case of a four-component system a given point in $\mathcal{M}_A$ (or $\mathcal{M}_C$) is associated with a plane in $\mathcal{M}_C$ (or $\mathcal{M}_A$) and so on.

For three-component systems this simultaneous representation of $\mathcal{M}_A$ and $\mathcal{M}_C$ together with a representation of the restrictions of the AFS by the complementarity theorem is a functionality of the **Complementarity & AFS** module of *FAC-PACK*. This module, its options and results are explained in the users' guide in Section 3.1.3. This module gives the user the opportunity to fix a number of up to three points in the AFS. These points correspond to known spectra or known concentration profiles. For example such information can be available for the reactants or for the final products of a chemical reaction. Typically any additional information on the reaction system can be used in order to reduce the rotational ambiguity for the remaining components. Within the AFS computation module the locking of spectra (or concentration profiles) results in smaller AFS segments for the remaining components, see Section 3.5.6. Here we focus on the simultaneous representation of the spectral and the concentrational AFS and the mutual restrictions of the AFS by adding partial information on the factors. Up to three points (for a three-component system) in an AFS can be fixed. These three points determine a triangle in this AFS and also a second triangle in the other AFS. These two triangles correspond to a factorization of D into nonnegative factors C and A. All this is explained in the users' guide in Section 3.6. The reader is invited to experiment with the MATLAB GUI of *FAC-PACK*, to move the vertices of the triangles through the AFS by using the mouse pointer and to watch the resulting changes for all spectra and concentration profiles. This interactive graphical representation of the system provides the user with the full control of the factors and supports the selection of a chemically meaningful factorization.

In Figure 4 the results of the application of the **Complementarity & AFS** module are shown for the sample problem from Section 2.1. The four rows of this figure show the following:

1. First row: A certain point in the AFS $\mathcal{M}_A$ is fixed and the associated spectrum $A(1,:)$ is shown. This point is associated with a straight line in the concentrational AFS $\mathcal{M}_C$. This line intersects two segments of $\mathcal{M}_C$. The complementary concentration profiles $C(:,2)$ and $C(:,3)$ are restricted to the two continua of the intersection. Up to now no concentration profile is uniquely determined.

2. Second row: A second point in $\mathcal{M}_A$ is fixed and this second spectrum $A(2,:)$ is shown. A second straight line is added to $\mathcal{M}_C$ by the complementarity theorem. The intersection of these two lines in $\mathcal{M}_C$ uniquely determines the complementary concentration $C(:,3)$.

3. Third row: Three points are fixed in $\mathcal{M}_A$ which uniquely determines the factor A. The triangle in $\mathcal{M}_A$ corresponds with a second triangle in $\mathcal{M}_C$. All this determines a nonnegative factorization $D = CA$.

4. Fourth row: A second nonnegative factorization is shown, which exactly reproduces the original components of the model problem from Section 2.1.

2.3.1. Soft Constraints and the AFS

Multivariate curve resolution methods often use soft-constraints in order to favor solutions which are particularly smooth, monotone, unimodal or that have other comparable properties. While moving vertices of triangle-factors interactively through the AFS the user might be interested to see where in the AFS solutions with certain properties can be found.

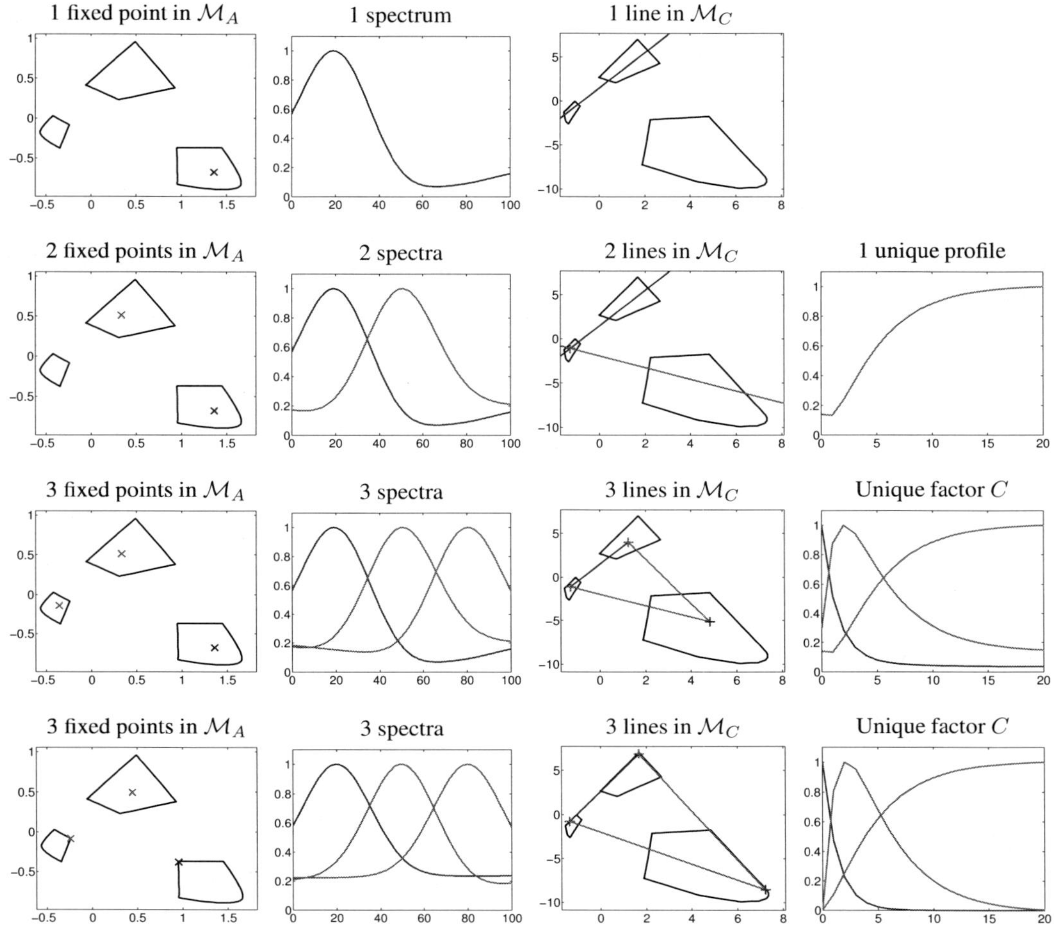

Figure 4. Successive reduction of the rotational ambiguity. First row: one fixed spectrum is associated with a straight line in $\mathcal{M}_C$. Second row: two fixed spectra in $\mathcal{M}_A$ result in two straight lines in $\mathcal{M}_C$. The point of intersection uniquely determines one concentration profile. Third row: three fixed points in $\mathcal{M}_A$ uniquely determine a complete factorization $D = CA$. Fourth row: This complete factorization reproduces the original factors from Section 2.1.

To this end the **Complementarity & AFS** module of *FAC-PACK* provides the option to plot the level set contours for certain constraint functions within the AFS window. Contour plots are available for estimating the

- monotonicity of the concentration profiles (in $\mathcal{M}_C$),
- the smoothness of spectra (in $\mathcal{M}_A$) or concentration profiles (in $\mathcal{M}_C$),
- the Euclidean norm of the spectra (in $\mathcal{M}_A$) in order to estimate the total absorbance of the spectra,

- and the closeness to an exponentially decaying function (in $\mathcal{M}_C$) in order to find concentration profiles of reactants which are degraded in a first order reaction.

In Figure 5 two contour plots are shown. First, in $\mathcal{M}_A$ the closeness of a concentration profile to an exponentially decaying function is shown. The smallest distance, which is shown by the lightest grey has the coordinates $(\alpha, \beta) \approx (7.2714, -8.5516)$. This point is the same which is shown in the fourth row and third column of Figure 4 and which can be identified with the reactant X in the model problem from Section 2.1. Second, the contours of the Euclidean norm of the spectra within the $\mathcal{M}_A$ window. These contours serve to estimate the total absorbance of the spectra and can for instance be used in order to identify spectra with few and narrow peeaks.

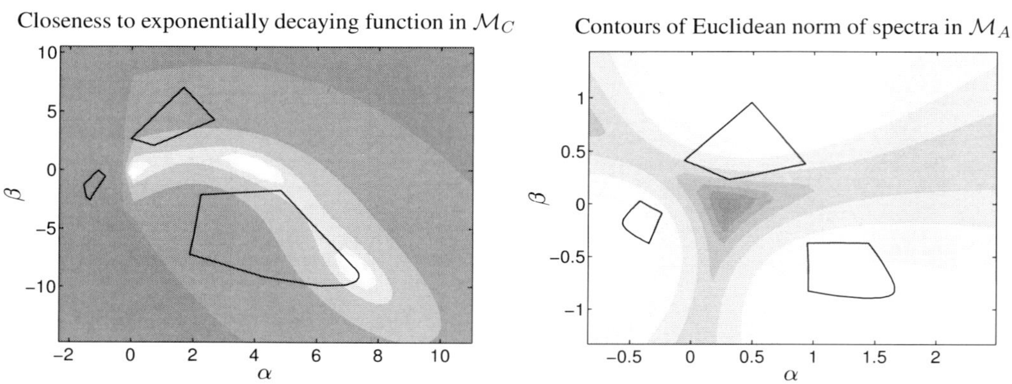

Figure 5. Contour plots within the (α, β)-plane close to the areas of feasible solutions $\mathcal{M}_C$ and $\mathcal{M}_A$ for the data from Section 2.1. Left: Contours of closeness of a concentration profile to an exponentially decaying function in the $\mathcal{M}_C$ window. Right: The contour plot of the Euclidean norm (total absorbance) of the pure component spectra within the $\mathcal{M}_A$ window.

3. Users' Guide to *FAC-PACK*

Here the users' guide on the revision 1.1 of *FAC-PACK* follows.

FAC-PACK is an easy-to-use software for the computation of nonnegative multi-component factorizations and for the numerical approximation of the area of feasible solutions (AFS). Revision 1.1 contains new functionalities for the reduction of the rotational ambiguity.

Important features of *FAC-PACK* are:

- a fast C program with a graphical user interface (GUI) in MATLAB,
- baseline correction of spectral data,
- computation of a low-rank approximation of the spectral data matrix and initial nonnegative matrix factorization,

- computation of the spectral and concentrational AFS by the polygon inflation method,
- applicable to two- and three-component systems,
- live-view mode and factor-locking for a visual exploration of all feasible solutions,
- reduction of the AFS by factor-locking,
- simultaneous representation of the spectral and concentrational AFS,
- reduction of the rotational ambiguity by the complementarity theorem.

3.1. Quick Start

Download the *FAC-PACK* software from

http://www.math.uni-rostock.de/facpack/

Then extract the software package facpack.zip, open a MATLAB desktop window and start facpack.m. The user can now select one of the three modules of *FAC-PACK*, see Figure 6.

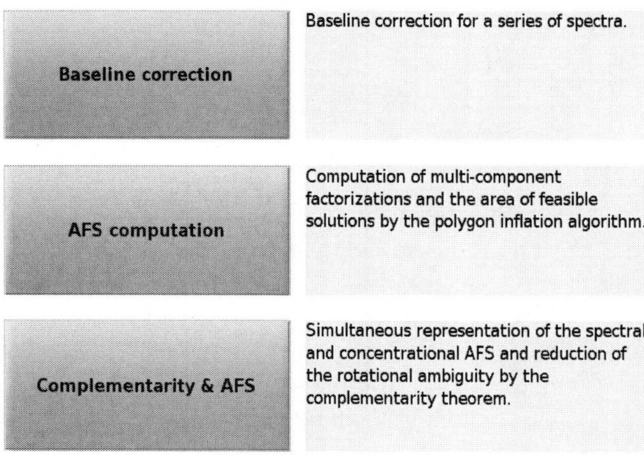

Figure 6. The start window of *FAC-PACK*.

3.1.1. Baseline Correction Module

Step 1: If the baseline needs a correction (e.g., if a background subtraction has partially turned the series of spectra negative), then the **Baseline correction** button can be pressed. This simple correction method only works for spectral data in which frequency windows with a distorted baseline are clearly separated from the main signal groups of the spectrum; see Figure 7.

Step 2: Load the spectral data matrix. Sample data shown above is example0.

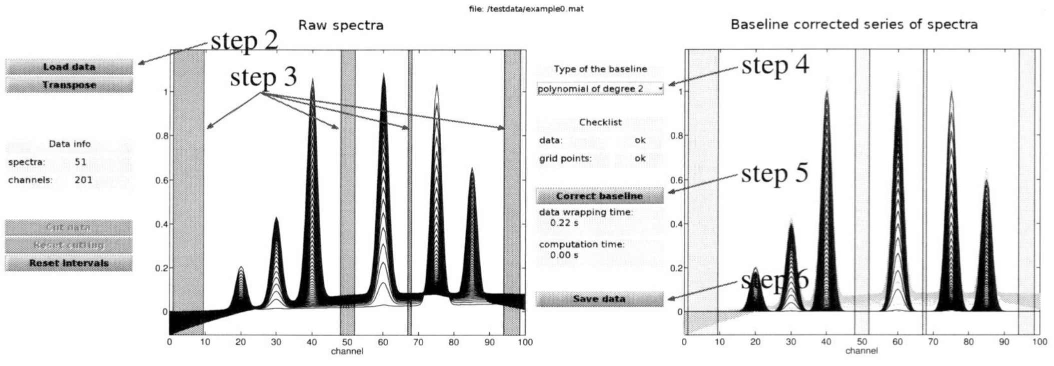

Figure 7. Baseline correction within 6 steps.

Step 3: One or more frequency windows should be selected in which more or less only the distorted background signal is present and which are clearly separated from the relevant spectral signals. In each of these frequency windows (marked by columns) the series of spectra is corrected towards zero. However, the correction is applied to the full spectrum. In order to select these frequency windows click the left mouse button within the raw data window, move the mouse pointer and release the mouse button. This procedure can be repeated.
Step 4: Select the correction method, e.g., polynomial degree 2.
Step 5: Compute the corrected baseline.
Step 6: Save the corrected data.

3.1.2. AFS Computation Module

Step 1: Press the Button **AFS computation** in order to compute the AFS.
Step 2: Select example2 as sample spectral data matrix.
Step 3: Select 3 as the number of components.
Step 4: Compute an initial nonnegative matrix factorization (NMF).
Step 5: Choose Polygon inflation.
Step 6: Compute the AFS which consists of three segments.
Step 7: Plot the range of spectral factors associated with segment 1 (leftmost segment) of the AFS.

A second test problem is shown in Figure 9:

Step 1: Select example3 as the data matrix.
Step 2: Select 3 as the number of components.
Step 3: Compute an initial nonnegative matrix factorization (NMF).
Step 4: Choose Inverse polygon inflation. The AFS consists of only one segment with a hole.
Step 5: Compute the AFS.
Step 6: Select live-view on.
Step 7: Move the mouse pointer through the AFS and watch the interactively computed solutions.

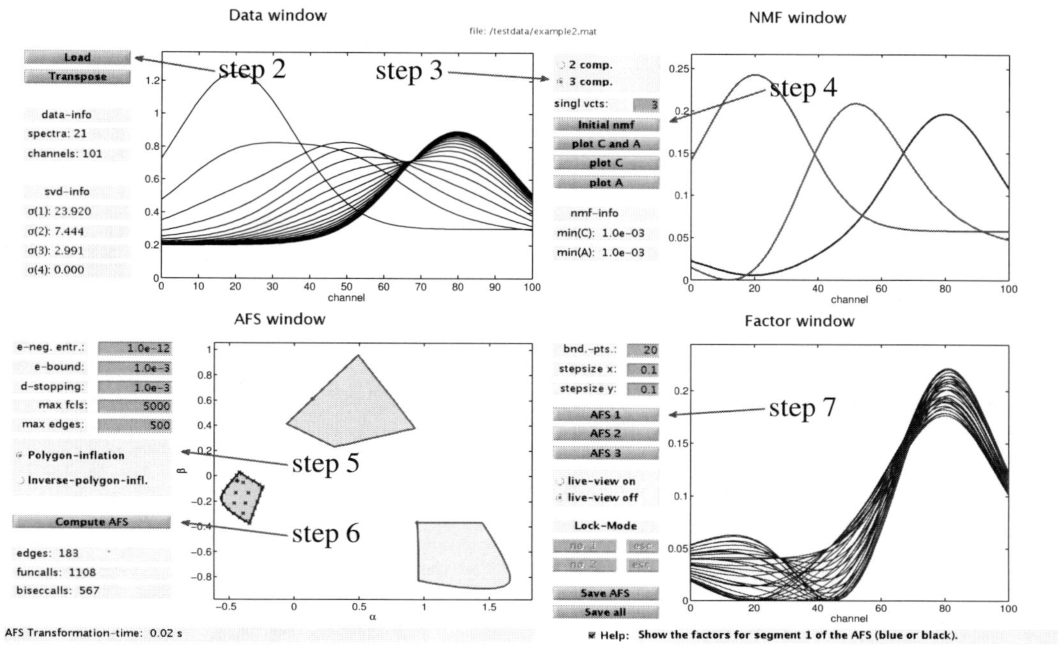

Figure 8. Quickstart in 6 steps.

3.1.3. Complementarity & AFS Module

Step 1: Press the button **Complementarity & AFS** to activate this GUI.

Step 2: The sample data example2 can be selected.

Step 3: Select FIRPOL to plot two supersets including the AFS for C and A and/or select full to show the AFS for the spectral and the AFS for concentrational factor. The sets FIRPOL can easily and quickly be computed; the computation of the AFS for C and for A may be time-consuming.

Step 4: Click the **first** button in the AFS live-view/factor A input block and move the mouse pointer through the spectral AFS (factor A). A first spectrum $A(1,:)$ can be locked by clicking in the AFS. The associated spectrum is shown in the spectral factor window (right-upper window).

Step 5: Click the **second** button and move the mouse pointer once again through the AFS for A. While moving the mouse pointer through the AFS a second spectrum $A(2,:)$ is shown in the spectral factor window together with the unique (by the complementarity theorem) concentration profile of the remaining third component. A certain second spectrum $A(2,:)$ can be locked by clicking in the AFS.

Step 6: Finally, click the **third** button and the spectrum $A(3,:)$ can be selected by moving the mouse pointer through the spectral AFS. The resulting predictions on the spectral factor are shown interactively. This third spectrum can also be locked.

Step 7: These last three buttons and also the buttons **first**, **second**, **third** in the AFS live-view/factor C block can be clicked and then a spectrum or concentration profile can be modified by moving the mouse pointer through the AFS. This allows to modify the two triangles in the spectral and concentrational AFS which uniquely determine

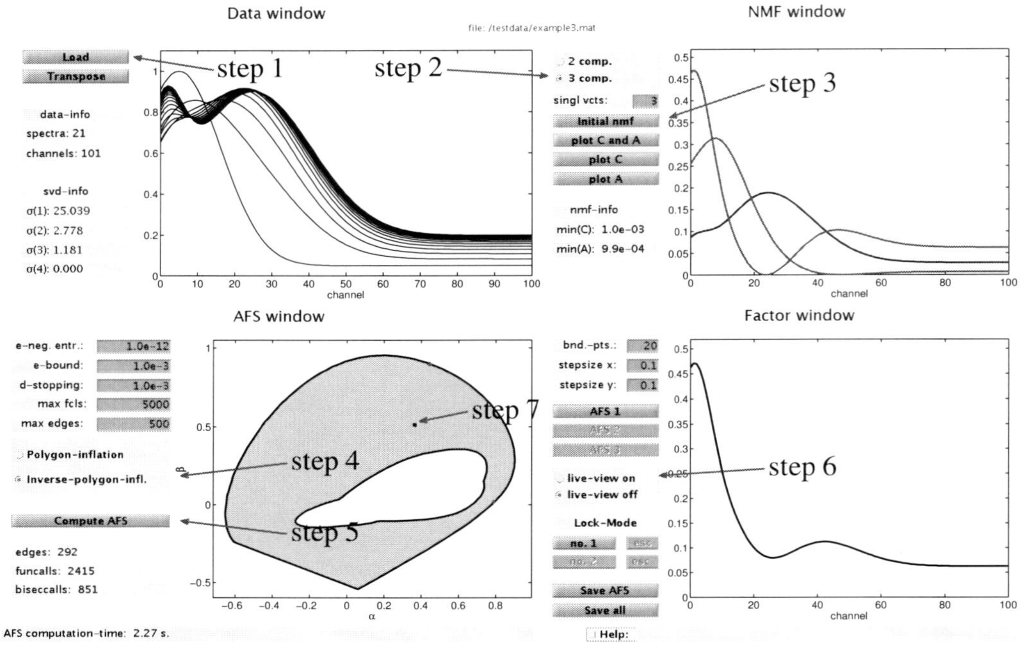

Figure 9. Quick start with live-view mode in 7 steps.

a feasible factorization of the given spectral data matrix.

In the following sections the functionalities of *FAC-PACK* are explained in detail.

3.2. Introduction to *FAC-PACK*

FAC-PACK is a software for the computation of nonnegative multi-component factorizations and for the numerical approximation of the area of feasible solutions (AFS). Currently, the software can be applied to systems with $s = 2$ or $s = 3$ components.

Given a nonnegative matrix $D \in \mathbb{R}^{k \times n}$, which may even be perturbed in a way that some of its entries are slightly negative, a multivariate curve resolution (MCR) technique can be used to find nonnegative matrix factors $C \in \mathbb{R}^{k \times s}$ and $A \in \mathbb{R}^{s \times n}$ so that

$$D \approx CA. \tag{1}$$

Some references on MCR techniques are [10, 7, 12, 11, 15]. Typically the factorization (1) does not result in unique nonnegative matrix factors C and A. Instead a continuum of possible solutions exists [12, 22, 8]; this non-uniqueness is called the rotational ambiguity of MCR solutions. Sometimes additional information can be used to reduce this rotational ambiguity, see [9, 16] for the use of kinetic models.

The most rigorous approach is to compute the complete continuum of nonnegative matrix factors (C, A) which satisfy (1). In 1985 Borgen and Kowalski [4] found an approach for a low dimensional representation of this continuum of solutions by the so-called area of feasible solutions (AFS). For instance, for a three-component system ($s = 3$) the AFS

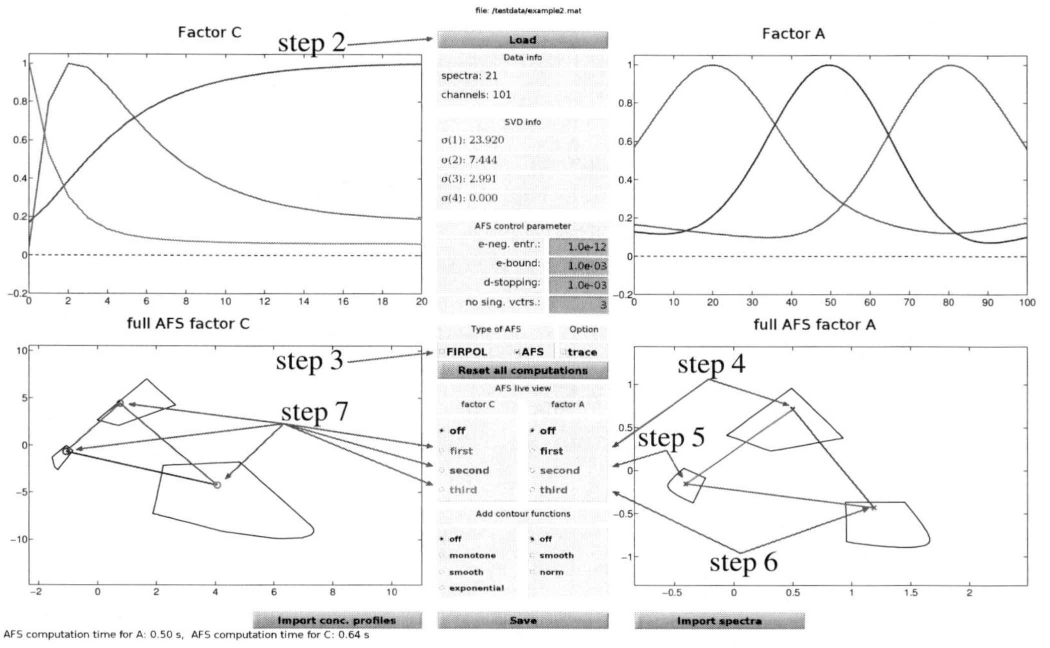

Figure 10. The complementarity & AFS module.

is a two-dimensional set. Further references on the AFS are [21, 14, 22, 1]. For a numerical computation of the AFS two methods have been developed: the triangle-boundary-enclosure scheme [5] and the polygon inflation method [18]. *FAC-PACK* uses the polygon inflation method.

FAC-PACK works as follows: First the data matrix D is loaded. The singular value decomposition (SVD) is used to determine the number s of independent components underlying the spectral data in D. *FAC-PACK* can be applied to systems with $s = 2$ or $s = 3$ predominating components. Noisy data is not really a problem for the algorithm as far as the SVD is successful in separating the characteristic system data (larger singular values and the associated singular vectors) from the smaller noise-dependent singular values. Then the SVD is used to compute a low rank approximation of D. After this an initial nonnegative matrix factorization (NMF) is computed from the low rank approximation of D. This NMF is the starting point for the polygon inflation algorithm since it supplies two or three points within the AFS. From these points an initial polygon can be constructed, which is a first coarse approximation of the AFS. The initial polygon is inflated to the AFS by means of an adaptive algorithm. This algorithm allows to compute all three segments of an AFS separately. Sometimes the AFS is a single topologically connected set with a hole. Then an "inverse polygon inflation" scheme is applied. The program allows to compute from the AFS the continuum of admissible spectra. The concentration profiles can be computed if the whole algorithm is applied to the transposed data matrix D^T. Alternatively, the spectral and the concentrational AFS can be computed simultaneously within the "Complementarity & AFS" graphical user interface (GUI).

FAC-PACK provides a live-view mode which allows an interactive visual inspection of the spectra (or concentration profiles) while moving the mouse pointer through the AFS. Within the live-view mode the user might want to lock a certain point of the AFS, for instance, if a known spectrum has been found. Then a reduced and smaller AFS can be computed, which makes use of the fact that one spectrum is locked. For a three-component system this locking-and-AFS-reduction can be applied to a second point of the AFS.

Within the GUI "Complementarity & AFS" the user can explore the complete factorization $D = CA$ simultaneously in the spectral and the concentrational AFS. The factorization $D = CA$ is represented by two triangles. The vertices of these triangles can be moved through the AFS and can be locked to appropriate solutions. During this procedure the program always shows the associated concentration profiles and spectra for all components. In this way *FAC-PACK* gives the user a complete and visual control on the factorization. It is even possible to import certain pure component spectra or concentration profiles in order to support this selection-and-AFS-reduction process within the "Complementarity & AFS" GUI.

3.3. Get Ready to Start

Please download *FAC-PACK* from

http://www.math.uni-rostock.de/facpack/

Then extract the file facpack.zip. Open a MATLAB desktop and run facpack.m.

This software is shareware that can be used by private and scientific users. In all other cases (e.g. commercial use) please contact the authors. We cannot guarantee that the software is free of errors and that it can successfully be used for a particular purpose.

3.3.1. Program Structure

The current revision of *FAC-PACK* consists of three graphical user interface (GUI) windows to solve the following problems in multivariate curve resolution:

1. Correction of the baseline for given spectroscopic data or pre-processed data, e.g., after background subtraction,

2. Computation of the area of feasible solutions for two- and three-component systems,

3. Simultaneous representation of the spectral and concentrational AFS and interactive reduction of the rotational ambiguity up to uniqueness by means of the complementarity theorem [17, 20].

Once *FAC-PACK* is started it is necessary to select one of the three GUIs "Baseline correction", "AFS computation" or "Complementarity & AFS". No additional MATLAB toolboxes are required to run the software.

Baseline correction This GUI has the following two windows:

1. Raw data window (Left window): The raw data window shows the k raw spectra which are the rows of the data matrix $D \in \mathbb{R}^{k \times n}$.

2. Baseline corrected data window (Right window): This window shows the control parameters for the baseline computation together with the corrected spectra. Seven simple functions (polynomials, Gaussian- and Lorentz curves) can be used to approximate the baseline.

AFS computation The GUI uses four windows:

1. Data window (Upper-left window): The data window shows the k rows (spectra) of the data matrix $D \in \mathbb{R}^{k \times n}$ in a 2D-plot. The number of spectra k is printed together with the number of spectral channels n. The four largest singular values of D are shown. By transposing the data matrix D it is possible to compute the AFS for the concentration factor C (instead of the AFS for the spectral factor A).
2. NMF window (Upper-right window): This window allows to set the number of components to $s = 2$ or $s = 3$ and to compute an initial NMF. The smallest minimal components of C and A are printed. The figure shows the profiles of the so-called abstract matrix factors. The buttons allow to compute the profiles of C and/or A.
3. AFS window (Lower-left window): Various control parameters allow to make certain settings for the AFS computation in order to control the approximation quality or maximal number of edges. Pressing the Compute AFS button starts the AFS computation. The live-view mode is active after the AFS has been drawn. Just move the mouse pointer to (and through) the AFS.
4. Factor window (Lower-right window): This window shows the spectral factors (or concentration profiles if the transpose option has been activated in the first window) for the grid points shown in the AFS window.

Complementarity & AFS The GUI is built around four windows and a central control bar:

1. Pure factor windows (Upper row): These windows show three concentration profiles and three spectra of a factorization $D = CA$ which is computed by a step-wise reduction of the rotational ambiguity for a three-component system.
2. C, A - AFS windows (Lower row): Here FIRPOL, a superset which includes the AFS, and/or the AFS can be shown for the factors C and A. A feasible factorization $D = CA$ is associated with two triangles whose vertices represent the concentration profiles and spectra.
3. Control bar (Center): Includes all control parameters for the computations, the live-view mode, the import of pure component spectra or concentration profiles as well as the control buttons for adding contour plots on certain soft constraints.

3.3.2. Importing Initial Data

Spectral data are imported to *FAC-PACK* by means of a MAT-file $*$.mat. This file must contain the matrix $D \in \mathbb{R}^{k \times n}$ whose rows are the k spectra. Each spectrum contains absorption values at n frequencies. The file may also contain a vector x with n components representing the spectral wavenumbers/frequencies and a time-coordinates vector $t \in \mathbb{R}^k$.

The spectral data matrix D is loaded by pressing one of the buttons **(1)**[1], **(18)** or **(64)**

[1]Red numbers refer to Section 3.7.

depending on the active GUI. The dimension parameters k and n are shown in the GUI. The four largest singular values are shown in the fields (**22**)-(**25**) or (**71**)-(**74**). These singular values allow to determine the numerical rank of the data matrix and are the basis for assigning the number of components s.

3.3.3. Sample Data

FAC-PACK provides some sample data matrices:

- example0.mat: $k = 51$ spectra, $n = 201$ channels, $s = 3$ components with a baseline turning negative to demonstrate the baseline correction.
- example1.mat: $k = 51$ spectra, $n = 101$ channels, $s = 2$ components.
- example2.mat: $k = 21$ spectra, $n = 101$ channels, $s = 3$ components, the AFS has three clearly separated segments.
- example3.mat: $k = 21$ spectra, $n = 101$ channels, $s = 3$ components, the AFS is one topologically connected set with a hole.
- example4.mat: some random noise has been added to data matrix D given in example2.mat, the AFS has three clearly separated segments.

3.3.4. External C-Routine

All time-consuming numerical computations are externalized to a C-routine in order to accelerate *FAC-PACK*. This C-routine is AFScomputationSYSTEMNAME.exe, wherein SYSTEMNAME stands for your system including the bit-version. For example AFScomputationWINDOWS64.exe is used on a 64 bit Windows system. The external routine is called if any of the buttons **Initial nmf** (**30**), **Compute AFS** (**43**), **no. 1** (**57**) or **no. 2** (**59**) is pressed. Pre-compiled versions of the C-routine for the following systems are parts of the distribution:

- Windows 32/64 bit,
- Unix 32/64 bit and
- Mac 64 bit.

The execution of the external routine can always be stopped by CTRL + C in the command window. If the C-routine takes too much computation time, the reason for this can be large values for k, n, max fcls (**40**) or max edges (**41**) or too small values for ε-bound (**38**) or δ-stopping (**39**).

3.3.5. Further Included Libraries

The C-routine AFScomputationSYSTEMNAME.exe includes the netlib library lapack and the ACM routine nl2sol. Any use of *FAC-PACK* must respect
the lapack license, see http://www.netlib.org/lapack/LICENSE.txt, and
the nl2sol license, see http://www.acm.org/publications/policies/softwarecrnotice.

3.4. Baseline Correction Module

AFS computations depend sensitively on distorted baselines and negative components in the spectral data. Such perturbing signals can pretend further components in the reaction system and negative components are not consistent with the nonnegative factorization problem. The **Baseline correction** GUI tries to correct the baseline and is accessible by pressing the button **Baseline correction** on the *FAC-PACK* start window, see a screen shot on page 129. Data loading is explained in Section 3.3.2. The **Transpose** button (**2**) serves to transpose the data matrix D.

The **Baseline correction** GUI is a simple tool for spectral data preprocessing. It can be applied to series of spectra in which the distorted baseline in some frequency windows is well separated from the relevant signal. We call such frequency intervals, which more or less show the baseline, zero-intervals. The idea is to remove the disturbing baseline from the series of spectra by adapting a global correction function within in the zero-intervals to the baseline. See Figure 11 where four zero-intervals have been marked by columns. The zero-intervals are selected in the raw data window (**8**) by clicking in the window, dragging a zero-interval and releasing the mouse button. The procedure can be repeated in order to define multiple zero-intervals. The **Reset intervals** button (**7**) can be used in order to reset the selection of all zero-intervals.

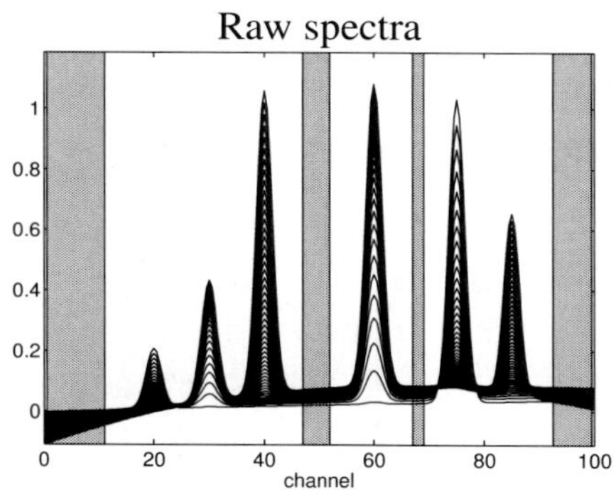

Figure 11. Four zero-intervals are marked by columns. In these intervals the spectral data is corrected towards zero. Data: example0.mat.

Mathematically every spectrum is treated separately. Let $a \in \mathbb{R}^n$ be a certain spectrum. Then the aim is to compute a baseline function $b \in \mathbb{R}^n$ so that the norm of the residuum $a - b$ is minimal within the selected intervals. The baseline-corrected spectrum is then $a_{\text{new}} = a - b$. Once again, this explains the naming zero-intervals, since there the spectrum is assumed to be zero.

The sum of the interval lengths of these zero-intervals should be as large as possible in order to ensure a reliable baseline approximation. However, the required number of frequency grid points depends on the degrees of freedom of the type of the baseline function. The text field (**11**) shows the information "ok"/"not ok" in order to indicate whether or not the selected intervals are large enough for the baseline approximation. Data cutting is a further functionality of the baseline correction GUI: If only a certain frequency subinterval is to be analyzed or the baseline correction is only to be applied to a subinterval, then this subinterval can be marked by mouse clicking and dragging and then the **Cut data**-button can be pressed. The user can always return to the initial data by pressing the **Reset cutting**-button (**6**).

3.4.1. Type of the Baseline

Four types of baselines are available:

- Polynomials of order zero up to order four,
- a Gauss curve,
- or a Lorentz curve.

The most recommended baselines are polynomials with the degrees 0, 1 or 2. A polynomial of degree 0 is a constant function so that a constant is added or subtracted for each spectral channel of a spectrum.

The baselines "Gauss curve" and "Lorentz curve" can be used to remove single isolated peaks from the series of spectra. Once again, the curve profile is fitted in the least-squares sense to the spectroscopic data within the selected frequency interval. Then the fitted profile is subtracted from the spectrum.

3.4.2. Program Execution and Data Export

The baseline correction by an external C-procedure is started by pressing the **Correct baseline**-button (**12**). The computation times for data wrapping and for solving the least-squares problem are shown in the text fields (**13**) and (**14**). The corrected series of spectra (in black) is shown in the window (**16**) together with the original data (in gray). The baseline-corrected data can be stored in a MATLAB file by pressing the **Save**-button (**15**).

By clicking the right mouse button in a figure a separate MATLAB figure opens. The figure can now be modified, printed or exported in the usual way. If for some reason (e.g. program runs too long) the external C-procedure for the baseline correction is to be stopped, then CTRL + C only works the MATLAB command window (and not in the GUI).

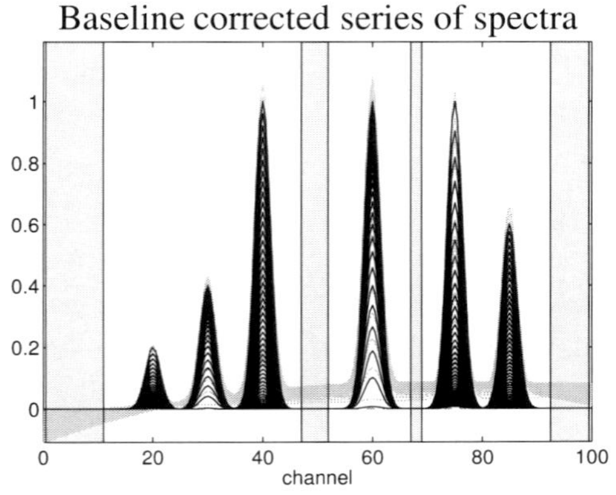

Figure 12. Sample problem example0.mat with corrected baseline by a 2nd order polynomial.

3.5. AFS Computation Module

This section describes how to compute the area of feasible solutions (AFS) for two- and three-component systems with *FAC-PACK*.

3.5.1. Initial Steps

The AFS computation module can be started by pressing the **AFS computation** button on the *FAC-PACK* start window. Data loading by the **Load** button (**18**) is explained in Section 3.3.2. The loading process is followed by a singular value decomposition (SVD) of the data matrix. The computing time is printed in the field **Computing time info** (**47**). The dimensions of D are shown in the data window together with the singular values $\sigma_1, \ldots, \sigma_4$. These singular values allow to determine the "numerical rank" of the data matrix and are the basis for assigning the number of components s in the NMF window.

FAC-PACK provides some sample data matrices. These sample data sets are introduced in Section 3.3.3. Figure 13 shows the series of spectra from example2.mat. Three clearly nonnegative singular values indicate a three-component system.

3.5.2. Transposing the Problem

FAC-PACK computes and displays the spectral matrix factor A, see (1), together with its AFS. In order to compute the first matrix factor C and its AFS, press the **Transpose** button (**19**) to transpose the matrix D and to interchange x and t.

3.5.3. Initial NMF

To run the polygon inflation algorithm an initial NMF is required. First the number of components (either $s = 2$ or $s = 3$) is to be assigned. In the case of noisy data additional singular vectors can be used for the decomposition by selecting a larger number of **singl vcts** (**29**). For details on this option see [13], wherein the variable z equals **singl vcts**.

The initial NMF is computed by pressing the button **Initial nmf** (**30**). The NMF uses a genetic algorithm and a least-squares minimizer. The smallest relative entries in the columns of C and rows of A, see [18] for the normalization of the columns of C and rows of A, are also shown in the NMF window, see (**34**) and (**35**). These quantities are used to define an appropriate noise-level ε; see Equation (6) in [18] for details.

The initial NMF provides a number of s interior points of the AFS; see Equations (3) and (5) in [18].

The buttons **plot C and A** (**31**), **plot C** (**32**) and **plot A** (**33**) serve to display the factors C and A together or separately. Note that these abstract factors are associated with the current NMF. After an NMF computation the A factor is displayed.

Figure 14 shows the factor A for an NMF for example2.mat. The relative minimal components in both factors are greater than zero ($9.9 \cdot 10^{-4}$ and 10^{-3}). So the noise-level **e-neg. entr.:** (**37**) can be set to the lower bound **e-neg. entr.:** $1 \cdot 10^{-12}$.

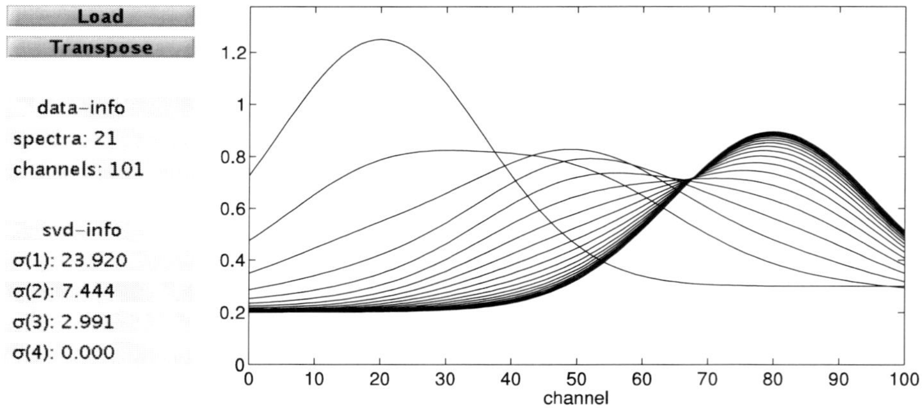

Figure 13. Test matrix example2.mat loaded by pressing the **Load** button (**18**).

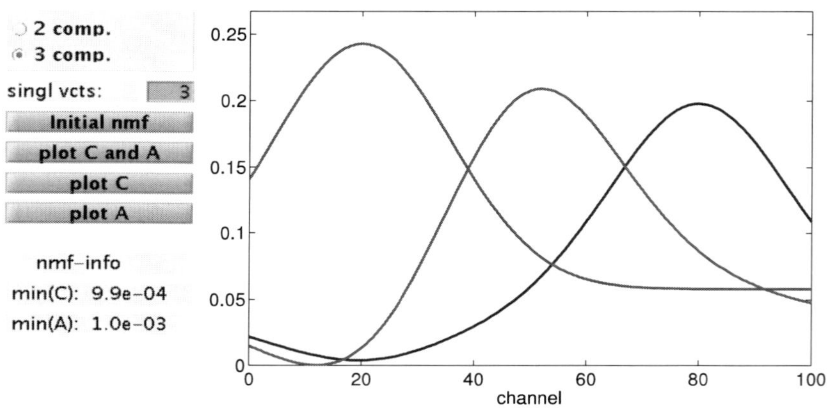

Figure 14. The initial NMF for example2.mat with $s = 3$ components.

3.5.4. Computation of the AFS

For two-component systems ($s = 2$) the AFS consists of two real intervals. *FAC-PACK* represents the associated AFS by the two orthogonal sides of a rectangle. The sides are just the intervals of admissible values for α and β where

$$T = \begin{pmatrix} 1 & \alpha \\ 1 & \beta \end{pmatrix}$$

is the transformation matrix which constitutes the rotational ambiguity. See [1, 2] for the AFS for the case $s = 2$. However in [1], see Equation (6), the entries of the second row of T are interchanged.

For three-component systems ($s = 3$) the AFS is formed by all points (α, β) so that

$$T = \begin{pmatrix} 1 & \alpha & \beta \\ 1 & s_{11} & s_{12} \\ 1 & s_{21} & s_{22} \end{pmatrix}$$

is a transformation which is associated with nonnegative factors C and A, see [18]. A permutation argument shows that with (α, β) the points (s_{11}, s_{12}) and (s_{21}, s_{22}) belong to the AFS, too.

The initial NMF is the starting point for the numerical computation of the AFS since it provides first α and β in the AFS.

Next various control parameters for the numerical computation of the AFS are explained. Further two ways of computing different kinds of the AFS are introduced.

Control parameters Five control parameters are used to steer the adaptive polygon inflation algorithm. *FAC-PACK* aims at the best possible approximation of the AFS with the smallest computational costs. Default values are pre-given for these control parameters.

- The parameter **e-neg. entr.** (**37**) is the noise level control parameter which is denoted by ε in [18]. Negative matrix elements of C and A do not contribute to the penalization functional if their relative magnitude is larger than $-\varepsilon$. In other words, an NMF with such slightly nonnegative matrix elements is accepted as valid.
- The parameter **e-bound** (**38**) controls the precision of the boundary approximation and is denoted by ε_b in [18]. Decreasing the value ε_b improves the accuracy of the positioning of new vertices of the polygon.
- The parameter **d-stopping** (**39**) controls the termination of the adaptive polygon inflation procedure. This parameter is denoted by δ in Section 3.5 of [18] and is an upper bound for change-of-area which can be gained by a further subdivision of any of the edges of the polygon.
- **max fcls.** (**40**) and **max edges** (**41**) are upper bounds for the number of function-calls and for the number of vertices of the polygon.

Type of polygon inflation *FAC-PACK* offers two possibilities to apply the polygon inflation method for systems with three components ($s = 3$). The user should select the proper method according to the following explanations:

1. The "classical" version of the polygon inflation algorithm is introduced in [18] and applies best to an AFS which consists of three clearly separated segments. Select **Polygon inflation** by pressing the upper button (**42**). A typical example is shown in Figure 15 for the test problem example2.mat. In each of the segments an initial polygon has been inflated from the interior. Interior points are accessible from an initial NMF.
2. Alternatively, the **Inverse polygon inflation** procedure is activated by pressing the lower button in (**42**). This should be done if the AFS is only one topologically connected set (with a hole) or if the isolated segments of AFS are in close neighborhood. See Figure 16 for an example.

   The inverse polygon inflation method is more expensive than the classical version. First the complement of the AFS is computed and then some superset of the AFS is computed. A set subtraction results in the desired AFS. The details are explained in [20].

The user should always try the second variant of the polygon inflation algorithm if the results of the first variant are not satisfying or if something appears to be doubtful.

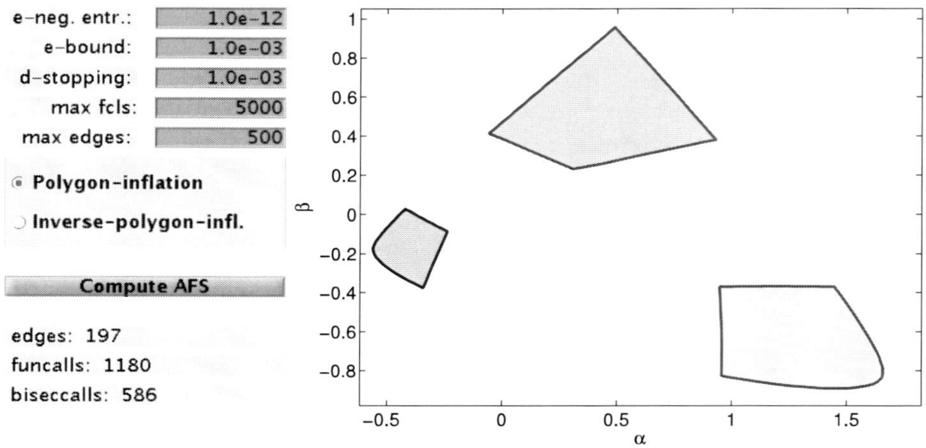

Figure 15. An AFS with three clearly separated segments as computed by the "classical" polygon inflation algorithm. Data: example2.mat.

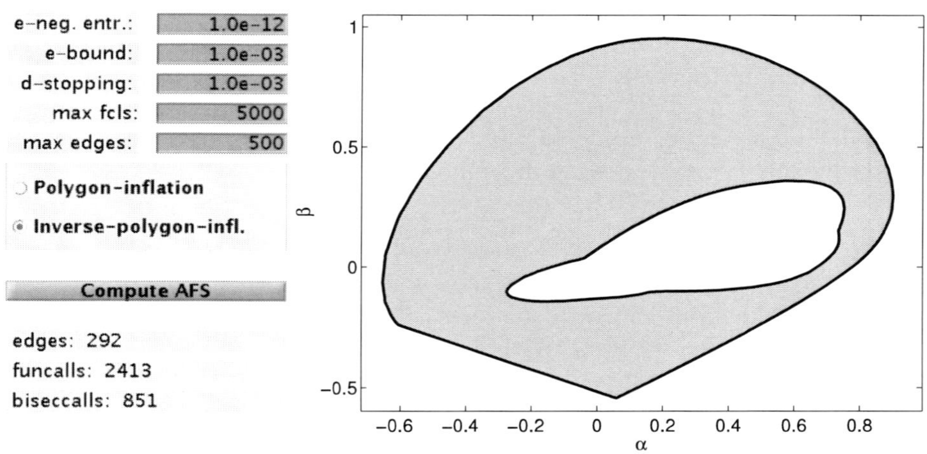

Figure 16. An AFS which consists of only one segment with a hole. For the computation the inverse polygon inflation algorithm has been applied to example3.mat.

3.5.5. Factor Representation & Live-View Mode

Next the plotting of the spectra and/or concentration profiles is described. These factors can be computed from the AFS by certain linear combinations of the right and/or left singular vectors. The live-view mode allows an interactive representation of the factors by moving the mouse pointer through the AFS.

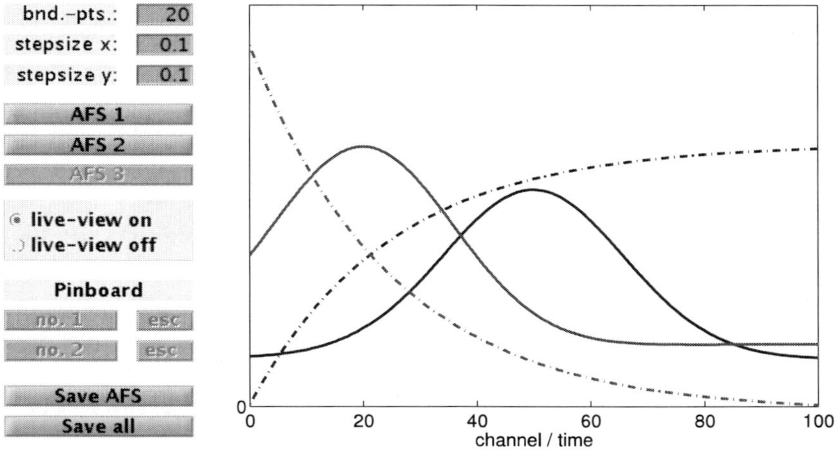

Figure 17. The live-view mode for the two-component system example1.mat. By moving the mouse pointer through the AFS the transformation to the factors C and A is computed interactively and all results are plotted in the factor window. The spectra are drawn by solid lines and the concentration profiles are represented by dash-dotted lines.

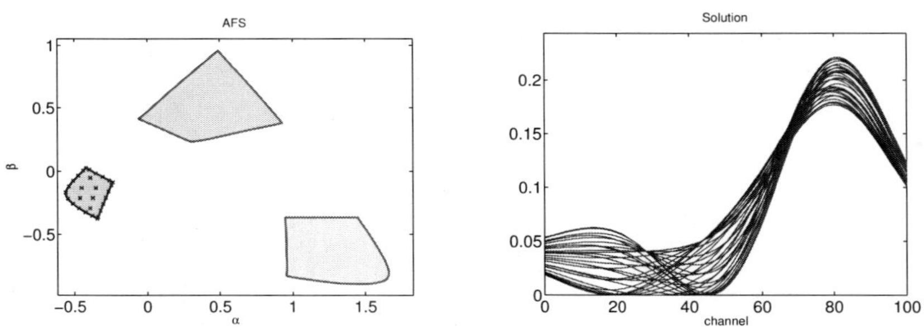

Figure 18. The AFS for the problem example2.mat consists of three separated segments. The leftmost segment of the AFS is covered with grid points. For each of these points the associated spectrum is plotted in the right figure.

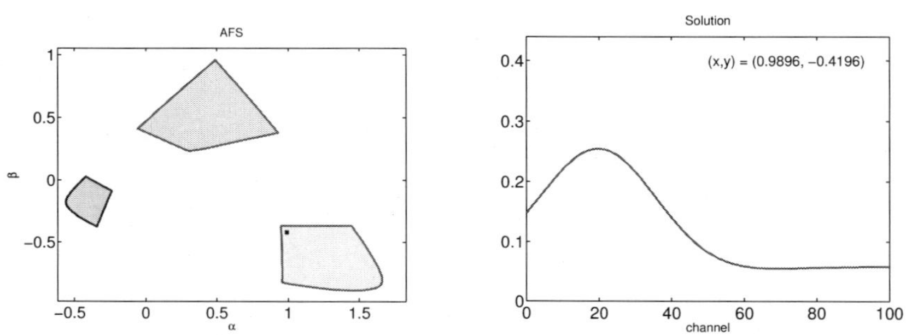

Figure 19. The live-view mode is active for the 3 component system example2.mat. In the rightmost AFS-segment the mouse pointer is positioned in the left upper corner at the coordinates $(0.9896, -0.4196)$. The associated spectrum is shown in the right figure together with the mouse pointer coordinates.

Two-component systems For two-component systems a full and simultaneous representation of the factors C and A is possible. In order to get an overview of the range of possible spectra one can discretize the AFS-intervals equidistantly. For each point of the resulting 2D discrete grid the associated spectrum is plotted in the factor window. The discretization parameter (subinterval length) is selected in the field (**50**) and/or (**51**). The plot of the spectra is activated by pressing button **AFS 1** (**52**) or button **AFS 2** (**53**).

Live-view mode: The live-view mode can be activated in the field (**55**). By moving the pointer through the AFS the associated solutions are shown interactively. Figure 17 shows a typical screen shot. The y-axis labels are turned off.

Three component systems For three-component systems the range of possible solutions is presented for each segment of the AFS separately. The user can select with the buttons **AFS 1** (**52**), **AFS 2** (**53**) and **AFS 3** (**54**) a specific AFS which is then covered by a grid. First the boundary of the AFS segment (which is a closed curve) is discretized by a number of **bnd-pts** nodes. The discretization parameters for the interior of this AFS segment are **step size x** (**50**) and **step size y** (**51**); the interior points are constructed line by line. The resulting nodes are shown in the AFS window by symbols $\times$ in the color of the AFS segment. For each of these nodes the associated spectrum is drawn in the factor window. For the test problem example2.mat the bundle of spectra is shown in Figure 18. The resolution can be refined by increasing the number of points on the boundary or by decreasing the step size in the direction of x or y.

If the admissible concentration profiles are to be printed, then the whole procedure is to be applied to the transposed data matrix D; activate the **Transpose** button (**19**) at the beginning and recompute everything.

Live-view mode: The live-view mode can be activated in the field (**55**). By moving the pointer through the AFS, the associated solutions are shown interactively. An example is shown in Figure 19.

Additionally, a certain point in the AFS can be locked by clicking the left mouse button; then the AFS for the remaining two components is re-computed. The resulting AFS is a smaller subset of the original AFS, which reflects the fact that some additional information is added by locking a certain point of the AFS. See Section 3.5.6 for further explanations.

3.5.6. Reduction of the Rotational Ambiguity

With the computation of the AFS *FAC-PACK* provides a continuum of admissible matrix factorizations. All these factorizations are mathematically correct in the sense that they represent nonnegative matrix factors, whose product reproduces the original data matrix D. However, the user aims at the *one* solution which is believed to represent the chemical system correctly. Additional information on the system can help to reduce the so-called rotational ambiguity. *FAC-PACK* supports the user in this way. If, for instance one spectrum within the continuum of possible spectra is detected, which can be associated with a known chemical compound, then this spectrum can be locked and the resulting restrictions can be used to reduce the AFS for the remaining components.

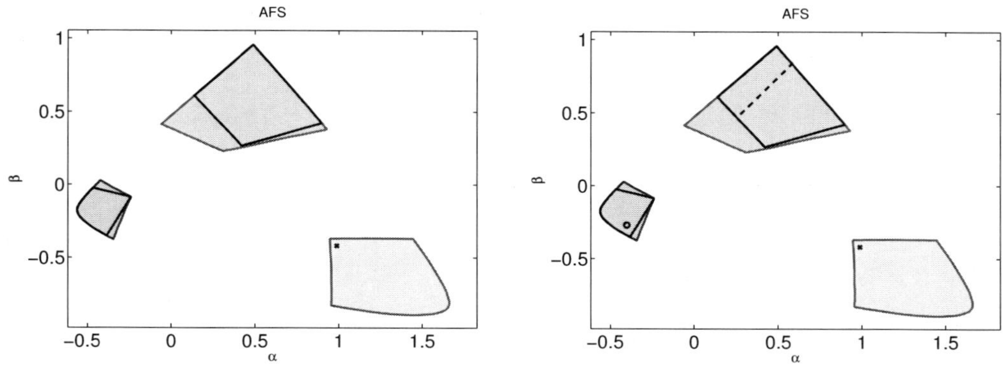

Figure 20. Reduction of the AFS for the test problem example2.mat. Left: the reduced AFS is shown by black solid lines after locking a first spectrum (marker $\times$). Right: The AFS of the remaining third component is shown by a black broken line.

Locking a first spectrum After computing the AFS, activate the **live-view** mode (**55**). While moving the mouse pointer through any segment of the AFS, the factor window shows the associated spectra. If a certain (known) spectrum is found, then the user can lock this spectrum by clicking the left mouse button. A $\times$ is plotted into the AFS and the button **no. 1** (**56**) becomes active. By pressing this button a smaller subset of the original AFS is drawn by solid black lines in the AFS window. This smaller AFS reflects the fact that some additional information on the system has been added. The button **esc** (**57**) can be used for unlocking any previously locked point.

Locking a second spectrum Having locked a first spectrum and having computed the reduced AFS the **live-view** mode can be reactivated. Then a second point within the reduced AFS can be locked. By pressing button **no. 2** (**58**) the AFS for the remaining component is reduced for a second time and is shown as a black broken line. Once again, the button **esc** (**59**) can be used for unlocking the second point.

Figure 20 shows the result of such a locking-and-AFS-reduction procedure for the test problem example2.mat. For this problem the AFS consists of three clearly separated segments and the final reduction of the AFS is shown by the black broken line in the uppermost segment of the AFS. The live-view mode allows to display the possible spectra along this line. The resulting restrictions on the complementary concentration factor C can be computed by the module AFS & Complementarity, see Section 3.6.

3.5.7. Save Data and Extract Axes

To save the results in a MATLAB *.mat file press either **Save AFS** (**60**) or **Save all** (**61**). A proper file name is suggested.

Save AFS saves the data D, the factors U, S, V of the singular value decomposition, the factors $Cinit$, $Ainit$ of the initial NMF and the AFS. The AFS has the data format of a structure which contains the following variables:

- For $s = 2$ the AFS consists of two segments: $AFS\{1\} \in \mathbb{R}^2$ and $AFS\{2\} \in \mathbb{R}^2$

with $\alpha \in [AFS\{1\}(1), AFS\{1\}(2)]$ and $\beta \in [AFS\{2\}(1), AFS\{2\}(2)]$ and vice versa.

- For $s = 3$ and an AFS consisting of three segments: $AFS\{i\} \in \mathbb{R}^{m_i \times 2}$ with $i = 1, 2, 3$ is a polygon whose x-coordinates are $AFS\{i\}(:, 1)$ and whose y-coordinates are $AFS\{i\}(:, 2)$.
- For $s = 3$ and an AFS consisting of one segment with a hole: $AFS\{1\} \in \mathbb{R}^{m \times 2}$ is the outer polygon ($AFS\{1\}(:, 1)$ the x-coordinates and $AFS\{1\}(:, 2)$ the y-coordinates) and $AFS\{2\} \in \mathbb{R}^{\ell \times 2}$ is the inner polygon surrounding the hole.

If the lock-mode has been used, see Section 3.5.6, then the results are saved in $AFSLockMode1$ and $LockPoint2$ (case of one locked spectrum) as well as $AFSLockMode2$ and $LockPoint1$ (case of two locked spectra).

The user can get access to all figures. Therefore click the right mouse button in the desired figure. Then a separate MATLAB figure opens. The figure can now be modified, printed or exported in the usual way.

3.5.8. Cancellation of the Program

Whenever the buttons **Initial nmf (30)**, **Compute AFS (43)**, **no. 1 (57)** or **no. 2 (59)** are pressed, an external C-routine is called, see Section 3.3.4. The GUI does not respond to any activities during the execution of the C-routine. Therefore, in case of any problems or in case of too long program runtimes, the C-routine cannot be canceled from the GUI. The termination can be enforced by typing CTRL + C in the MATLAB command window. Time consuming processes can be avoided if all the parameters are adjusted to reasonable (default) values.

3.5.9. How to Det Help

If the **Help** box **(62)** is checked and the mouse pointer is moved over a button, a text field or an axis, then a short explanation appears right next to the **Help** field. Further, a small pop-up window opens if the mouse pointer rests for more than one second on a button.

3.6. Complementarity & AFS Module

This *FAC-PACK* module simultaneously shows the AFS for the spectral factor and the AFS for the concentration factor. It demonstrates how the rotational ambiguity for a three-component system can be reduced by means of the complementarity theory. For the complementarity theorem see [19, 20] and for comparable results see [3]. Solutions can be selected and modified within a live-view mode. Any changes in the spectral AFS are immediately mapped to changes in the concentrational AFS and vice versa. This gives the user the full control over and a visualization of a the complete factorization $D = CA$.

3.6.1. Initial Steps

The Complementarity & AFS module module can be started by pressing the **Complementarity & AFS** button on the *FAC-PACK* start window. Data loading by the **Load** button

(**18**) is explained in Section 3.3.2. The loading process is followed by a singular value decomposition (SVD) of the data matrix. The dimensions of D are shown in the fields (**69**) and (**70**) together with the four largest singular values in the fields (**71**-**74**). For this module the fourth singular value should be "small" compared to the first three singular values so that the numerical rank of D equals 3. For noisy data one might try to use a larger number of vectors in the field (**78**), see also Section 3.5.3.

In Figure 13 we consider the sample data example2.mat, cf. Section 3.3.3. The parameters in the fields (**75**)-(**77**) control the AFS computation. The meaning of these parameters is explained in Section 3.5.4.

3.6.2. AFS Selection

In this module both the spectral AFS $\mathcal{M}_A$ and the concentrational AFS $\mathcal{M}_C$ are computed. It is also possible to compute FIRPOL which includes the AFS. The user can select to compute

1. the two FIRPOL sets which contain $\mathcal{M}_A$ and $\mathcal{M}_C$ as subsets. For the spectral factor FIRPOL is the set $\mathcal{M}_A^+ = \{(\alpha, \beta) : (1, \alpha, \beta) \cdot V^T \geq 0\}$; see also Equation (6) in [19]. Points in $\mathcal{M}_A^+$ represent only one nonnegative spectrum $A(1, :)$. The concentrational set FIRPOL $\mathcal{M}_C^+$ can be described in a similar way. The FIRPOL computation is computationally cheaper and faster since no optimization problems have to be solved.

2. the spectral AFS $\mathcal{M}_A$ and the concentrational AFS $\mathcal{M}_C$. These computations are often time-consuming.

FIRPOL as well as the AFS are computed by the polygon inflation method [18, 19]. Numerical calculations are done by the C-routine AFScomputationSYSTEMNAME, see Section 3.3.4, immediately after a selection is made in the checkbox (**79**). The results are shown in the windows (**65**) and (**67**).

Figure 21 shows FIRPOL $\mathcal{M}_A^+$ and the AFS $\mathcal{M}_A$ for the sample data **example2**. The trace of D, see Section 3.6.2, is also shown. The triangle in $\mathcal{M}_A$ represents a complete feasible factor A.

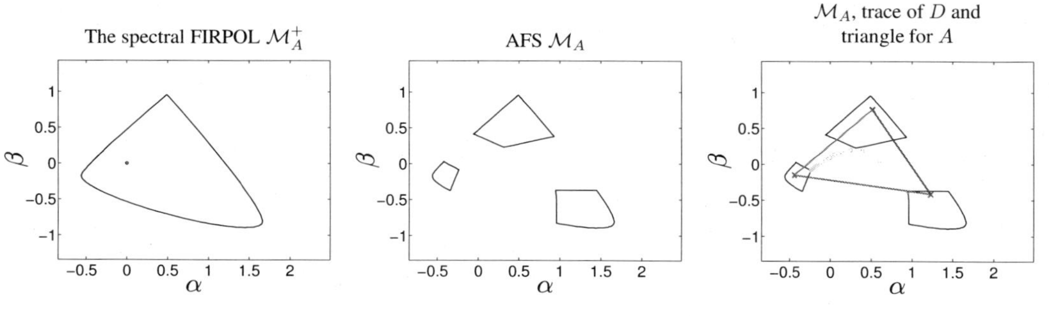

Figure 21. Sample data **example2**. Left: FIRPOL $\mathcal{M}_A^+$. Center: $\mathcal{M}_A$. Right: $\mathcal{M}_A$ with the trace of D marked by gray circles and a triangle representing a feasible factor A.

Traces of the spectral data matrix D The traces of spectral data matrix D can be drawn in $\mathcal{M}_A$ and $\mathcal{M}_C$. The traces in $\mathcal{M}_A$ are the normalized expansion coefficients of the rows of D with respect to the right singular vectors $V(:,2)$ and $V(:,3)$. The normalization is that the expansion coefficient for $V(:,1)$ equals 1. Thus the trace $\mathcal{M}_A$ is given by the k points

$$w_i = \left(\frac{(DV)_{i2}}{(DV)_{i1}}, \frac{(DV)_{i3}}{(DV)_{i1}}\right) = \left(\frac{D(i,:)\cdot V(:,2)}{D(i,:)\cdot V(:,1)}, \frac{D(i,:)\cdot V(:,3)}{D(i,:)\cdot V(:,1)}\right), \quad i = 1,\ldots,k.$$

Analogously, the traces of D in $\mathcal{M}_C$ are the normalized expansion coefficients of the columns of D with respect to the scaled left singular vectors $\sigma_2^{-1}U(:,2)$ and $\sigma_3^{-1}U(:,3)$

$$u_j = \left(\frac{(\Sigma^{-1}U^TD)_{2j}}{(\Sigma^{-1}U^TD)_{1j}}, \frac{(\Sigma^{-1}U^TD)_{3j}}{(\Sigma^{-1}U^TD)_{1j}}\right) = \left(\frac{\sigma_1 U(:,2)^T\cdot D(:,j)}{\sigma_2 U(:,1)^T\cdot D(:,j)}, \frac{\sigma_1 U(:,3)^T\cdot D(:,j)}{\sigma_3 U(:,1)^T\cdot D(:,j)}\right)$$

for $j = 1,\ldots,n$. If the checkbox **trace** (**80**) is activated, then the traces of D in $\mathcal{M}_A$ and $\mathcal{M}_C$ are plotted by small gray circles.

3.6.3. Live-View Mode

The AFS can be explored in a live-view mode which allows an interactive representation of the factors by moving the mouse pointer through the AFS. Points in $\mathcal{M}_A$ and or $\mathcal{M}_C$ can be locked by clicking the left mouse button in the AFS. Any points can be unlocked and afterwards moved in the AFS. The modification of a certain point in one AFS results in changes for the components in the other AFS. All this is shown interactively. A unique factor C or A is represented by a triangle in the AFS. The vertices of these triangles can be unlocked by the buttons (**82**)-(**84**) and (**86**)-(**88**). By a step-by-step selection and modification of three vertices in the AFS for A (or three vertices in the AFS for C) the user can construct two triangles and thus a complete factorization $D = CA$. Next an initial triangle is constructed in $\mathcal{M}_A$; the whole procedure can also be applied to $\mathcal{M}_C$.

Initialization: Selection of a first point By activating the **first**-radio button (**83**) a first spectrum/point of the spectral AFS can be selected. If the mouse pointer is moved through the AFS window (**67**), then the associated spectra are shown in the window (**66**) for the factor A. The complementary concentration profiles are restricted by the complementarity theorem to a one-dimensional affine space; the associated straight line is shown in the concentrational AFS simultaneously. By pressing the left mouse button a certain point in the AFS can be locked. All this is illustrated in Figure 22.

Initialization: Selection of a second point Next a second point can be selected in $\mathcal{M}_A$. Therefore the second-radio button (**84**) can be activated and the mouse pointer can be moved through the AFS $\mathcal{M}_A$. The following items are plotted simultaneously:

- the two selected points in the AFS $\mathcal{M}_A$,
- the two associated spectra in the factor window (**66**),
- the two complementary straight lines and the point of intersection in the AFS $\mathcal{M}_C$ (**65**),

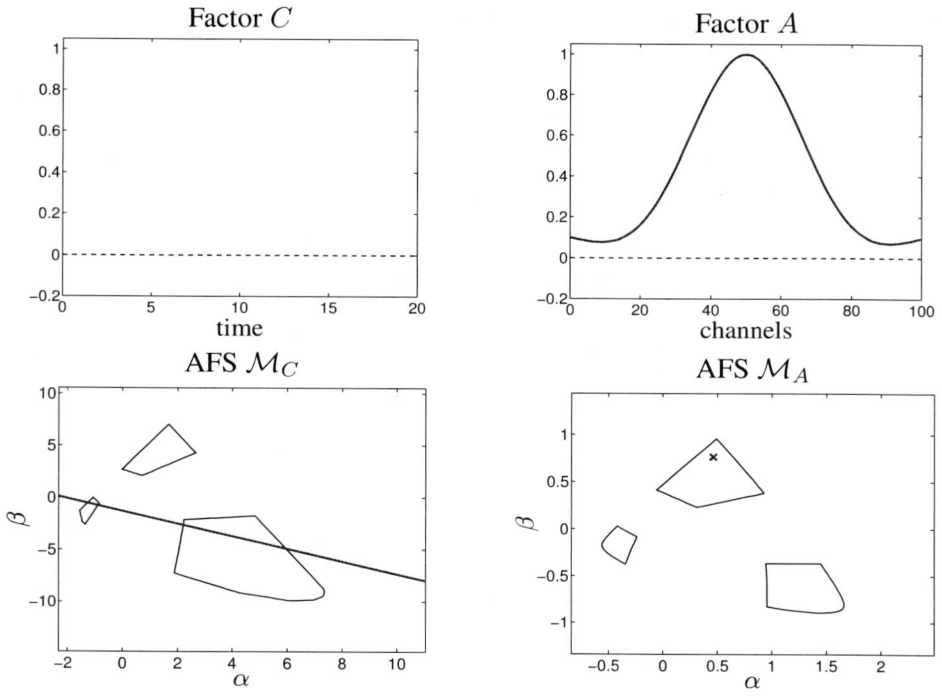

Figure 22. A first point, marked by $\times$, is locked in the AFS $\mathcal{M}_A$ (lower-right). This results in: Upper-left: no concentration profile has uniquely been defined up to now. Upper-right: a single pure component spectrum has been fixed. Lower-left: the straight line is a restriction on the complementary concentration profiles $C(:, 2:3)$ in $\mathcal{M}_C$.

- the associated concentration profile for the uniquely determined complementary concentration profile in the factor window (**64**).

By pressing the left mouse-button a certain point can be locked. The result is illustrated in Figure 23.

Initialization: Selection of a third point A third point can be added by activating the red third-radio button (**85**) and moving the mouse pointer through the AFS $\mathcal{M}_A$. The following items are plotted simultaneously:

- a triangle in the AFS $\mathcal{M}_A$ whose vertices uniquely determine the factor A,
- the three associated spectra in the factor window (**66**),
- the three complementary straight lines which define a triangle in the AFS $\mathcal{M}_C$ (**65**),
- the associated concentration profiles in the factor window C (**64**).

Once again a third point can be locked by clicking the left mouse button. Figure 24 illustrates this.

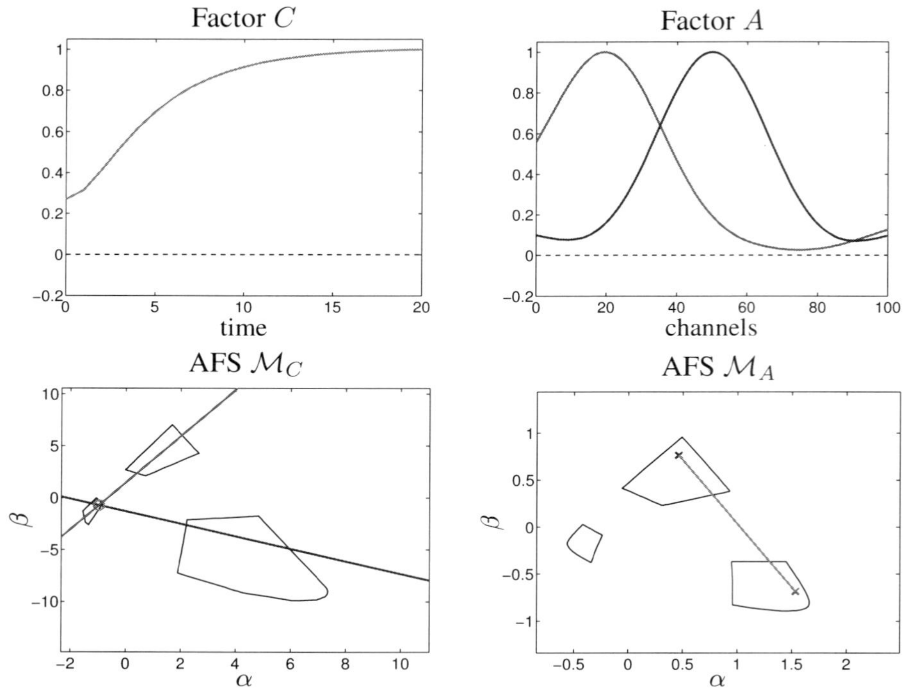

Figure 23. A second point, marked by $\times$, is locked in the AFS $\mathcal{M}_A$ (lower-right). This results in: Upper-left: one complementary concentration profile is uniquely determined. Upper-right: the two locked pure component spectra. Lower-left: the falling straight line restricts the complementary concentration profiles $C(:, 2:3)$ and the rising straight line restricts the complementary concentration profiles $C(:, 1)$ and $C(:, 3)$ in $\mathcal{M}_C$. One concentration profile marked by $\circ$ is now uniquely determined.

3.6.4. Modification of the Solution

Once an initial solution has been selected, one can now modify the vertices of the triangles. This can be done by activating the buttons

- **(80)**-**(82)** to modify the triangle in $\mathcal{M}_C$,
- **(84)**-**(86)** to modify the triangle in $\mathcal{M}_A$.

The selection and re-positioning of the vertices has to follow the rule that all six vertices are within the segments of $\mathcal{M}_A$ and $\mathcal{M}_C$. Otherwise, negative components can be seen in the factor windows (**64**) and (**66**). In general a vertex in one AFS is associated with a line segment in the other AFS; see [17] for the complementarity theorem and [3, 19, 20] for applications to three-component systems. Any vertex in one AFS and the associated line segment in the other AFS are plotted in the same color in order to express their relationship. By changing one vertex of a triangle, two vertices of the other triangle are affected.

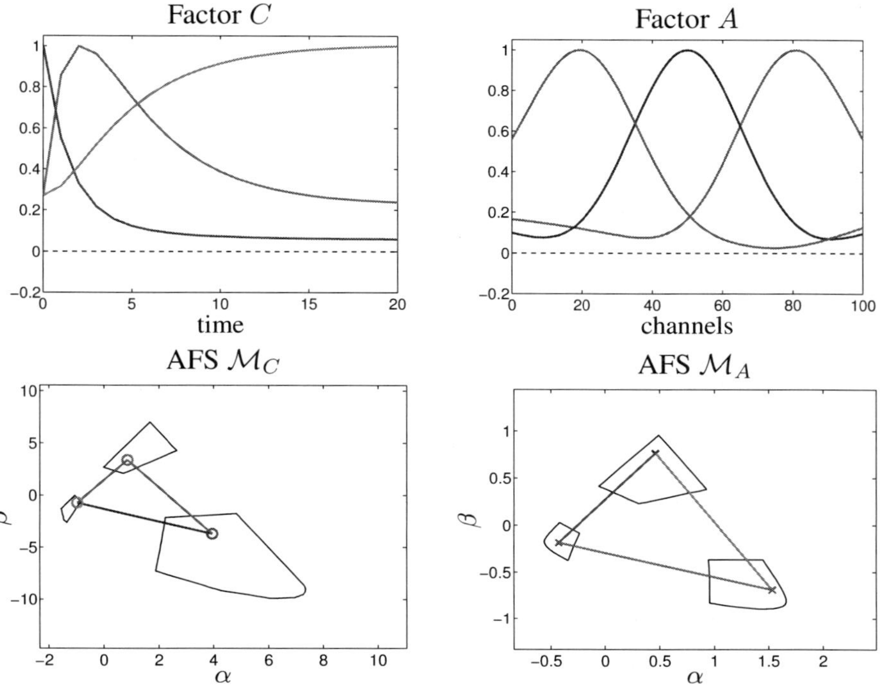

Figure 24. Three locked points in $\mathcal{M}_A$ uniquely define the factor A represented by a feasible triangle (lower-right). This results in: Upper-left: three complementary concentration profiles. Upper-right: three pure component spectra. Lower-left: a triangle in the AFS $\mathcal{M}_C$.

3.6.5. Contour Plots on Factor Properties

The Complementarity & AFS module can help the user to select pure components in the AFS which have certain desired properties. Sometimes it might help to know where in an AFS monotone concentration profiles or exponentially decaying profiles (of a reactant) can be found. Smoothness of a pure component profile or a small integral of a spectrum might also be valuable selection criteria. In the Complementarity & AFS module any point in the (α, β)-plane can be evaluated with respect to various constraint functions which estimate the smoothness, monotonicity, exponential decay and so on. The result in form of a contour plot is then shown in the AFS.

By using the radio buttons (**90**)-(**92**) for the factor C and by using the radio buttons (**94**) and (**95**) for the factor A the contour plots are shown. These plots can be removed from the AFS by clicking the buttons (**89**) and (**93**). The numerical evaluation of the constraint function may take some time.

Contour plots for the following constraint functions are available:

- **monotone** concentration profiles (**90**): monotonously increasing or decreasing profiles are favored and are shown in light gray or white in the contour plot,
- **smooth** concentration profiles (**91**): profiles with a small Euclidean norm of the discrete second derivative are favored (**91**) (shown white in the contour plot),

- **exponentially** decaying concentration profiles (**92**): for every point of a grid in the (α, β)-plane the approximation error for an exponentially decaying function (with an optimized decay constant) is calculated. This allows to favor reactants decaying exponentially, which can be found in the light gray or white areas in $\mathcal{M}_C$,

- **smooth** spectra (**94**): spectra with a small Euclidean norm of the discrete second derivative are favored,

- small **norm** spectra: spectra with a small Euclidean norm of the representing vector are favored. Thus spectra with few isolated and narrow peaks are shown in light gray compared to those with wide absorbing peaks (shown in darker gray).

Figure 25 shows two examples.

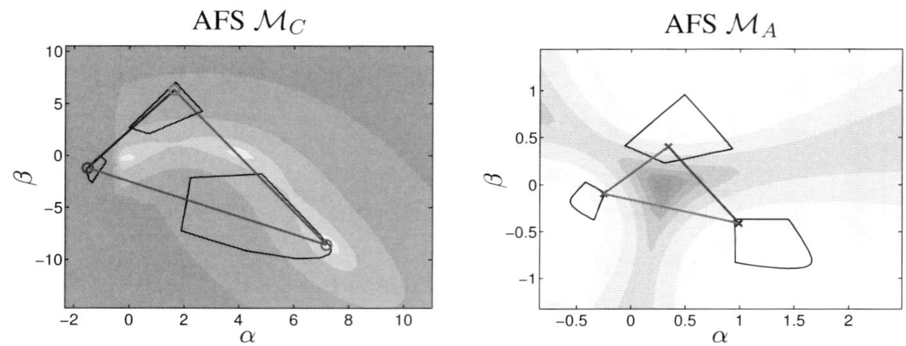

Figure 25. Contour plots for two soft constraints. Left: Contour evaluating the closeness to an exponentially decaying function. The right lower vertex (marked by ∘) represents an exponentially decaying concentration profile. Right: Contour plot for the Euclidean norm of the spectrum for example2.mat.

3.6.6. Import of Spectra and Concentration Profiles

If the user can provide one or more pure component spectra or certain concentration profiles for the reaction system under consideration, then these data can be loaded into the Complementarity & AFS module. After pressing the **Import spectra**-button (**98**) one, two or three pure component spectra can be loaded. These should be stored in a MATLAB variable named a being an $\ell \times n$ vector in case of ℓ pre-given spectra. (Linear interpolation should be used if a known spectrum does not fit the dimension n.) Any loaded spectrum is marked in the spectral AFS $\mathcal{M}_A$ by a $*$.

Analogously, known concentration profiles can be loaded by pressing the **Import concentration profiles** (**96**) button. The data must be provided in a $*$.mat file containing the variable c; columns with k components determine pre-given concentration profiles.

3.6.7. Save Data and Extract Axes

In order to export the solutions press the **Save**-button (**97**). Then

- the variables determining the geometry of FIRPOL and of the AFS,
- the initial spectral data, the SVD,
- and, if available, the locked points in AFS $\mathcal{M}_A$ and $\mathcal{M}_C$, the associated spectra plus concentration profiles and transformation matrix T determining a unique factorization CA

are saved in a MATLAB file.

Any figure window in the GUI can be accessed by clicking the right mouse button in the desired figure. Then a separate MATLAB figure opens. The figure can now be modified, printed or exported in the usual way.

3.6.8. Program Abortion

The AFS & complementarity module calls an external C-routine if FIRPOL or the AFS is computed through the check fields (**79**). The GUI does not respond to any commands during the execution of the C-routine. In case of any problems or in case of too long program runtimes, the C-routine cannot be canceled from the GUI. Program abortion can be enforced by typing CTRL + C in the MATLAB command window.

3.7. Screen Shots & Description of Operation Panels

I: Screen shot of the **Baseline-correction** module:

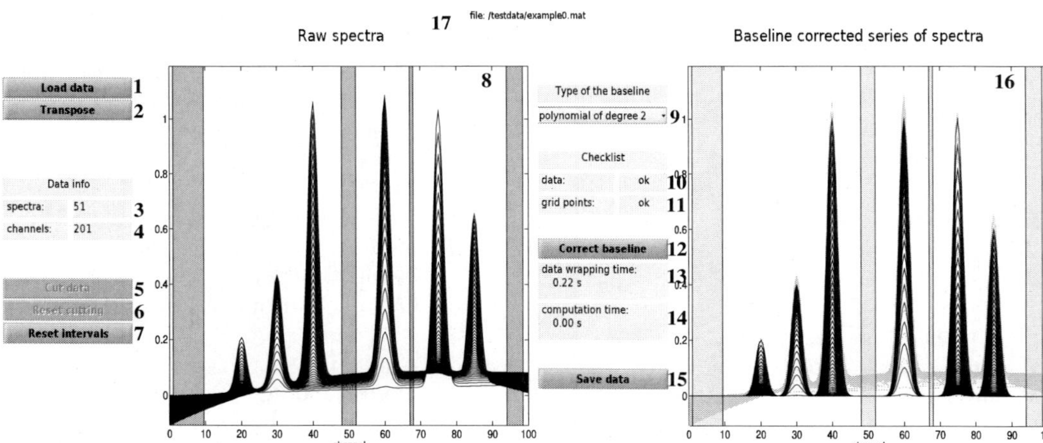

The GUI buttons, data fields and figures:

1 Load data matrix D
2 Transpose D
3 # rows of D
4 # columns of D
5 Cut spectra as marked by mouse
6 Reset spectral cutting
7 Reset selected intervals
8 Raw spectra window
9 Select type of baseline correction
10 Status of data consistency
11 Status of band selection consistency
12 Apply baseline correction
13 Computing time for data wrapping
14 Computing time for baseline correction
15 Save button
16 Corrected spectra window
17 Name of data file

II: Screen shot of the **AFS computation** module:

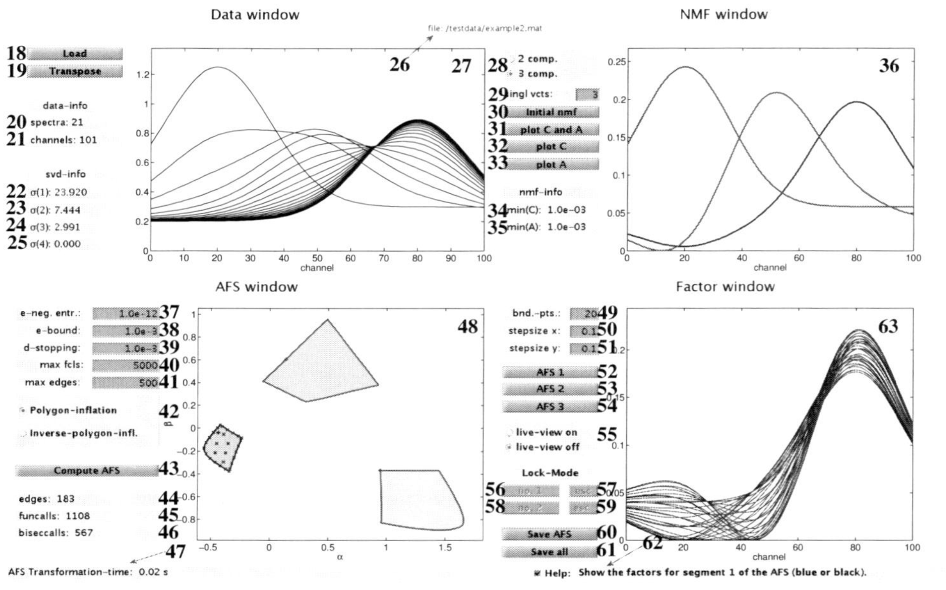

The GUI buttons, data fields and figures:

18 Load button
19 Transpose button
20 # rows of data matrix D
21 # columns in D
22 Singular value 1 of D
23 Singular value 2 of D
24 Singular value 3 of D
25 Singular value 4 of D
26 File name
27 Initial spectra window
28 Input # components
29 # of singular values used for initial NMF
30 Compute NMF
31 Plot factors C and A
32 Plot factor C
33 Plot factor A
34 Minimal matrix elements of C in the NMF
35 Minimal matrix elements of A in the NMF
36 NMF window
37 Acceptable negativeness for the NMF
38 Precision at the boundary
39 Stopping precision
40 Maximal # calls of the target function
41 Maximal # of edges
42 Select type of polygon inflation
43 Compute AFS
44 # edges
45 # calls of target function
46 No. used function calls for the boundary-precision
47 Output on the numerical computation
48 AFS window
49 # of boundary points for the factor representation
50 Step size x direction for the factor representation
51 Step size y direction for the factor representation
52 Plot factors for segment 1 of the AFS
53 Plot factors for segment 2 of the AFS
54 Plot factors for segment 3 of the AFS

55 Activate live-view mode
56 Compute AFS for 1 selected point
57 Reset 1. restricted AFS
58 Compute AFS for 2 selected points
59 Reset 2. restricted AFS
60 Save button (AFS)
61 Save button (all)
62 Help field
63 Factor window

III: Screen shot of the **Complementarity & AFS** module:

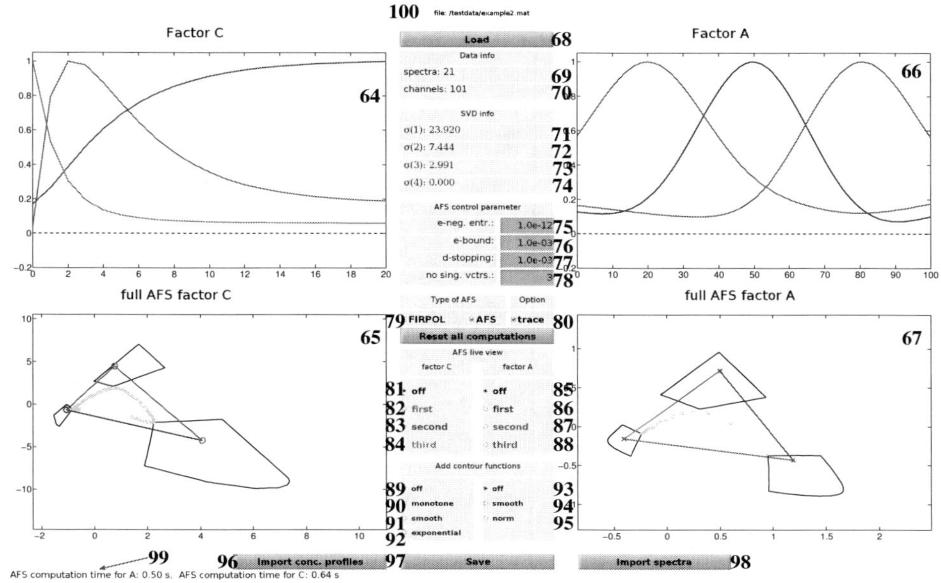

The GUI buttons, data fields and figures:

64 Factor C
65 AFS for C
66 Factor A
67 AFS for A
68 Load spectral data matrix D
69 # rows of D
70 # columns of D
71 Singular value 1 of D
72 Singular value 2 of D
73 Singular value 3 of D
74 Singular value 4 of D
75 Acceptable negativeness for the NMF
76 Precision at the boundary
77 Stopping precision
78 # of singular vectors used for initial NMF
79 Select AFS (FIRPOL and/or full AFS)
93 Contour mode for A off
94 Contour mode for A - smooth spectra
95 Contour mode for A - spectra with a small norm
96 Import external concentration profiles
80 Mark mixed spectra (rows of D) in the AFS
81 Live-view mode C off
82 Live-view mode $C(:,1)$ on
83 Live-view mode $C(:,2)$ on
84 Live-view mode $C(:,3)$ on
85 Live-view mode A off
86 Live-view mode $A(1,:)$ on
87 Live-view mode $A(2,:)$ on
88 Live-view mode $A(3,:)$ on
89 Contour mode C off
90 Contour mode for C - monotone profiles
91 Contour mode for C - smooth profiles
92 Contour mode for C - exponential decay
97 Save the results
98 Import external spectra
99 Computing time
100 Name of data file

References

[1] H. Abdollahi, M. Maeder, and R. Tauler, Calculation and Meaning of Feasible Band Boundaries in Multivariate Curve Resolution of a Two-Component System, *Anal. Chem.* 81 (2009), no. 6, 2115–2122.

[2] H. Abdollahi and R. Tauler, Uniqueness and rotation ambiguities in Multivariate Curve Resolution methods, *Chemom. Intell. Lab. Syst.* 108 (2011), no. 2, 100–111.

[3] S. Beyramysoltan, R. Rajkó, and H. Abdollahi, Investigation of the equality constraint effect on the reduction of the rotational ambiguity in three-component system using a novel grid search method, *Anal. Chim. Acta* 791 (2013), no. 0, 25–35.

[4] O.S. Borgen and B.R. Kowalski, An extension of the multivariate component-resolution method to three components, *Anal. Chim. Acta* 174 (1985), 1–26.

[5] A. Golshan, H. Abdollahi, and M. Maeder, Resolution of Rotational Ambiguity for Three-Component Systems, *Anal. Chem.* 83 (2011), no. 3, 836–841.

[6] A. Golshan, M. Maeder, and H. Abdollahi, Determination and visualization of rotational ambiguity in four-component systems, *Anal. Chim. Acta* 796 (2013), no. 0, 20–26.

[7] J.C. Hamilton and P.J. Gemperline, Mixture analysis using factor analysis. II: Self-modeling curve resolution, *J. Chemom.* 4 (1990), no. 1, 1–13.

[8] J. Jaumot and R. Tauler, MCR-BANDS: A user friendly MATLAB program for the evaluation of rotation ambiguities in Multivariate Curve Resolution, *Chemom. Intell. Lab. Syst.* 103 (2010), no. 2, 96–107.

[9] A. Juan, M. Maeder, M. Martínez, and R. Tauler, Combining hard and soft-modelling to solve kinetic problems, *Chemometr. Intell. Lab.* 54 (2000), 123–141.

[10] W.H. Lawton and E.A. Sylvestre, Self modelling curve resolution, *Technometrics* 13 (1971), 617–633.

[11] M. Maeder and Y.M. Neuhold, *Practical data analysis in chemistry*, Elsevier, Amsterdam, 2007.

[12] E. Malinowski, *Factor analysis in chemistry*, Wiley, New York, 2002.

[13] K. Neymeyr, M. Sawall, and D. Hess, Pure component spectral recovery and constrained matrix factorizations: Concepts and applications, *J. Chemom.* 24 (2010), 67–74.

[14] R. Rajkó and K. István, Analytical solution for determining feasible regions of self-modeling curve resolution (SMCR) method based on computational geometry, *J. Chemom.* 19 (2005), no. 8, 448–463.

[15] C. Ruckebusch and L. Blanchet, Multivariate curve resolution: A review of advanced and tailored applications and challenges, *Anal. Chim. Acta* 765 (2013), 28–36.

[16] M. Sawall, A. Börner, C. Kubis, D. Selent, R. Ludwig, and K. Neymeyr, Model-free multivariate curve resolution combined with model-based kinetics: Algorithm and applications, *J. Chemom.* 26 (2012), 538–548.

[17] M. Sawall, C. Fischer, D. Heller, and K. Neymeyr, Reduction of the rotational ambiguity of curve resolution technqiues under partial knowledge of the factors. Complementarity and coupling theorems, *J. Chemom.* 26 (2012), 526–537.

[18] M. Sawall, C. Kubis, D. Selent, A. Börner, and K. Neymeyr, A fast polygon inflation algorithm to compute the area of feasible solutions for three-component systems. I: Concepts and applications, *J. Chemom.* 27 (2013), 106–116.

[19] M. Sawall and K. Neymeyr, A fast polygon inflation algorithm to compute the area of feasible solutions for three-component systems. II: Theoretical foundation, inverse polygon inflation, and FAC-PACK implementation, *J. Chemom.* 28 (2014), 633–644.

[20] ———, On the area of feasible solutions and its reduction by the complementarity theorem, *Anal. Chim. Acta* 828 (2014), 17–26.

[21] R. Tauler, Calculation of maximum and minimum band boundaries of feasible solutions for species profiles obtained by multivariate curve resolution, *J. Chemom.* 15 (2001), no. 8, 627–646.

[22] M. Vosough, C. Mason, R. Tauler, M. Jalali-Heravi, and M. Maeder, On rotational ambiguity in model-free analyses of multivariate data, *J. Chemom.* 20 (2006), no. 6-7, 302–310.

In: Current Applications of Chemometrics
Editor: Mohammadreza Khanmohammadi

ISBN: 978-1-63463-117-4

Chapter 7

MULTIWAY CALIBRATION APPROACHES TO HANDLE PROBLEMS LINKED TO THE DETERMINATION OF EMERGENT CONTAMINANTS IN WATERS

Mirta R. Alcaráz, Romina Brasca, María S. Cámara, María J. Culzoni, Agustina V. Schenone, Carla M. Teglia, Luciana Vera-Candioti and Héctor C. Goicoechea*

Laboratorio de Desarrollo Analítico y Quimiometría (LADAQ),
Cátedra de Química Analítica I,
Facultad de Bioquímica y Ciencias Biológicas,
Universidad Nacional del Litoral-CONICET,
Ciudad Universitaria, Santa Fe, Argentina

ABSTRACT

This chapter includes an overview of current chemometric methodologies using the second-order advantage to solve problems related to the analysis of emergent contaminants in water samples. Discussions are focus on pharmaceuticals and other compounds in tap, river and other sort of water samples. Second-order data generation through several techniques such as fluorescence spectroscopy, high performance liquid chromatography and capillary electrophoresis is commented. This work describes the most frequently used algorithms, such as the standard approaches for second-order data analysis: PARAFAC (parallel factor analysis) and MCR-ALS (multivariate curve resolution alternating least squares), and the recently implemented U-PLS/RBL (unfolded partial least squares/residual bilinearization) as well. This chapter deals with the nature of the drawbacks related to their implementation, as well as other problems that are inherent to the involved analytical techniques and samples.

* Corresponding Author: E-mail: hgoico@fbcb.unl.edu.ar

INTRODUCTION

Up to date, a great variety of synthetic and naturally occurring chemical compounds have been found in groundwater, surface water (i.e., river, lake, sea, wetland), tap water, raw and treated wastewater (domestic, industrial and municipal), sediments, soil, air, food, etc [1-10]. The so-called "emerging contaminants" cover a great variety of man-made or naturally occurring chemicals as well as their metabolites or degradation products, and microorganisms that are present in the environment, but are not usually monitored due to lack of regulations, and cause known or suspected adverse ecological and/or human health effects [11, 12].

Generally, this kind of compounds is also referred to as "contaminants of emerging concern" because of the potential risks for the environment and human health associated with their massive environmental release. The presence of this kind of compounds in the environment generates an emerging concern in many countries. In that way, environmental and health protection agencies have established policies for their regulation and monitoring to protect freshwater, the aquatic environment and wildlife. For example, the Environmental Protection Agency of the United States (www.epa.gov.ar/waterscience) periodically develops a list of contaminants, i.e., the Contaminant Candidate List, and decides which contaminants to regulate, which is called Regulatory Determinations. Therefore, this list of unregulated contaminants is useful to prioritize research and data collection efforts and determine whether a specific contaminant should be regulated in the future. The Network of References Laboratories for Monitoring of Emerging Environmental Pollutants (NORMAN) (http://www.norman-network.net) in Europe is also an important association which enhances the exchange of information on emerging environmental substances.

Emerging contaminants typically include pharmaceuticals (i.e., analgesics, anti-inflammatories, antimicrobials, β-blockers, lipid regulators, X-ray contrast media, veterinary drugs), personal care products (e.g., acetophenone, polycyclic musks, nitro musks), artificial sweeteners (e.g., sucralose, acesulfame, saccharin, cyclamate, aspartame), stimulants (e.g., caffeine), illicit drugs and drugs of abuse (e.g., opiates, cocaine, cannabis, amphetamines), hormones and steroids (e.g., 17β-estradiol, estrone, estriol, 17α-ethinyloestradiol, mestranol), surfactants (e.g., alkylphenols), pesticides (e.g., chlorpyrifos, chlordane, malathion, diuron, prodiamine, phenylphenol, carbendazim), plasticizers (e.g., bisphenol A, bisphenol A diglycidyl ether, diethyl phthalate, dibutylphthalate), flame retardants (e.g., polybrominateddiphenyl ethers), anticorrosives (e.g., benzotriazole, totyltriazole), and nanomaterials (e.g., fullerenes, nanotubes, quantum dots, metal oxanes, nanoparticles: TiO_2, silver, gold) [12, 13]. Products which contained these contaminants are consumed or used worldwide and became indispensable for modern civilizations [12]. Therefore, occurrence of emerging contaminants is reported in a variety of environments such as influents and effluents from wastewater treatment plants, surface water, groundwater, sediments and even tap water. The environmental concentration in water, which commonly range from ng/L to μg/L, can change significantly depending on the season (consumption patterns are different in summer and winter), climatic conditions (e.g., temperature, rainfalls), bedrock geology, soil type, rate of industrial production, human metabolism (e.g., excretion rate) and distance to primary sources, i.e., industrial sewage treatment plants, landfilling areas, hospitals, municipal sewage systems or private septic tanks, among others [14].

It is clear that emerging contaminants are present in complex matrixes at trace levels, a combination that places considerable demands on the analytical methods used for their determination. Monitoring of emerging contaminants requires specialized sampling and analytical techniques. In general, the analytical methods involve sample clean up to remove possible interferences from the sample matrix, and pre-concentration step to increase analyte concentrations. The most relevant techniques for sample clean up and pre-concentration are solid-phase extraction (SPE) and liquid-liquid extraction (LLE) [15-20]. Emerging contaminants are usually analyzed by liquid chromatography tandem mass spectrometry (LC-MS/MS), gas chromatography coupled to mass spectrometry (GC-MS), and ultrahigh performance liquid chromatography mass spectrometry (UHPLC-MS/MS) [21-27].

One disadvantage of the cited analytical methods is associated with the sample preparation steps due to the fact that sample clean up represents a tedious and time-consuming procedure for reducing matrix effects and removing potential interferences [21, 28]. For this reason, the development of simple, high selective and sensitive new analytical methods is required. In that way, second-order calibration allows for the analysis of the target analytes in the presence of unexpected interferences and matrix effects, but avoiding sample preprocessing.

SECOND ORDER CALIBRATION

Second-Order Data Generation

In multivariate calibration the term 'order' is usually employed to denote the number of modes for the data array which is recorded for a single sample. The term 'way', on the other hand, is reserved for the number of modes of the mathematical object which is built by joining data arrays measured for a group of samples [29]. When second-order data for a group of samples are joined into a single three-dimensional array, the resulting object is known as three-way array, and these data are usually known as three-way data [30].

Second-order data can be generated in two ways: (1) using a single instrument, such as a spectrofluorimeter registering excitation-emission matrices (EEMs) or a diode-array spectrophotometer following the kinetics of a chemical reaction, or (2) coupling two 'hyphenated' first-order instruments, as in GC-MS, bi-dimensional GC (GC-GC), MS-MS, etc. [30].

In the implementation of multi-way analysis, the analytical community has found a way of improving the quality of the results when developing analytical methods to be applied for the quantitation of target analytes in complex matrices. Using modern instrumentation, analytical laboratories can generate a variety of second-order instrumental data [31].

Separative techniques such as chromatography and electrophoresis coupled to any detection system, i.e., diode array detector (DAD), fluorescence detector (FLD), Fourier transform infrared spectroscopy (FTIR) or MS, generate matrix data which have the retention/elution times in one mode and the spectral sensor on the other mode. The same kind of data is obtained with non separative methods which register the temporal signals, such as flow injection analysis (FIA) and sequential injection analysis (SIA) coupled to the above mentioned detectors. On the other hand, when the data is recorded using a spectrofluorimeter, the matrix will have excitation and emission signals in the excitation and emission dimensions, respectively. Another way of obtaining fluorescence second-order data is by measuring synchronous fluorescence spectra (SFS) at different offsets ($\Delta\lambda$). Moreover, electrochemical methods provide multi-way data by changing an instrumental parameter, i.e., pulse time in differential pulse voltammetry (DPV). All these data can be arranged into a data table or matrix, where columns and rows correspond to each data dimension.

An important second-order data property is the trilinearity. Multi-linearity can be defined as the possibility of mathematically expressing a generic element of a multi-way data array as a linear function of component concentrations and profiles in the data modes [32, 33]. Second-order data are trilinear when each compound in all experiments treated together can be described by a triad of invariant pure profiles, and the spectral shape of a component is not modified by changes in the other two modes (bilinearity property). However, some three-way arrays that are considered non-trilinear can be conveniently modeled, i.e., those having shifted retention times in different chromatographic runs. On the other hand, total synchronous fluorescence spectra do not fulfill the bilinearity property due to the fact that the shape of the SF for each analyte varies with the change in $\Delta\lambda$ [34].

As was previously commented, the chemical data for a single sample in the second-order domain comprise a matrix, which can be visualized as a two-dimensional surface or "landscape". Figure 1 illustrates one example of second-order data: a chromatographic run between 0-2 min which was registered with a DAD in the wavelength range of 200-360 nm. As can be appreciated in this figure, the time variation was represented at λ = 268 nm on the x-axis, while on the other, the wavelength variation was displayed at time = 1 min. A multi-way array can be modeled with an appropriate algorithm to achieve selectivity by mathematical means as will be developed in the next section.

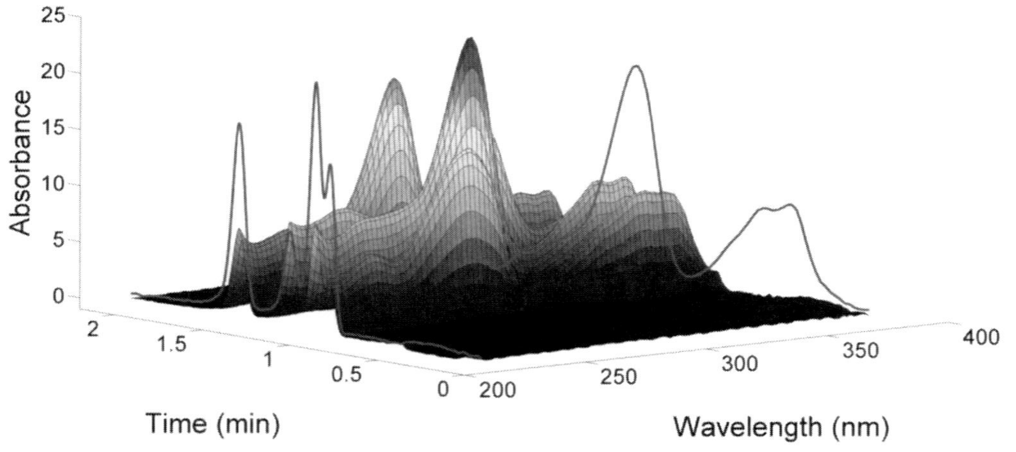

Figure 1. Landscape obtained by plotting the chromatographic elution time registered with a DAD for a given sample. The time variation is represented on the x-axis at λ = 268 nm (purple line), and the wavelength variation at time = 1 min (green line) on the y-axis.

Second-Order Algorithms

Several algorithms can be cited among those involving the second-order advantage: Generalized Rank Annihilation (GRAM) [35], Direct Trilinear Decomposition (DTLD) [36, 37], Self-Weighted Alternating Trilinear Decomposition (SWATLD) [38], Alternating Penalty Trilinear Decomposition (APTLD) [39], Parallel Factor Analysis (PARAFAC) [40], Multivariate Curve Resolution Alternating Least Squares (MCR-ALS) [41], Bilinear Least Squares (BLLS) [42], N-way and Unfolded -Partial Least Squares /Residual Bilinearization (N- and U-PLS/RBL) [43, 44, 45] and Artificial Neural Networks followed by Residual Bilinearization (ANN/RBL) [46]. A description of the three most used algorithms, i.e., PARAFAC, MCR-ALS and U-PLS/RBL, will be presented below.

PARAFAC

PARAFAC decomposes a three-way array $\underline{\mathbf{X}}$, obtained by joining second-order data from the calibration and test samples according to the following equation [40]:

$$\underline{\mathbf{X}} = \sum_{n=1}^{N} \mathbf{a}_n \otimes \mathbf{b}_n \otimes \mathbf{c}_n + \underline{\mathbf{E}} \quad (1)$$

where $\otimes$ indicates the Kronecker product, N is the total number of responsive components, the column vectors a_n, b_n and c_n are usually collected into the score matrix **A** and the loading matrices **B** and **C**; and $\underline{\mathbf{E}}$ is a residual error term of the same dimensions as $\underline{\mathbf{X}}[(I+1)\times J\times K]$. I is the number of calibration samples and, hence, the first dimension of $\underline{\mathbf{X}}$ is $(I+1)$, since the test sample is also included in $\underline{\mathbf{X}}$. J and K are the number of digitized wavelengths (in the case of EEM data).

The application of the PARAFAC model to two-way data sets calls for:(1) initializing and/or constraining the algorithm, (2) establishing the number of components, (3) identifying specific components from the information provided by the model, and(4) calibrating the model in order to obtain absolute concentrations for a particular component in an unknown sample. Initialization of the least-squares fit can be carried out with profiles obtained by several procedures, the most usual being Direct Trilinear Decomposition (DTLD) [36]. The number of components (N) can be estimated by considering the internal PARAFAC parameter (known as core consistency) [40] or by analyzing the univariate calibration line obtained by regressing the PARAFAC relative concentration values from the training samples against their standard concentrations [48]. Identification of the chemical constituent under investigation is accomplished with the aid of the profiles B and C, as extracted by PARAFAC, and comparing them with those for a standard solution of the analyte of interest. This is required, since the components obtained by decomposition of $\underline{\mathrm{X}}$ are sorted according to their contribution to the overall spectral variance, and this order is not necessarily maintained when the unknown sample is changed. Finally, absolute analyte concentrations are obtained after proper calibration, since only relative values (A) are provided by decomposing the three-way data array (see Figure 2). Experimentally, this is done by preparing a data set of standards of known composition and regressing the first I elements of column A against known standard concentrations y of analyte n.

MCR-ALS

This algorithm, instead of forming a three-dimensional data array like PARAFAC, arranges the matrices along the mode that is suspected of breaking the trilinear structure (i.e., if a matrix-to-matrix variation of profiles occurs along the column direction, a column-wise augmented matrix is created). The bilinear decomposition of the augmented matrix **D** is performed according to the expression [33, 41]:

$$\mathbf{D} = \mathbf{C}\mathbf{S}^{T} + \mathbf{E} \tag{2}$$

where the rows of **D** contain, in the case of chromatographic data registered with DAD, the absorption spectra measured as a function of time, the columns of **C** contain the time profiles of the compounds involved in the process, the columns of **S** contain their related spectra, and **E** is a matrix of residuals not fitted by the model. Appropriate dimensions of **D**, **C**, $\mathbf{S}^T$ and **E** are thus$(1+I)J\times K$, $(1+I)\ J\times N$, $N\times K$ and $(1+I)J\times K$, respectively (I = number of training samples, J = number of elution times, K = number of digitized wavelengths and N = number of responsive components).

Decomposition of **D** is achieved by iterative least-squares minimization of $\|\mathbf{E}\|$ under suitable constraining conditions (i.e., non-negativity in spectral profiles, unimodality and non-negativity in concentration profiles), and also closure relations between reagents and products of a given reaction.

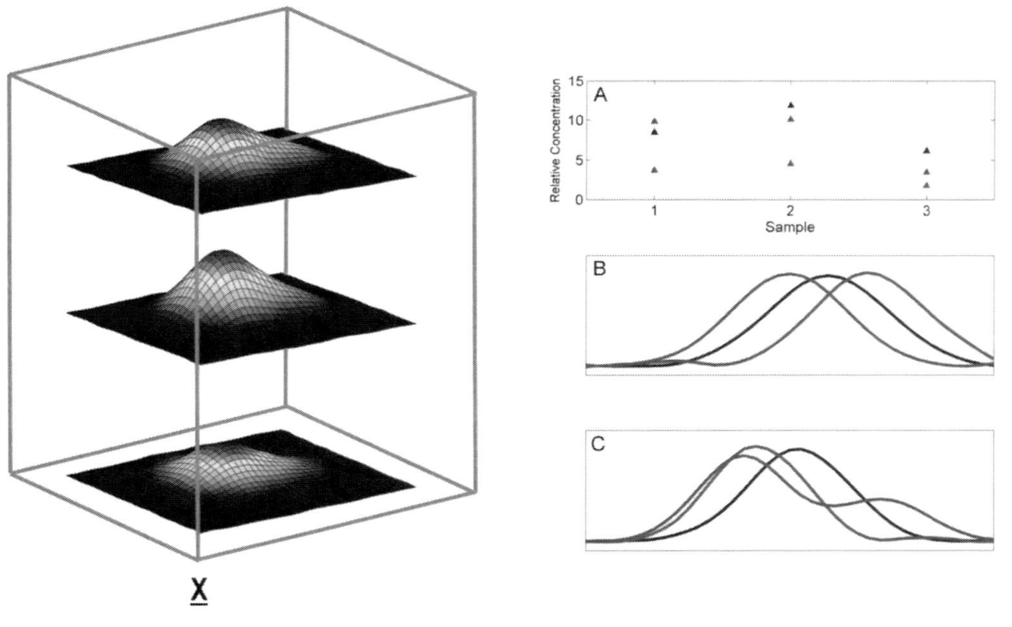

Figure 2. Schematic representation of a three-component PARAFAC model: (B) emission, (C) excitation and (A) relative concentration loadings.

The pure spectra of the compounds should be the same in all experiments, but the profiles in the different **C** sub-matrices need not share a common shape. This is the reason why kinetic experiments performed in different conditions (e.g., pH and temperature) and, hence, showing different kinetic profiles can be analyzed together, as long as the spectra of the compounds involved in the process remain invariant (absorption spectra usually show limited temperature dependence). Similar considerations can be made for variations in chromatogram peaks.

In the present context, we need to point out that MCR-ALS requires initialization with system parameters as close as possible to the final results [41]. In the column-wise augmentation mode, the species spectra are required, as obtained from either pure analyte standards or the analysis of purest spectra.

Figure 3 represents an array built with three HPLC-DAD matrices recorded for three-component mixtures with different concentrations of each analyte (74 times × 81 wavelengths). As can be seen, the **D** column-wise augmented matrix (222 times × 81 wavelengths) is decomposed into two matrices: the augmented **C** matrix of size 222 × 3 (three components) containing the evolution during the chromatographic run of every component in each individual matrix, and the $\mathbf{S}^T$ matrix of size 3 × 81 containing the spectra of each component. The area under the evolution curve for each component can be used to build the univariate calibration graph with a quantitative purpose.

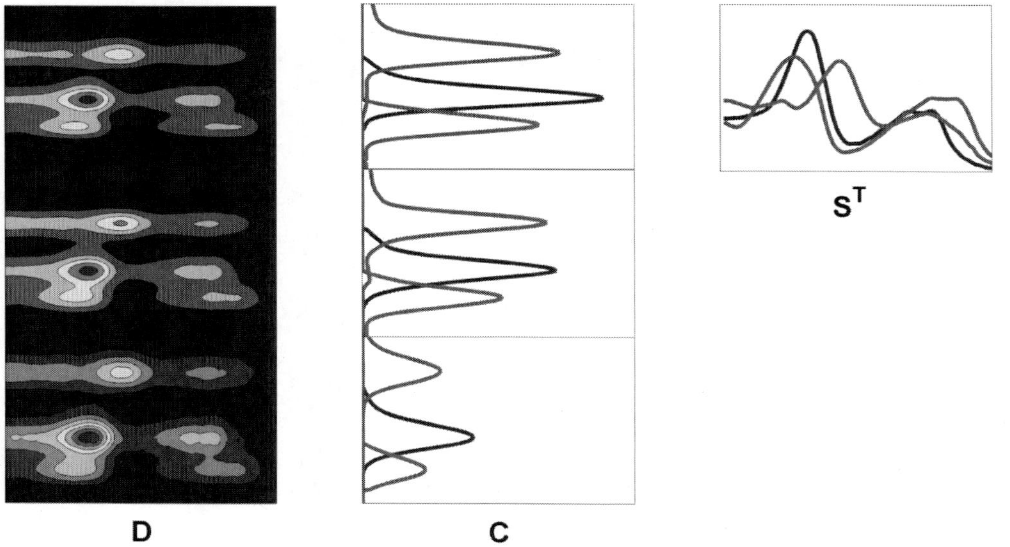

Figure 3. Schematic representation of an MCR-ALS modeling. A column-wise augmented matrix (D), and the profiles retrieved after the MCR-ALS decomposition, i.e., temporal profiles (C) and spectral profile (S^T), for the three compounds are represented.

U-PLS/RBL

This algorithm involves a first calibration step in which information regarding concentrations is employed, but without including data for the unknown sample [43].The I calibration data matrices $\mathbf{X}_{c,i}$ (size $J \times K$, where J and K are the number of channels in each dimension) are vectorized (unfolded), and a usual U-PLS model is calibrated with these data and the vector of calibration concentrations y ($N_c \times 1$, where N_c is the number of calibrated analytes). This provides a set of loadings **P** and weight loadings **W** (both of size $JK \times A$, where A is the number of latent factors), as well as regression coefficients v (size $A \times 1$). Techniques such as leave-one-out cross-validation allow for the acquisition of the parameter A [49]. If no unexpected interferences occur in the test sample, v can be employed to estimate the analyte concentration:

$$y_u = \mathbf{t}_u^T \mathbf{v} \tag{3}$$

where t_u is the test sample score, obtained by projection of the (unfolded) data for the test sample $\mathbf{X}_u$ onto the space of the A latent factors:

$$\mathbf{t}_u = (\mathbf{W}^T \mathbf{P})^{-1} \mathbf{W}^T \text{vec}(\mathbf{X}_u) \tag{4}$$

When unexpected constituents occur in $\mathbf{X}_u$, then the sample scores given by equation (4) are not suitable for analyte prediction using equation (3). In this case, the residuals of the U-PLS prediction step [s_p, see equation (5)] will be abnormally large in comparison with the typical instrumental noise, which is easily assessed by replicate measurements:

$$\begin{aligned} s_p &= \| \mathbf{e}_p \| / (JK-A)^{1/2} = \| \text{vec}(\mathbf{X}_u) - \mathbf{P} (\mathbf{W}^T \mathbf{P})^{-1} \mathbf{W}^T \text{vec}(\mathbf{X}_u) \| / (JK-A)^{1/2} = \\ &= \| \text{vec}(\mathbf{X}_u) - \mathbf{P}\mathbf{t}_u \| / (JK-A)^{1/2} \end{aligned} \tag{5}$$

where $\| \cdot \|$ indicates the Euclidean norm.

Interestingly, if unexpected constituents occur in the test sample, the situation can be handled by RBL [44], which is based on Singular Value Decomposition (SVD) modeling of the interferent effects. RBL aims at minimizing the norm of the residual vector e_u, computed while fitting the sample data to the sum of the relevant contributions to the sample signal. For a single interferent:

$$\text{vec}(\mathbf{X}_u) = \mathbf{P}\mathbf{t}_u + \text{vec}[g_{int} \mathbf{b}_{int} (\mathbf{c}_{int})^T] + \mathbf{e}_u \tag{6}$$

where $\mathbf{b}_{int}$ and $\mathbf{c}_{int}$ are the left and right eigenvectors of $\mathbf{E}_p$ and g_{int} is a scaling factor:

$$(g_{int}, \mathbf{b}_{int}, \mathbf{c}_{int}) = \text{SVD}_1(\mathbf{E}_p) \tag{7}$$

where $\mathbf{E}_p$ is the $J \times K$ matrix obtained after reshaping the $JK \times 1$ e_p vector of equation (3), and SVD_1 indicates taking the first principal component.

During this RBL procedure, **P** is kept constant at the calibration values and t_u is varied until $\|\mathbf{e}_u\|$ is minimized. The minimization can be carried out using a Gauss-Newton (GN) procedure starting with $\mathbf{t}_u$ from equation (3). Once $\| \mathbf{e}_u \|$ is minimized in equation (4), the analyte concentrations are provided by equation (1), by introducing the final t_u vector found by the RBL procedure.

The number of unexpected constituents N_{unx} can be assessed by comparing the final residuals s_u with the instrumental noise level:

$$s_u = \| \mathbf{e}_u \| / [JK - (N_c + N_{unx})]^{1/2} \tag{8}$$

where $\mathbf{e}_u$ is from equation (4). A plot of s_u computed for trial number of components will show decreasing values, starting at s_p when the number of components is equal to A (the number of latent variables used to described the calibration data), until it stabilizes at a value compatible with the experimental noise, allowing to locate the correct number of components. It should be noticed that for $N_{unx}> 1$, the profiles provided by the SVD analysis of $\mathbf{E}_p$ unfortunately no longer resemble the true interferent profiles, due to the fact that the principal components are restricted to be orthonormal.

Software

Finally, the software freely available on the Internet to implement second-order calibration should be mentioned. They are in the form of MATLAB codes [50], including several useful graphical user interfaces (GUI) [51, 52, 53]. Among the most popular GUI it can be mentioned: a) N-way toolbox (http://www.models.life.ku.dk/nwaytoolbox/download) for PARAFAC, PARAFAC2, GRAM, DTLD, U-PLS and N-PLS, b) MCR graphical interface (http://www.mcrals.info/) for MCR-ALS, and c) MVC2 graphical interface (www.iquir-conicet.gov.ar/descargas/mvc2.rar) for PARAFAC, APTLD, SWATLD, BLLS/RBL, U-PLS/RBL, N-PLS/RBL and MCR-LS.

Literature Examples

In this section, an up-to-date literature search on the use of chemometric tools for solving problems related to the determination of emerging contaminants in water samples is presented (Table 1). The considerable relevance of chemometric algorithms in the development of methods for the determination of emerging contaminants in a wide range of water samples is highlighted in many reviews [30, 54, 55, 56, 57, 58].

In general terms, the main purpose of the works was to solve problems that arise from the analysis of samples in which the target analytes are present at very low concentrations and embedded in complex matrices, which contains species that interfere with the determination. To overcome these drawbacks, different kinds of strategies, which in many cases comprise the application of chemometric approaches, were followed. In this sense, most of the works describe the application of second-order chemometric algorithms in the simultaneous determination of several emerging contaminants by separation methods such as HPLC coupled to DAD or MS [59, 60, 61, 62, 63]. The use of the chemometric algorithms allowed solving problems such as lack of chromatographic resolution due to coeluting analytes and/or presence of interferences in the sample matrix. Besides, several works demonstrate the capability of the use of EEMs for the determination of emerging contaminants in complex matrices without any prior sample pretreatment [64, 65, 66, 67].

In general terms, MCR-ALS, PARAFAC and U-PLS/RBL are the preferred algorithms to process this kind of data, but their implementation is related to the data intrinsic characteristics. Maggio et al. [68] observed distorting effects on the chromatograms, i.e., the analyte retention time shift from run to run and the shift magnitude increases with the presence of potential interferents in some samples, making it impossible to align the chromatograms in the time dimension, in order to restore the trilinearity required by most second-order multivariate algorithms. Consequently, MCR-ALS was the preferred algorithm, since it is able to handle second-order data deviating from trilinearity. On the other hand, PARAFAC has been mainly applied to resolved data involving EEMs [64, 66, 67, 69, 70, 72], and there are also several applications related to the determination of different analytes by chromatographic methods such as GC [73] and LC [63]. However, PARAFAC requires second-order data following trilinear structures.

When working with separative methods, the appearance of baseline drift and coelution problem between analytes, and also with the matrix components, led to the fact that neither identification nor quantification could be performed using classical univariate calibration. In fact, erroneous determinations of starting and ending points of each peak, in case of baseline drift, have a drastic effect on the uncertainty of the predicted concentrations. In consequence, handling baseline drifts or background contributions before applying second-order calibration algorithms has been a critical preprocessing step in many chromatographic analyses [60, 63, 74, 79, 80]. An approach is to set the background as a systematic part of the model and treat it as a chemical component. Another way to solve this problem is exhibited by De Zan et al. [75], whom proposed a method for the determination of eight tetracycline antibiotics in effluent wastewater by LC-DAD. Because of the high complexity of the analytical problem under study, a considerable number of matrix compounds coeluted with the target analytes leading to strong overlapping and baseline drift. To solve these drawbacks, the asymmetric least-squared algorithm developed by Eilers [76] was employed to eliminate the chromatogram baseline. Then, to eliminate unexpected interferences and sensitivity changes, a strategy involving standard addition calibration in combination with U-PLS/RBL was implemented. The same procedures were conducted by García et al. [74] in the analysis of seven non-steroidal anti-inflammatory drugs and the anticonvulsant carbamazepine, by Martinez Galera et al. for the quantitation of nine β-blockers [60], and by Vosough et al. for the quantification of sulfonamids, metronidazole and chloramphenicol [73], all of themin river water and modeling LC-DAD data with MCR-ALS. Yu et al. [63] employed the same aforementioned strategy but using another algorithm, i.e., the chromatographic background drift was completely eliminated by orthogonal spectral signal projection (OSSP) to improve the quantitative performance of PARAFAC. On the other hand, a novel method for the determination of carbaryl in effluents using excitation-emission-kinetic fluorescence data was proposed by Zhu et al. [67]. The followed strategy comprised the application of quadrilinear PARAFAC to eliminate the strong and broad native fluorescence of different chromophores present in the matrix effluent and estimate the carbaryl concentration. Moreover, the effects of both Rayleigh and Raman scattering could be avoided and weakened. Kumar et al. [34] employed MCR-ALS for the simultaneous extraction of the SFS for the pure components at various wavelength offsets, without any prior knowledge of them. The principal aim of this work was to create "fingerprints" for the identification of the fluorophores benzo[*a*]pyrene, perylene, and pyrene in groundwater. In this case, the methodology based on the subtraction of the blank solvent spectra was used to correct the Raman scattering.

Another interesting consideration when analyzing complex environmental samples is the matrix effect caused by the presence of interferents or by the pretreatment steps. In this sense, several works were conducted to reduce the number of calibration samples by using transference of calibration with piecewise direct standardization (PDS) [77], especially when the standard addition method should be implemented [73, 75, 78]. Very recently, Schenone et al. [71] presented a methodology based on the use of EEM's modeled with PARAFAC combined with PDS to predict the phenyleprine concentration in waters when significant inner filter effects occur, even in the presence of unexpected sample components. Results were compared with those obtained after the application of the classical standard addition method combined with PARAFAC, carrying out five additions to each sample, in triplicate, and it was showed that the methodology constitutes a simple and low-cost method for the determination of emerging contaminants in water samples.

Moreover, some researchers used different strategies to observe and/or quantitate the degradation products of certain emerging contaminants. In order to quantitate the analytes involved in the photodegradation of phenol, Bosco et al. [64] applied PARAFAC to resolve EEMs recorded throughout the reaction. Besides, Razuc et al. [79] used hybrid hard-soft modeling multivariate curve resolution (HS-MCR) to study the kinetic behavior of the photodegradation reaction of ciprofloxacin carried out in a flow system with UV–Vis monitoring. The UV–Vis spectra recorded for a given sample were arranged in matrices and subsequently analyzed using HS-MCR. The spectral overlap and lack of selectivity inherent to the use of spectrophotometric detection in the analysis of mixtures was overcome by using this algorithm.

In some cases of highly complex systems, a strategy that involves data partitioning into small regions should be applied to troubleshoot issues such as rank deficiencies due to complete spectral overlap between analytes. In this sense, García et al. [80] developed an HPLC-DAD method coupled to MCR-ALS and U-PLS/RBL to simultaneously determine eight tetracyclines in wastewater. In order to simplify the analysis, the total chromatographic data registered for each sample was partitioned in eight regions, each of them corresponding to each analyte and the associated interferences. Another example is discussed in the work done by Alcaráz et al., in which an electrophoretic method coupled to MCR-ALS involving two compounds that share the same UV spectra was presented [81]. The resolution strategy consisted of considering them as if they were a single analyte, but with two electrophoretic peaks. In this sense, they had to be completely separated to be satisfactorily quantitated. Therefore, in order to obtain the individual contribution for each analyte, the vector profile was split into two parts after resolution by MCR-ALS.

On some occasions, researchers applied different algorithms to compare their predictive ability, and then select the most suitable for the intended application. For example, in the work of Culzoni et al. [65] experimental data was modeled with PARAFAC, U-PLS/RBL, and N-PLS/RBL. The quantitative analysis was carried out by registering EEMs of a variety of water samples containing galantamine and additional pharmaceuticals as interferences, and further processing through the chosen algorithms. In order to compare their performance on modeling second-order data, the statistical comparison of the prediction results was made *via* the Bilinear Least Squares (BLS) regression method and the Elliptical Joint Confidence Region test. In this work, while the ellipses corresponding to PARAFAC and U-PLS include the theoretically expected point (1,0) suggesting a high-quality prediction, the one corresponding to N-PLS does not include it, indicating a deficient accuracy.

Table 1. Applications of chemometric algorithm used to the resolution of second-order instrumental data aiming at the determination of emerging contaminants in water samples

Analytes	Sample	Instrumentation	Challenges	Algorithm	LOD	Year [Ref.]
Methomyl, seethylsimazine, deethylatrazine, carbendazim, carbofuran, simazine, atrazine, alachlor, chlorpyrifos-oxon, terbutryn, chlorfenvinphos pirimiphosmethyl, chlorpyrifos	Wastewater, sediment	HPLC-DAD	Chromatographic coelution	MCR-ALS	–	2004 [86]
Resorcinol, phenol, oxamyl, methomyl	River water, wastewater	HPLC-DAD	Baseline drift Chromatographic coelution	GRAM PARAFAC MCR–ALS	–	2004 [59]
Methomyl, deethylsimazinedeethylatrazine, carbendazimcarbofuran, simazine, atrazine, alachlor, chlorpyrifosoxon, terbutryn, chlorfenvinphos, pirimiphos-methyl, chlorpyrifos	Wastewater, sediment	LC-MS	Baseline drift Chromatographic coelution Matrix interferences	MCR-ALS	–	2005 [90]
Carbendazim, thiabendazole	Tap, mineral and river water	SPE-EEM	Matrix interferences	PARAFAC PLS/RBL	MBC: 1×10^{-3} µg mL^{-1} TBZ: 2×10^{-4} µg mL^{-1}	2006 [69]
Phenol, hydroquinone, catechol and resorcinol	Wastewater	EEM	Spectral overlapping Matrix interferences	PARAFAC	–	2006 [64]
Tetracycline, chlortetracycline and oxytetracycline	Surface water	Photochemically induced fluorescence(PIF)-EEM	Spectral overlapping Matrix interferences	N-PLS PARAFAC BLLS	–	2006 [78]
Alachlor, atrazine, carbendazimcarbofuran, chlorfenvinphos, chlorpyrifoschlorpyrifos-oxondeethylatrazinedeethylsimazine, methomylpirimiphos-methylsimazine, terbutryn	Wastewater, sediment	HPLC-DAD/MS	Chromatographic coelution Matrix interferences	MCR-ALS	–	2006 [85]
Carbaryl and its degradation product 1-naphthol	River and tapwater	EEM	Matrix interferences	PARAFAC	–	2008 [70]
Tetracycline, oxytetracycline, meclocycline, minocycline, metacycline, chlortetracycline, demeclocycline and doxycycline	Wastewater	HPLC-DAD	Matrix interferences	U-PLS/RBL MCR-ALS	0.02–0.03 mg L^{-1}	2008 [80]
Benzo[*a*]pyrene and dibenzo[*a,h*]anthracene	Natural water, wastewater	Solid phase EEM	Spectral overlapping Matrix interferences	PARAFAC U-PLS/RBL	0.14 g L^{-1}	2008 [91]

Analytes	Sample	Instrumentation	Challenges	Algorithm	LOD	Year [Ref.]
Atrazine, simazine,chlorpyrifos, diazinon, lindane, endosulphan I, endosulphan II, endosulphan, sulphate, pyrene, fluoranthene, phenanthrene, nonylphenol,etoxylate, t-octylphenol)	River water	Data sets from monitoring and simulation studies	Identification of contamination patterns	MCR-ALS	–	2009 [88]
Carbaryl	Effluent	EEM	Spectral overlapping Matrix interferences	PARAFAC	3.58 μg L^{-1}	2009 [67]
Naproxen, ketoprofen, diclofenac, piroxicam, indomethacin, sulindac, diflunisal, carbamazepine	River water, wastewater	SPME-HPLC-DAD	Chromatographic coelution Spectral overlapping Matrix interferences	MCR-ALS		2009 [74]
Sotalolatenolol, nadolol, pindolol, metoprolol, timolol, bisoprolol, propanolol and betaxolol, paracetamol and phenazone	River water	HPLC-DAD	Chromatographic coelution Spectral overlapping Matrix interferences	MCR-ALS	0.2 – 0.5 μg L^{-1}	2010 [60]
Organochlorinated compounds, polycyclic aromatic hydrocarbons, pesticides and alkylphenols	River water, sediment and soil	Data sets from monitoring and simulation studies	Identification of contamination patterns	MCR-ALS	–	2010 [92]
Galantamine	River, tap and well water	EEM	Spectral overlapping	PARAFAC U-PLS/RBL N-PLS/RBL	–	2010 [65]
Benzophenone-3,4-methylbenzilidene camphor, octocrylene, 1-(4-tert-butylphenyl)-3-(4methyoxyphenyl)1,3-propanedione), ethylhexylmethoxycinnamate, ethylhexylsalicylate, homosalate	Wastewater	HPLC-DAD	Chromatographic coelution Spectral overlapping Matrix interferences	MCR-ALS	–	2011 [84]
Bentazone	River water	EEM	Spectral overlapping Matrix interferences	PARAFAC	5 ng mL^{-1}	2011 [66]
Aldicarbsulfoxide, oxamyl, aldicarbsulfone, methomyl, 3-hydroxy-carbofuran, aldicarb, propoxur, carbofuran, carbaryl, 1-naphthol and methiocarb	Surface and tap water	HPLC-DAD	Chromatographic coelution Spectral overlapping Matrix interferences	MCR-ALS	0.1 – 2.3 μg mL^{-1}	2011 [68]
Simazin, atrazine, propazine	Stream, well, and river water	GC-MS	Chromatographic coelution Matrix interferences	PARAFAC	0.15 –1.34 μg L^{-1}	2012 [72]
Benzo[*a*]pyrene, perylene, and pyrene	Groundwater	TSFS	Spectral overlapping	MCR–ALS	–	2012 [34]
Ciprofloxacin	Aqueous solution	Flow system-UV-Vis	Monitoring of photodegradation reaction	HS-MCR	–	2013 [79]

Table 1. (Continued)

Analytes	Sample	Instrumentation	Challenges	Algorithm	LOD	Year [Ref.]
Sulfacetamide, sulfamerazine, sulfamethoxazole, tetracycline, pipemidicacid, pefloxacin, danofloxacin, lomefloxacin, metronidazole,ornidazole, andoxytetracycline	Tap water	HPLC-DAD	Baseline drift Chromatographic coelution Matrix interferences	OSSP PARAFAC	–	2013 [63]
Sulfamethoxazole, metronidazole, chloramphenicol, sulfadizine and sulfamerazine	Wastewater	SPE-HPLC-DAD	Baseline drift Chromatographic coelution Matrix interferences	MCR/ALS	19 – 40 ng mL^{-1}	2013 [73]
Carbamazepine	Tap, ground and river water	PIF-EEM	Matrix interferences	PARAFAC U- and N-PLS MCR-ALS	0.2 ng mL^{-1}	2013 [82]
Ofloxacin and ciprofloxacin	Tap water	HPLC-EEM	Chromatographic coelution Matrix interferences	PARAFAC U-PLS/RTL MCR-ALS	0.4 – 3.2 μg L^{-1}	2014 [31]
Flumenique, enoxacin, ofloxacin, cinoxacin, enrofloxacin and ciprofloxacin	Tap water	CE-DAD	Electrophoretic coelution Matrix interferences	MCR-ALS	5 – 18 μg L^{-1}	2014 [81]

This latter fact demonstrates that in the system under study it is not appropriate to maintain the tridimensional nature of the data for the quantitative analysis in order to successfully quantitate the analyte. Another example related to the application of BLS to compare the performance of different algorithms can be found in the work of Lozano et al. [82].

An interesting aspect to point out is the calculation of figures of merit for the different chemometric methods.

The traditional calibration curves (signal as a function of analyte concentration) could not be employed to inform figures of merits when applying multivariate methods to process higher-order data. In the last years, the net analyte signal (NAS) emerged as a concept that helps to improve the multivariate calibration theory. It allows separating the analyte specific information from the whole signal, and can be used for estimating important figures of merit. In addition, NAS values of each sample can be used to represent multivariate models as pseudo-univariate curve, an easier and simpler manner to interpret them in routine analyses [60, 68, 73, 74, 79, 80, 82, 84, 85, 85]. García et al. [80] showed that, in general terms, lower LODs are obtained when performing univariate calibration, because multivariate figures of merit are highly affected by the presence of unexpected components in real samples. Basically, computation and report of figures of merit is fundamental to establish the method efficiency. In this sense, most of the works under study do not report or calculate detection and quantitation limits [59, 74, 78, 83, 84, 85, 86], or those in which they are mentioned were calculated using univariate curves [73]. It is pertinent to remark that sensitivity (SEN) can be considered as one of the most relevant figures of merit in the field of analytical chemistry due to the fact that it is a decisive factor in estimating others, such as LOD and LOQ. SEN can be defined as the variation in the net response for a given change in analyte concentration [31]. For more information related to the calculation of SEN with multivariate quantitative purposes the reader is referred to the work of Bauza et al. [87].

Regarding the abundance of conducted research, pesticides (in particular, insecticides) are the most studied analytes [67, 72, 82, 88], followed by drugs (i.e., antibiotics) [78, 79, 81], phenolic compounds [59, 64], and others [34, 62, 89]. With respect to the analysis of pesticides, Peré et al. [91] developed an LC-MS method coupled to MCR-ALS for the analysis of biocides in complex environmental samples, including methomyl, deethylsimazine, deethylatrazine, carbendazim, carbofuran, simazine, atrazine, alachlor, chlorpyrifos-oxon, terbutryn, chlorfenvinphos, pirimiphosmethyl and chlorpyrifos. With regard to the analysis of hydrocarbons, a method for the simultaneous determination of two polycyclic aromatic hydrocarbons, i.e., benzo[*a*]pyrene and dibenzo[*a,h*]anthacene, which consisted in EEMs registration of samples deposited on nylon membranes and their resolution by applying PARAFAC was developed by Bortolato et al. [91]. An example related to the assessment of drug traces in highly complex wastewater samples can be found in the work conducted by Vosough et al. [73]. In this work, five important antibiotics, i.e., sulfamethoxazole, metronidazole, chloramphenicol, sulfadizine and sulfamerazine, were analysed using SPE-HPLC-DAD. Due to the matrix interferences and the resulting sensitivity changes, a strategy implementing standard addition calibration in combination with MCR-ALS was performed.

The versatility of MCR-ALS was also exemplified in the work of Terrado et al. [92]. An interesting discussion related to the use of MCR-ALS as a powerful chemometric tool for the analysis of environmental monitoring data sets was carried out. The proposed strategy allowed for the investigation, resolution, identification, and description of pollution patterns distributed over a particular geographical area. The results obtained by MCR-ALS analysis of surface water, groundwater, sediment and soil data sets obtained in a 3-year extensive monitoring were used to perform an integrated interpretation of the main features that characterize pollution patterns of organic contaminants affecting the Ebro River basin (Catalonia, NE Spain). Then, concentrations of four different families of compounds, i.e., organochlorinated compounds, polycyclic aromatic hydrocarbons, pesticides and alkylphenols were analysed. This example showed the capability of chemometric methods in relation to their field of action, i.e., they could be employed not only to estimate analyte concentrations but also to evaluate analyte patterns or behaviors in different environmental locations.

Conclusion

In this chapter a complete overview of the applications of second-order algorithms for the determination of emerging contaminants in a variety of complex environmental water samples was carried out. The selected examples, which include diverse approaches, show the capability of the different algorithms to enhance data resolution and reduce the tedious workbench associated with this kind of analytes which are embedded in complex matrices. As a consequence, the analytical chemist should be encouraged to exploit more frequently these tools, originated by the association of second-order data generation and modeling with convenient algorithms, which are nowadays easily available on the Internet.

Acknowledgments

The authors are grateful to Universidad Nacional del Litoral, CONICET (Consejo Nacional de Investigaciones Científicas y Técnicas) and ANPCyT (Agencia Nacional de Promoción Científica y Tecnológica) for financial support.

References

[1] Bayen S., Yi X., Segovia E., Zhou Z. Kelly B.C. (2014) *Journal of Chromatography A*, 1338, 38-43.

[2] Deblonde T., Cossu-Leguille C., Hartemann P. (2011) *International Journal of Hygiene and Environmental Health*, 214, 442-448.

[3] Ferrer I., Thurman E.M. (2013) In: *Comprehensive Analytical Chemistry*. Elsevier. 91-128.

[4] Kopperi M., Ruiz-Jiménez J., Hukkinen J.I., Riekkola M.-L. (2013) *Analytica Chimica Acta*, 761, 217-226.

[5] Lima Gomes P.C.F., Barnes B.B., Santos-Neto Á.J., Lancas F.M., Snow N.H. (2013) *Journal of Chromatography A*, 1299, 126-130.

[6] Lisboa N.S., Fahning C.S., Cotrim G., Dos Anjos J.P., De Andrade J.B., Hatje V. Da Rocha G.O. (2013) *Talanta*, 117, 168-175.
[7] Montesdeoca-Esponda S., Vega-Morales T., Sosa-Ferrera Z., Santana-Rodríguez J.J. (2013) *TrAC Trends in Analytical Chemistry*, 51, 23-32.
[8] Patrolecco L., Ademollo N., Grenni P., Tolomei A., Barra Caracciolo A., Capri S. (2013) *Microchemical Journal*, 107, 165-171.
[9] Snow D.D., Cassada D.A., Bartelt-Hunt S.L., Li X., Zhang Y., Zhang Y., Yuan Q., Sallach J.B. (2012) *Water Environment Research*, 84, 764-785.
[10] Farré M., Barceló D., Barceló D. (2013) *TrAC Trends in Analytical Chemistry*, 43, 240-253.
[11] Smital T. (2008) 105. In: *Emerging Contaminants from Industrial and Municipal Waste*. Springer Berlin Heidelberg. 105-142.
[12] Thomaidis N.S., Asimakopoulos A.G., Bletsou A. (2012) *Global NEST Journal* 14, 72-79.
[13] Murray K.E., Thomas S.M., Bodour A.A. (2010) *Environmental Pollution*, 158, 3462-3471.
[14] Luo Y., Guo W., Ngo H.H., Nghiem L.D., Hai F.I., Zhang J., Liang S., Wang X.C. (2014) *Science of The Total Environment*, 473–474, 619-641.
[15] Davankov V., Tsyurupa M. (2011) In: *Comprehensive Analytical Chemistry*. Elsevier. 523-565.
[16] Jiménez J.J. (2013) *Talanta*, 116, 678-687.
[17] Robles-Molina J., Gilbert-López B., García-Reyes J.F., Molina-Díaz A. (2013) *Talanta*, 117, 382-391.
[18] Weigel S., Bester K., Hühnerfuss H. (2001) *Journal of Chromatography A*, 912, 151-161.
[19] Zapf A., Heyer R., Stan H.-J. (1995) *Journal of Chromatography A*, 694, 453-461.
[20] Robles-Molina J., Lara-Ortega F.J., Gilbert-López B., García-Reyes J.F., Molina-Díaz A. *Journal of Chromatography A*,
[21] Jiang J.-Q., Zhou Z., Sharma V.K. (2013) *Microchemical Journal*, 110, 292-300.
[22] Loos R., Carvalho R., António D.C., Comero S., Locoro G., Tavazzi S., Paracchini B., Ghiani M., Lettieri T., Blaha L., Jarosova B., Voorspoels S., Servaes K., Haglund P., Fick J., Lindberg R.H., Schwesig D., Gawlik B.M. (2013) *Water Research*, 47, 6475-6487.
[23] Bourdat-Deschamps M., Leang S., Bernet N., Daudin J.-J., Nélieu S. n. *Journal of Chromatography A*,
[24] Farré M., Kantiani L., Petrovic M., Pérez S., Barceló D. (2012) *Journal of Chromatography A*, 1259, 86-99. *Journal of Chromatography A*, 1218, 6712-6719.
[25] Portolés T., Mol J.G.J., Sancho J.V., Hernández F. (2014) *Journal of Chromatography A*, 1339, 145-153.
[26] Wilson W.B., Hewitt U., Miller M., Campiglia A.D. (2014) *Journal of Chromatography A*, 1345, 1-8.
[27] Guitart C., Readman J.W., Bayona J.M. (2012) In: *Comprehensive Sampling and Sample Preparation*. Academic Press. Oxford. 297-316.
[28] Arancibia J.A., Damiani P.C., Escandar G.M., Ibañez G.A., Olivieri A.C. (2012) *Journal of Chromatography B*, 910, 22-30.

[29] Escandar G.M., Goicoechea H.C., Muñoz De La Peña A., Olivieri A.C. (2014) *Analytica Chimica Acta*, 806, 8-26.
[30] Escandar G.M., Olivieri A.C., Faber N.M., Goicoechea H.C., Muñoz De La Peña A., Poppi R.J. (2007) *TrAC Trends in Analytical Chemistry*, 26, 752-765.
[31] Alcaráz M.R., Siano G.G., Culzoni M.J., De La Peña A.M., Goicoechea H.C. (2014) *Analytica Chimica Acta*, 809, 37-46.
[32] De Juan A., Tauler R. (2003) *Analytica Chimica Acta*, 500, 195-210.
[33] De Juan A., Tauler R. (2001) *Journal of Chemometrics*, 15, 749-771.
[34] Kumar K., Mishra A.K. (2012) *Chemometrics and Intelligent Laboratory Systems*, 116, 78-86.
[35] Sanchez E., Kowalski B.R. (1986) *Analytical Chemistry*, 58, 496-499.
[36] Sanchez E., Kowalski B.R. (1990) *Journal of Chemometrics*, 4, 29-45.
[37] Wentzell P.D., Nair S.S., Guy R.D. (2001) *Analytical Chemistry*, 73, 1408-1415.
[38] Chen Z.-P., Wu H.-L., Jiang J.-H., Li Y., Yu R.-Q. (2000) *Chemometrics and Intelligent Laboratory Systems*, 52, 75-86.
[39] Xia A.L., Wu H.-L., Fang D.-M., Ding Y.-J., Hu L.-Q., Yu R.-Q. (2005) *Journal of Chemometrics*, 19, 65-76.
[40] Bro R. (1997) *Chemometrics and Intelligent Laboratory Systems*, 38, 149-171.
[41] Tauler R. (1995) *Chemometrics and Intelligent Laboratory Systems*, 30, 133-146.
[42] Linder M., Sundberg R. (1998) *Chemometrics and Intelligent Laboratory Systems*, 42, 159-178.
[43] Wold S., Geladi P., Esbensen K., Öhman J. (1987) *Journal of Chemometrics*, 1, 41-56.
[44] Olivieri A.C. (2005) *Journal of Chemometrics*, 19, 253-265.
[45] Lozano V.A., Ibañez G.A., Olivieri A.C. (2008) *Analytica Chimica Acta*, 610, 186-195.
[46] Olivieri A.C. (2005) *Journal of Chemometrics*, 19, 615-624.
[47] Smilde A., Bro R., Geladi P. (2005) In: *Multi-Way Analysis with Applications in the Chemical Sciences*. John Wiley Sons, Ltd. 257-349.
[48] Muñoz De La Peña A., Espinosa Mansilla A., González Gómez D., Olivieri A.C., Goicoechea H.C. (2003) *Analytical Chemistry*, 75, 2640-2646.
[49] Haaland D.M., Thomas E.V. (1988) *Analytical Chemistry*, 60, 1193-1202.
[50] Mathworks T. (2010) MATLAB.
[51] Jaumot J., Gargallo R., De Juan A., Tauler R. (2005) *Chemometrics and Intelligent Laboratory Systems*, 76, 101-110.
[52] Gemperline P.J., Cash E. (2003) *Analytical Chemistry*, 75, 4236-4243.
[53] Olivieri A.C., Wu H.-L., Yu R.-Q. (2009) *Chemometrics and Intelligent Laboratory Systems*, 96, 246-251.
[54] Escandar G.M., Damiani P.C., Goicoechea H.C., Olivieri A.C. (2006) *Microchemical Journal*, 82, 29-42.
[55] Galera M.M., García M.D.G., Goicoechea H.C. (2007) *TrAC Trends in Analytical Chemistry*, 26, 1032-1042.
[56] Goicoechea H.C., Culzoni M.J., García M.D.G., Galera M.M. (2011) *Talanta*, 83, 1098-1107.
[57] Mas S., De Juan A., Tauler R., Olivieri A.C., Escandar G.M. (2010) *Talanta*, 80, 1052-1067.
[58] Ballesteros-Gómez A., Rubio S. (2011) *Analytical Chemistry*, 83, 4579-4613.

[59] Comas E., Gimeno R.A., Ferré J., Marcé R.M., Borrull F., Rius F.X. (2004) *Journal of Chromatography A*, 1035, 195-202.
[60] Martínez Galera M., Gil García M.D., Culzoni M.J., Goicoechea H.C. (2010) *Journal of Chromatography A*, 1217, 2042-2049.
[61] Peré-Trepat E., Petrovic M., Barceló D., Tauler R. (2004) *Analytical and Bioanalytical Chemistry*, 378, 642-654.
[62] Selli E., Zaccaria C., Sena F., Tomasi G., Bidoglio G. (2004) *Water Research*, 38, 2269-2276.
[63] Yu Y. J., Wu H. L., Fu H. Y., Zhao J., Li Y. N., Li S. F., Kang C., Yu R. Q. (2013) *Journal of Chromatography A*, 1302, 72-80.
[64] Bosco M.V., Garrido M., Larrechi M.S. (2006) *Analytica Chimica Acta*, 559, 240-247.
[65] Culzoni M.J., Aucelio R.Q., Escandar G.M. (2010) *Talanta*, 82, 325-332.
[66] Porini J.A., Escandar G.M. (2011) *Analytical Methods*, 3, 1494-1500.
[67] Zhu S. H., Wu H. L., Xia A.L., Nie J. F., Bian Y. C., Cai C. B., Yu R. Q. (2009) *Talanta*, 77, 1640-1646.
[68] Maggio R.M., Damiani P.C., Olivieri A.C. (2011) *Talanta*, 83, 1173-1180.
[69] Piccirilli G.N., Escandar G.M. (2010) *Analyst*, 135, 1299-1308.
[70] Zhu S.-H., Wu H.-L., Xia A.L., Han Q.-J., Zhang Y., Yu R.-Q. (2008) *Talanta*, 74, 1579-1585.
[71] Schenone A.V., Culzoni M.J., Martínez Galera M., Goicoechea H.C. (2013) *Talanta*, 109, 107-115.
[72] Real B.D., Ortiz M.C., Sarabia L.A. (2012) *Journal of Chromatography B*, 910, 122-137.
[73] Vosough M., Mashhadiabbas Esfahani H. (2013) *Talanta*, 113, 68-75.
[74] García M.D.G., Cañada F.C., Culzoni M.J., Vera-Candioti L., Siano G.G., Goicoechea H.C., Galera M.M. (2009) *Journal of Chromatography A*, 1216, 5489-5496.
[75] De Zan M.M., Gil García M.D., Culzoni M.J., Siano R.G., Goicoechea H.C., Martínez Galera M. (2008) *Journal of Chromatography A*, 1179, 106-114.
[76] Eilers P.H.C. (2003) *Analytical Chemistry*, 76, 404-411.
[77] Wang Y., Veltkamp D.J., Kowalski B.R. (1991) *Analytical Chemistry*, 63, 2750-2756.
[78] Valverde R.S., García M.D.G., Galera M.M., Goicoechea H.C. (2006) *Talanta*, 70, 774-783.
[79] Razuc M., Garrido M., Caro Y.S., Teglia C.M., Goicoechea H.C., Fernández Band B.S. (2013) *Spectrochimica Acta Part A: Molecular and Biomolecular Spectroscopy*, 106, 146-154.
[80] García M.D.G., Culzoni M.J., De Zan M.M., Valverde R.S., Galera M.M., Goicoechea H.C. (2008) *Journal of Chromatography A*, 1179, 115-124.
[81] Alcaráz M.R., Vera-Candioti L., Culzoni M.J., Goicoechea H.C. (2014) *Analytical and Bioanalytical Chemistry*, 406, 2571-2580.
[82] Lozano V.A., Escandar G.M. (2013) *Analytica Chimica Acta*, 782, 37-45.
[83] Muñoz De La Pena A., Mora Diez N., Bohoyo Gil D., Cano Carranza E. (2009) *J. Fluoresc*, 19, 345-52.
[84] Vosough M., Mojdehi N.R. (2011) *Talanta*, 85, 2175-2181.
[85] Peré-Trepat E., Tauler R. (2006) *Journal of Chromatography A*, 1131, 85-96.
[86] Peré-Trepat E., Hildebrandt A., Barceló D., Lacorte S., Tauler R. (2004) *Chemometrics and Intelligent Laboratory Systems*, 74, 293-303.

[87] Bauza M.C., Ibañez G.A., Tauler R., Olivieri A.C. (2012) *Analytical Chemistry*, 84, 8697-8706.

[88] Terrado M., Lavigne M.-P., Tremblay S., Duchesne S., Villeneuve J.-P., Rousseau A.N., Barceló D., Tauler R. (2009) *Journal of Hydrology*, 369, 416-426.

[89] Bortolato S.A., Arancibia J.A., Escandar G.M. (2010) *Environmental Science Technology*, 45, 1513-1520.

[90] Peré-Trepat E., Lacorte S., Tauler R. (2005) *Journal of Chromatography A*, 1096, 111-122.

[91] Bortolato S.A., Arancibia J.A., Escandar G.M. (2008) *Anal. Chem*, 80, 8276-86.

[92] Terrado M., Barceló D., Tauler R. (2010) *Analytica Chimica Acta*, 657, 19-27.

In: Current Applications of Chemometrics
Editor: Mohammadreza Khanmohammadi

ISBN: 978-1-63463-117-4

Chapter 8

QUANTIFICATION OF PROTON MAGNETIC RESONANCE SPECTROSCOPY (^{1}H-MRS)

Mohammad Ali Parto Dezfouli[1,2] ***and Hamidreza Saligheh Rad***[1, 2*]

[1] Quantitative MR Imaging and Spectroscopy Group,
Research Center for Molecular and Cellular Imaging,
Tehran University of Medical Sciences, Tehran, Iran
[2] Department of Biomedical Engineering and Medical Physics,
Tehran University of Medical Sciences, Tehran, Iran

ABSTRACT

Proton magnetic resonance spectroscopy (^{1}H-MRS) is a non-invasive diagnostic tool for measuring biochemical changes in the human body. It is providing information on the chemical structure of complex molecules and capable of the quantitative analysis of complex mixtures. Magnetic resonance spectroscopy refers to the absorption and release of radio frequent energy by a nucleus in a magnetic field. While MRS provides spectral or chemical information, MRI gives anatomical information. In medical applications, MRS is applied to identify substances at specific locations in the body and to obtain an indication of their concentration. A disease condition can be indicated on a MRS spectrum by either a change in concentration of a metabolite, which is normally present in the brain, or by the appearance of a resonance, which is not normally detectable. Metabolites are the intermediates and products of metabolism, usually restricted to small molecules. The aim of quantification methods is to non-invasively determine the absolute concentration of metabolites. As the metabolite concentrations change with a range of disease conditions, quantification is important in a medical context to detect these changes. Accurate and robust quantification of proton magnetic resonance spectroscopy is to precisely synthesize the biochemical compounds of the tissue, and by means of correct computation of different metabolites' concentrations detected in the associated spectra. However, overlapping nature of metabolites of interests, existing static field (B0) inhomogeneity as well as low-SNR regime hinders the performance of such quantification methods. Here briefly describes principle of advanced quantitative data analysis and discuss biomedical applications of MRS.

[*] Corresponding author: h-salighehrad@tums.ac.ir

INTRODUCTION

Proton magnetic resonance spectroscopy (^{1}H-MRS) is a non-invasive technique for monitoring biochemical changes in a living tissue [1]. ^{1}H-MRS of the brain has information consisting from the molecular content of tissues, which made it a useful tool for monitoring human brain tumours. Metabolite concentrations are most often presented as ratios or absolute concentrations. Important methodological aspects in an absolute quantification strategy are needed to address, including radiofrequency coil properties, calibration procedures, spectral fitting methods, cerebrospinal fluid content correction, macromolecule suppression, and spectral editing.

The signal frequency that is detected in nuclear magnetic resonance (NMR) spectroscopy is proportional to the magnetic field applied to the nucleus. This would be a precisely determined frequency if the only magnetic field acting on the nucleus was the externally applied field. But the response of the atomic electrons to that externally applied magnetic field is such that their motions produce a small magnetic field at the nucleus which usually acts in opposition to the externally applied field. This change in the effective field on the nuclear spin causes the NMR signal frequency to shift. The magnitude of the shift depends upon the type of nucleus and the details of the electron motion in the nearby atoms and molecules. It is called a "chemical shift". The precision of NMR spectroscopy allows this chemical shift to be measured, and the study of chemical shifts has produced a large store of information about the chemical bonds and the structure of molecules [2].

The MRS signal is made up of several peaks in specific frequencies which are related to the chemical environment of the metabolites. The amplitude of the time-domain coefficients of each metabolite, which is equivalent to the peak area in the frequency-domain, is proportional to the amount of that metabolite's concentration [3, 4].

Concentration of metabolites may lead to accurate monitoring of the human brain tumours only if they are well quantified [5]. An accurate and robust quantification of ^{1}H-MRS is to precisely synthesize the biochemical compounds of tissues and by means of correct computation of different metabolites' abundance. The performance of such methods are limited due to the overlapping nature of metabolites of interests and the static field (B_0) inhomogeneity in the actual low signal-to-noise ratio (SNR) regime. Furthermore, a large and broad background baseline signal, which comes from macromolecules hampers the accuracy of MRS quantification.

Quantification methods to estimate the abundance of metabolites of interests are implemented either in time-domain [6] or in frequency-domain [7]. Also number of simultaneous dual domains techniques have been proposed for spectral line estimation in MRS [8-12].

BASICS OF MAGNETIC RESONANCE

One of the non-invasive diagnostic techniques used for detection of cancer and metabolic diseases is Nuclear Magnetic Resonance (NMR). NMR is based on the magnetic properties that certain nuclei experiment when exposed to an external magnetic field. These nuclei have a unique resonance frequency related to the radio waves emitted. Only certain atomic nuclei have a nuclear magnetic moment, for instance, Proton (^{1}H), Phosphorus (^{31}P), Carbon (^{13}C), Fluor (^{19}F), Sodium (^{23}Na), etc. Due to its abundance in the human body, the most common nucleus used in MR is ^{1}H.

The proton MRS signal is obtained by placing the sample under investigation into a static external magnetic field B_0 to which the angular momentum of the nuclei is aligned (Figure 1).

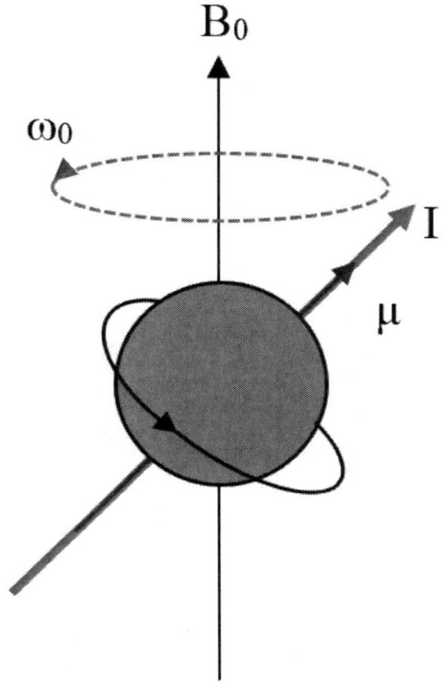

Figure 1. Spin rotation around its own axis and around the magnetic.

Spins rotate about the Z axis with a specific frequency called the Larmor frequency (f_0) proportional to the magnetic field B_0. At the Larmor frequency the nuclei absorb energy causing the proton to change its alignment and to be detected in a magnetic field. This Larmor frequency is defined by:

$$f_0 = \frac{\gamma B_0}{2\pi} \tag{1}$$

Since the gyromagnetic ratio (γ) is a constant of each nuclear species, the spin frequency of a certain nuclei (f_0) depends on the external magnetic field (B_0) and the local microenvironment. The electric shell interactions of these nuclei with the surrounding molecules cause a change in the local magnetic field leading to a change on the spin frequency of the atom (a phenomenon called chemical shift). The value of this difference in resonance frequency gives information about the molecular group carrying ^{1}H and is expressed in parts per million (ppm).

The effective magnetic field at the nucleus can be expressed in terms of the externally applied field B0 by the expression:

$$B = B_0(1 - \sigma) \tag{2}$$

where σ is called the shielding factor or screening factor. The factor s is small typically 10-5 for protons and <10-3 for other nuclei [13]. In practice the chemical shift is usually indicated by a symbol d which is defined in terms of a standard reference.

$$\delta = \frac{(v_s - v) \times 10^6}{v_R} \tag{3}$$

The signal shift is very small, parts per million, but the great precision with which frequencies can be measured permits the determination of chemical shift to three or more significant figures. The chemical shift position of a nucleus is ideally expressed in ppm because it is independent of the field strength.

$$ppm = \frac{f_{Hz} \times 10^6}{f_s} + ppm_{ref} \tag{4}$$

There are different field strengths clinically used for conventional MRI, ranging from 0.2 to 3T. Since the main objective of MRS is to detect weak signals from metabolites, a higher strength field is required (1.5T or more). Higher field strength units have the advantage of higher signal-to-noise ratio (SNR), better resolution and shorter acquisition times making the technique useful in sick patients and others that cannot hold still for long periods of time.

Magnetic Resonance Spectroscopy (MRS) Signals

MRS signals are measured in the time domain and they are exponentially decaying sine wave, which is termed free-induction decay (FID). FID signal contains the sum of the response from all metabolites, macromolecule and noise, which can be represented as a sum of complex-damped exponentials.

They can be measured from a specific anatomical region, or from an entire organ overlaid by a grid of multiple voxels (MRSI). In order to observe the contribution of individual metabolites, MRS signals are transformed to the frequency domain using the Fast Fourier Transform (FFT), producing a spectrum where the metabolite resonances can be visualized.

Some metabolites such as lactate have doublets, triplets or multiples instead of single peaks. These peaks are broken down into more complex peaks and are explained by J-coupling, also named spin-spin coupling. The J-coupling phenomenon occurs when the molecular structure of a metabolite is such that protons are found in different atomic groups. These groups have a slightly different local magnetic fields, thus each ^{1}H resonates at a frequency characteristic of its position in the molecule resulting in a multiple peak.

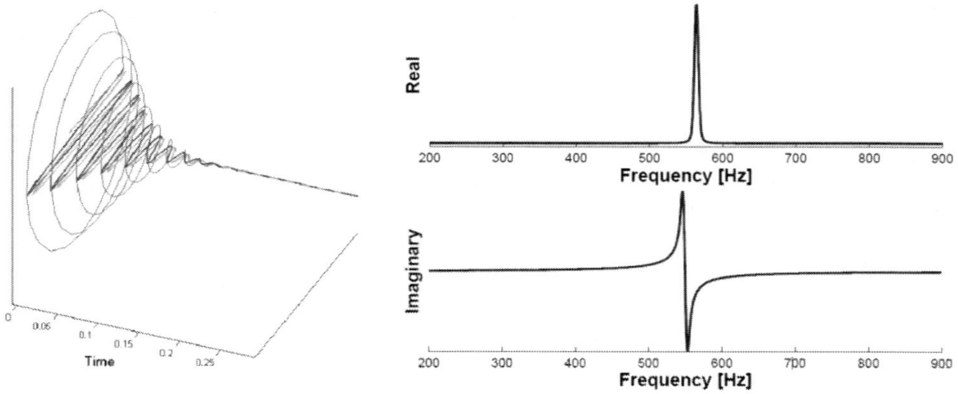

Figure 2. Left: Time domain signal representation in two planes corresponding to the real and imaginary parts. Right: Real and imaginary parts of the Fourier Transformed signal. Adapted from [14].

SIGNAL MODEL

¹H-MRS signal *s(t)* is contained three main terms, that can be modeled by:

$$s(t) = met(t) + b(t) + n(t) \tag{5}$$

The first term; *met(t)* represents the metabolites signal whose model function is known. The main goal is to obtain reliable estimation of the metabolite parameters. The function *met(t)* can be defined as:

$$\text{met}(t) = \sum_{k=1}^{K} a_k e^{(j\varphi_k)} e^{(-d_k t + 2\pi j f_k t)} m_k(t) \tag{6}$$

where K is the number of metabolites ($k = 1, ..., K$), $m_k(t)$ the profile of metabolite , a_k the amplitude, φ_k the phase shift, d_k the damping correction, f_k the frequency shift due to B_0 inhomogeneity and $j = \sqrt{-1}$. Although the distortions are not exactly known, but they have shown specified boundaries. The second term; *b(t)* represents the baseline signal which its model is not known exactly. In the frequency-domain, it is shown broad and smooth compared to the resonance signals and in the time-domain, it decays quickly and exists in few early samples. These two features of baseline are used in our dictionary construction algorithm to overcome the baseline problem. The third term; *n(t)* denotes the white Gaussian-distributed noise.

Technique

The ¹H-MRS acquisition usually starts with anatomical images, which are used to select a volume of interest (VOI), where the spectrum will be acquired. For the spectrum acquisition, different techniques may be used including single- and multi-voxel imaging.

Single-Voxel Spectroscopy

In the single voxel spectroscopy (SVS) the signal is obtained from a voxel previously selected. This voxel is acquired from a combination of slice-selective excitations in three dimensions in space, achieved when a RF pulse is applied while a field gradient is switched on. It results in three orthogonal planes and their intersection corresponds to VOI (Figure 3).

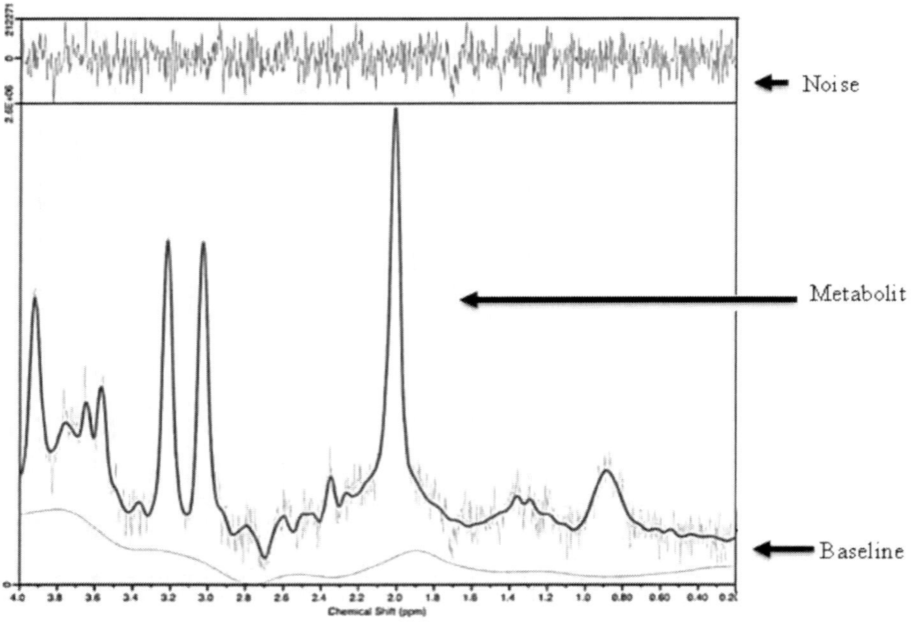

Figure 3. MRS signal [15].

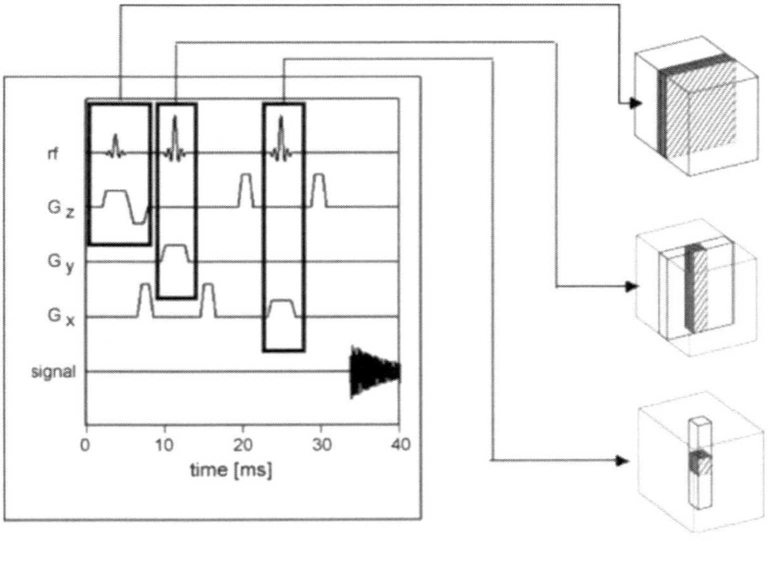

Figure 4. Left: The intersection of the orthogonal planes, given by slice selection and phase gradients, results in the VOI.

MAGENTIC RESONANCE SPECTROSCOPY IMAGING (MRSI)

Magnetic resonance spectroscopy imaging (MRSI) is multi voxel technique. The main objective of MRSI is to obtain simultaneously many voxels and a spatial distribution of the metabolites within a single sequence. Thus, this ^{1}H-MRS technique uses phase-encoding gradients to encode spatial information after the RF pulses and the gradient of slice selection.

MRSI is acquired using only slice selection and phase encoding gradients, besides the spoiler gradients. Instead of the anatomical information given by the conventional MRI signal, the MRS signal results in a spectrum of metabolites with different frequencies.

The main difference between MRSI and SVS is that, after the RF pulse, phase encoding gradients are used in one, two or three dimensions (1D, 2D or 3D) to sample the k-space. In a 1D sequence, the phase encoding has a single direction, in 2D has two orthogonal directions and, in 3D three orthogonal directions. The result of a 2D MRSI is a matrix, called a spectroscopy grid.

The size of this grid corresponds to the field of view (FOV) previously determined. In the 3D sequence, many grids are acquired within one FOV. The number of partitions (or voxels) of the grids is directly proportional to the number of phase encoding steps.

The spatial resolution is also proportional to the number of voxels in a determined FOV (more voxels give a better spatial resolution). However, for a larger number of voxels, more phase encoding steps are needed and this implies a longer time for acquisition. Spatial resolution is also determined by the FOV size (smaller FOV gives better spatial resolution) and by point of spread function (PSF).

ACQUISITION PARAMETERS

Pulse sequence: Mainly, two techniques are used for acquisition of SVS ^{1}H-MRS spectra: pointed-resolved spectroscopy (PRESS) and stimulated echo acquisition mode (STEAM). In the PRESS sequence, the spectrum is acquired using one 90° pulse followed by two 180° pulses. Each of them is applied at the same time as a different field gradient. Thus, the signal emitted by the VOI is a spin echo.

The first 180° pulse is applied after a time TE1/2 from the first pulse (90° pulse) and the second 180o is applied after a time TE1/2+TE. The signal occurs after a time 2TE (Figure 3). To restrict the acquired sign to the VOI selected, spoiler gradients are needed. Spoiler gradients de-phase the nuclei outside the VOI and reduce their signal.

STEAM is the second most commonly used SVS technique. In this sequence all three pulses applied are 90° pulses. As in PRESS, they are all simultaneous with a different field gradient. After a time TE1/2 from the first pulse, a second 90° is applied. The time elapsed between the second and the third is conventionally called “mixing time” (MT) and is shorter than TE1/2.

The signal is finally achieved after a time TE+MT from the first pulse (Figure 3). Thus, the total time for STEAM technique is shorter than PRESS. Spoiler gradients are also needed to reduce signal from regions outside the VOI.

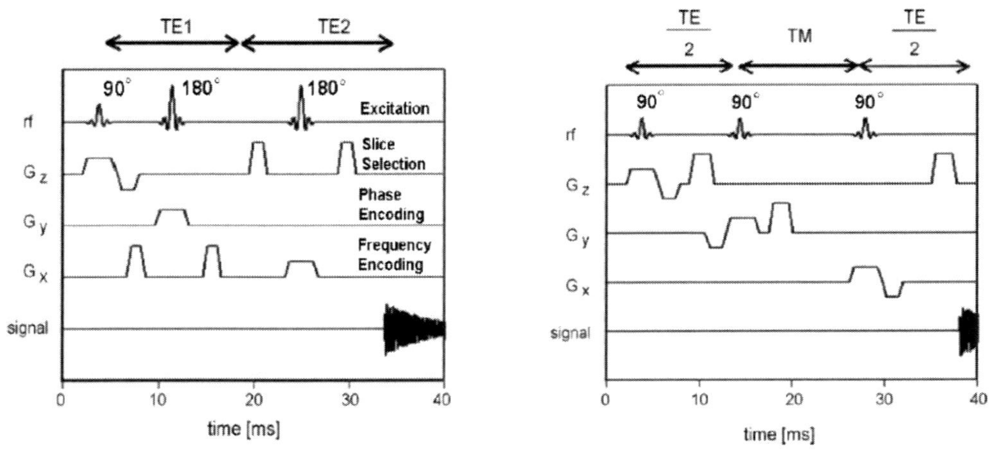

Figure 5. Left: PRESS sequence with three RF pulses applied simultaneously with field gradients along the main axes of the magnet. Right: STEAM sequence with the refocusing gradients positioned before the second RF pulse and after the third RF pulse. Only the first part of the data acquisition time is shown as the first part of the FID. Figures from [16].

Echo Time (TE)

Echo time represents the time in milliseconds (ms) between the application of the first pulse and the beginning of the data acquisition. MRS signals can be acquired at long and short TE, where long TE signals provide a good observation of slowly decaying components, while short TE signals provide more metabolic information. Typical long TE signals are in the range between 60 and 300ms and short TE signals are in the range between 1 and 50ms.

Repetition Time (TR)

Repetition time is the time between successive pulse sequences applied to the same slice.

T1, T2 and T2*

T1 is also called the spin-lattice relaxation time and refers to the time to reduce the difference between the longitudinal magnetization after the RF pulse and its equilibrium value by a factor of e^1, where e is Euler's number e = 2.71828. Thus, we have:

$$M_z(t) = M_0(1 - e^{-\frac{t}{T1}}) \quad (7)$$

where $M_z(t)$ is the longitudinal net magnetization at time t and M_0 the initial net magnetization aligned with B_0. T2 is also called the spin-spin relaxation time and refers to the time constant which describes the return to equilibrium of the transverse magnetization M_{xy} as:

$$M_{xy}(t) = M_{xy0}(1 - e^{-\frac{t}{T2}}) \quad (8)$$

where M_{xy0} is the net transversal magnetization at time 0. T2 is always ≤ T1. The net magnetization in the xy-plane goes to zero and then the longitudinal magnetization still grows until the net magnetization is reached along the z-axis. $T2^*$ is also known as the effective transverse relaxation time, thus, the time constant for the observed decay of the FID. It is a combination of transverse relaxation and magnetic field inhomogeneity effects. In the presence of a homogeneous magnetic field $T2^* = T2$, however, in an inhomogeneous field $T2^* < T2$. We have the following relation between time and frequency domain parameters:

$$T2^* = \frac{1}{\pi v} \quad (9)$$

where v corresponds to the linewidth1 of the Fourier transformed signal. (i.e., the Full Width at Half Maximum (FWHM)).

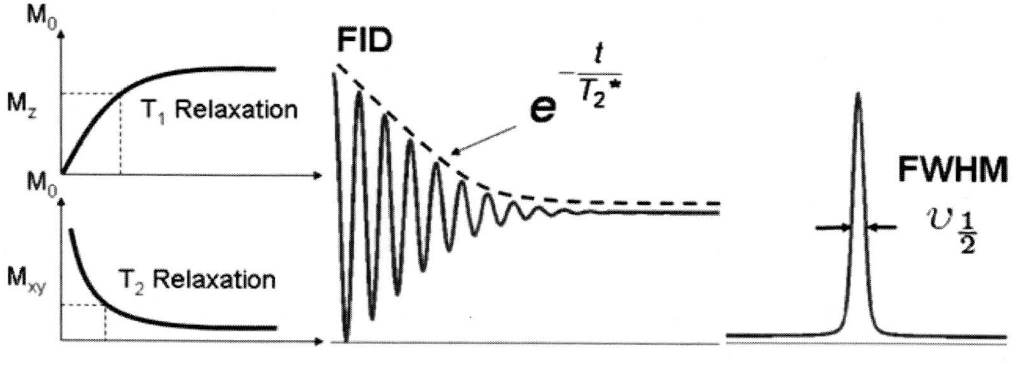

Figure 6. Left: Graphical representation of the T1 and T2 relaxation times along the magnetization M0. The time constants T1 and T2 are measured in seconds (or ms) and refer to the exponential nature of the two relaxation processes. Middle: real part of the FID during relaxation and its envelope, which is related to the $T2^*$ relaxation time. Right: Real part of FFT of the FID showing the linewidth of the spectrum, which corresponds to the Full Width at Half Maximum (FWHM).[17]

Single or Multi-Voxel

In the single voxel spectroscopy (SVS) the signal is obtained from a voxel previously selected. This voxel is acquired from a combination of slice-selective excitations in three dimensions in space, achieved when a RF pulse is applied while a field gradient is switched on. It results in three orthogonal planes and their intersection corresponds to VOI. In other hand, multi-voxel MRS measures signals from a grid of multiple volumes.

Number of averages is the number of acquisitions performed in order to increase the SNR. In addition the number of points (NDP) is the number of time samples used in the acquisition.

Proton MRS is the most common spectroscopy technique used in brain studies, where up to twenty metabolites can be distinguished and quantified. The metabolites commonly observed in 1H MRS are: N-acetyl-L-aspartate (NAA), Creatine (Cr), Choline (Cho), Myo-inositol (m-Ins), Lactate (Lac), Glutamate (Glu) and Glutamine (Gln).

Water Suppresion

MRS-visible brain metabolites have a low concentration in brain tissues. Water is the most abundant and thus its signal in MRS spectrum is much higher than that of other metabolites (the signal of water is 100.000 times greater than that of other metabolites). To avoid this high peak from water to be superimpose on the signal of other brain metabolites, water suppression techniques are needed. The most commonly used technique is chemical shift selective water suppression (CHESS) which pre-saturates water signal using frequency selective 90° pulses before the localizing pulse sequence. Other techniques sometimes used are VAriable Pulse power and Optimized Relaxation Delays (VAPOR) and Water suppression Enhanced through T1 effects.

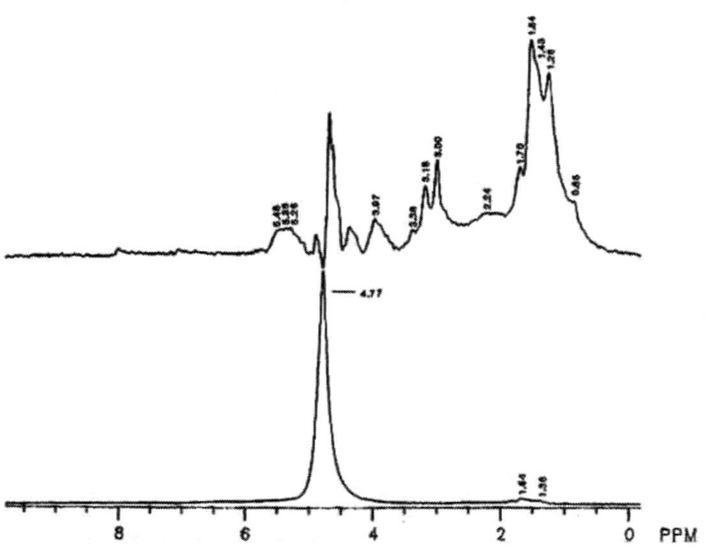

Figure 7. STEAM spectrum of normal human muscle with (top) and without water suppression (bottom). [18].

Higher Fields ^{1}H-MRS

Higher field MRI (3T and above) is used in many centers mostly for research purposes. On the past decade, 3T MRI started to be routinely used for clinical examinations and it results in better SNR and faster acquisitions factor which are important in sick patients that cannot hold still. ^{1}H-MRS performed at 3T MRI has a higher SNR and a reduced acquisition time compared to 1.5T. It was believed that SNR would increase linearly with the strength of the magnetic field but SNR does not double with 3T ^{1}H-MRS because others factors are also responsible for the SNR, including metabolite relaxation time and magnetic field homogeneity.

Spectral resolution is improved with higher magnetic field. A better spatial resolution increases the distance between peaks making it easier to distinguish between them. However, the metabolites linewidth also increases at higher magnetic field due to a markedly increase T2 relaxation time. Thus, short TE is more commonly used with 3T. The difference of T1 relaxation time from 1.5T to3T depends on the brain region studied.

3T ^{1}H-MRS is more sensitive to magnetic field inhomogeneity and some artifacts are more pronounced with it particularly susceptibility and eddy currents ones. Chemical shift displacement is also larger at 3T and this artifact increases linearly with the magnetic field.

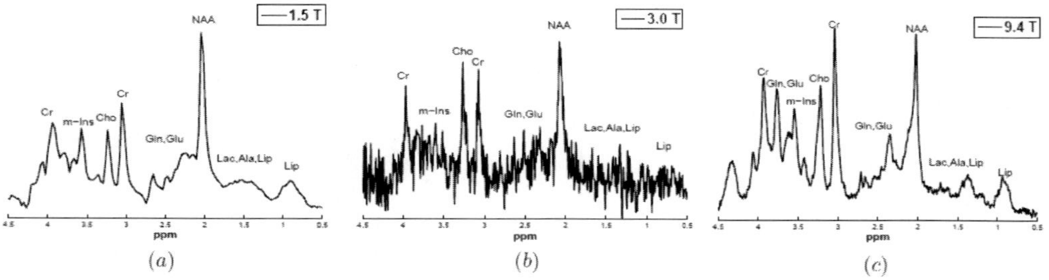

Figure 8. Real part of the spectra of three in vivo signals acquired at: (a) 1.5 T with acquisition parameters: PRESS pulse sequence, TR=6 s, TE=23 ms, SW=1 KHz, NDP=512 points, 64 averages. (b) 3.0 T with acquisition parameters: PRESS pulse sequence, TR=2 s, TE=35 ms, SW=2 KHz, NDP=2048 points, 1 average. (c) 9.4 T with acquisition parameters: PRESS pulse sequence, TR=4 s, TE=12 ms, SW=4 KHz, NDP=2048 points and 256 averages.

Metabolites in Proton MR Spectroscopy

Only a limited number of molecules with protons are observable in ^{1}H-MRS. In ^{1}H-MRS, the principle molecules that can be analyzed are:

- N-acetyl-asparate (NAA) (molecule present in healthy neurones) at 2.0 ppm
- Creatine/phosphocreatine (Cr) (energy metabolism molecules) at 3.0 ppm
- Choline compounds (Cho) (marker in the synthesis and breakdown of cell membranes) at 3.2 ppm
- Myo-inositol (mI) (only found in glial tissue) at 3.5 ppm
- Glutamine-Glutamate-GABA complex (Glx) (neurotransmitters) between 2.1 and 2.5 ppm
- Lactate (Lac) (anaerobic metabolism): doublet at 1.35 ppm
- Free lipids (Lip): wide resonance, doublet at 1.3 and 0.9 ppm

The number of discernable metabolites will vary according to the TE of the spectroscopy sequence: the longer the TE (135 or 270 ms), the more long T2 metabolites are selected. With short TE (15 to 20 ms), the spectrum will be more complex because of the greater number of superimposed peaks, producing a number of problems for quantification and interpretation.

N-Acetylaspartate (NAA)

Peak of NAA is the highest peak in normal brain. This peak is assigned at 2.02 ppm. NAA is synthesized in the mitochondria of neurons then transported into neuronal cytoplasm and along axons. NAA is exclusively found in the nervous system (peripheral and central) and is detected in both grey and white matter. It is a marker of neuronal and axonal viability and density. NAA can be found in immature oligodendrocytes and astrocyte progenitor cells, as well. NAA also plays a role as a cerebral osmolyte.

Absence or decreased concentration of NAA is a sign of neuronal loss or degradation. Neuronal destruction from malignant neoplasms and many white matter diseases result in decreased concentration of NAA. In contrast, increased NAA is nearly specific for Canavan disease. NAA is not demonstrated in extra-axial lesions such as meningiomas or intra-axial ones originating from outside of the brain such as metastases.

Creatine (Cr)

The peak of Cr spectrum is assigned at 3.02 ppm. This peak represents a combination of molecules containing creatine and phosphocreatine. Cr is a marker of energetic systems and intracellular metabolism. Concentration of Cr is relatively constant and it is considered a most stable cerebral metabolite. Therefore it is used as an internal reference for calculating metabolite ratios. However, there are regional and individual variability in Cr concentrations.

In brain tumors, there is a reduced Cr signal (see details below). On the other hand, gliosis may cause minimally increased Cr due to increased density of glial cells (glial proliferation). Creatine and phosphocreatine are metabolized to creatinine then the creatinine is excreted via kidneys (Hajek Dezortova, 2008). Systemic disease (e.g., renal disease) may also affect Cr levels in the brain (Soares Law, 2009).

Choline (Cho)

Its peak is assigned at 3.22 ppm and represents the sum of choline and choline-containing compounds (e.g., phosphocholine). Cho is a marker of cellular membrane turnover (phospholipids synthesis and degradation) reflecting cellular proliferation. Intumors, Cho levels correlate with degree of malignancy reflecting of cellularity. Increase Cho may be seen in infarction (from gliosis or ischemic damage to myelin) or inflammation (glial proliferation) hence elevated Cho is nonspecific.

Lactate (Lac)

Peak of Lac is not seen or is hardly visualized in the normal brain. The peak of Lac is a doublet at 1.33 ppm which projects above the baseline on short/long TE acquisition and inverts below the baseline at TE of 135-144 msec.

A small peak of Lac can be visible in some physiological states such as newborn brains during the first hours of life (Mullins, 2006). Lac is a product of anaerobic glycolysis so its concentration increases under anaerobic metabolism such as cerebral hypoxia, ischemia, seizures and metabolic disorders (especially mitochondrial ones). Increased Lac signals also occur with macrophage accumulation (e.g., acute inflammation). Lac also accumulates in tissues with poor washout such as cysts, normal pressure hydrocephalus, and necrotic and cystic tumors (Soares Law, 2009).

Lipids (Lip)

Lipids are components of cell membranes not visualized on long TE because of their very short relaxation time. There are two peaks of lipids: methylene protons at 1.3 ppm and methyl protons at 0.9 ppm (van der Graaf, 2010). These peaks are absent in the normal brain, but presence of lipids may result from improper voxel selection causing voxel contamination from adjacent fatty tissues (e.g., fat in subcutaneous tissue, scalp and diploic space).

Lipid peak scan be seen when there is cellular membrane breakdown or necrosis such as in metastases or primary malignant tumors.

Myoinositol (Myo)

Myo is a simple sugar assigned at 3.56 ppm. Myo is considered a glial marker because it is primarily synthesized in glial cells, almost only in astrocytes. It is also the most important osmolyte in astrocytes. Myo may represent a product of myelin degradation. Elevated Myo occurs with proliferation of glial cells or with increased glial-cell size as found in inflammation. Myo is elevated in gliosis, astrocytosis and in Alzheimer's disease (Soares Law, 2009; van der Graaf, 2010).

Alanine (Ala)

Ala is an amino acid that has a doublet centered at 1.48 ppm. This peak is located above the baseline in spectra obtained with short/long TE and inverts below the baseline on acquisition using TE= 135-144 msec .

Its peak may be obscured by Lac (at 1.33 ppm). The function of Ala is uncertain but it plays a role in the citric acid cycle (Soares Law, 2009). Increased concentration of Ala may occur in oxidative metabolism defects (van der Graaf, 2010). In tumors, elevated level of Ala is specific for meningiomas.

Glutamate-Glutamine (Glx)

Glx is a complex peaks from glutamate (Glu), Glutamine (Gln) and gamma-aminobutyric acid (GABA) assigned at 2.05-2.50 ppm.

These metabolite peaks are difficult to separate at 1.5 T. Glu is an important excitatory neurotransmitter and also plays a role in the redox cycle (Soares Law, 2009; van der Graaf, 2010). Elevated concentration of Gln is found in a few diseases such as hepatic encephalopathy (Fayed et al., 2006; van der Graaf, 2010).

Table 1. The metabolites and their information

Metabolite		Freq. (ppm)	short T2	long T2	Role	Anomalies	
mI	Myo-Inositol	3,6	•		Glial marker	↑ : gliomas, MS reactional gliosis ↓ : herpetic encephalitis	
Cho	Choline	3,2	•	•	Cell membrane metabolism marker	↑ : tumors, demyelinization	
Cr Pcr	Creatine Phosphocreatine	3,0	•	•	Energy metabolism marker, serves as reference peak as it is ~ constant		
Glx	GABA, Glutamate Glutamine	2,1-2,5	•		Intracellular neurotransmitter marker	↑ : hepatic encephalopathy	
NAA	N-Acetyl-Aspartate	2,0	•	•	Healthy neuron marker	↑ : Canavan's disease ↓ : neuronal distress	
Succ	Succinate	2,4	•	•	DISEASES	Pyogenic abscess	
Ac	Acetate	1,9	•	•		Abscess	
Ala	Alanine	1,5 (doublet)	•	•		Meningioma, Abscess	
Lac	Lactate	1,3 (doublet)	• +	• –		Ischemia, convulsions, tumors, mitochondrial cytopathies	↑ : anaerobic metabolism
Lip	Free lipids	0,9 1,4	•	•		Necrotic tumor (high grade)	

Pre-Processing

Due the signal acquisition procedure, many distortions, noise and shift are applied to signal. The pre-processing steps enhanced the signal or more processing. This steps be more important when absolute value need to extract from the signal. The pre-processing can be done in several parts as follow:

Frequency Alignment

Spectral alignment is essential for most of the quantification methods. In this preprocessing step the FID is transformed to the frequency domain and the spectra are shifted such that some recognizable peaks reach the desired frequency locations. In ^{1}H-MRS, the resonance frequency of observable metabolites is known and, therefore, they can be used for shifting the full spectra.

In some particular cases, the chemical shifts of in vivo metabolites can vary depending on the tissue, temperature and pH. These variations affect the accurate estimation of metabolite resonances, especially when fitting a set of metabolite profiles to the in vivo signal. Therefore, advanced alignment using for instance quantum mechanics, may be necessary if displacement of individual spectral peaks in a metabolite are detected.

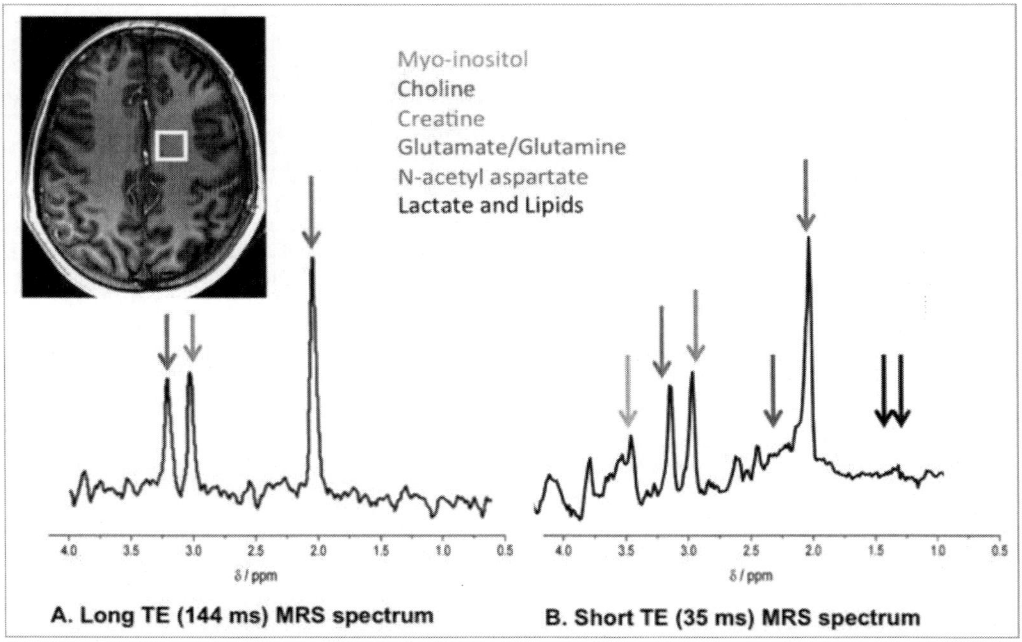

Figure 9. MRS of white matter in a normal brain. (A) Long TE spectra have less baseline distortion and are easy to process and analyze but show fewer metabolites than short TE spectra. Also, the lactate peaks are inverted, which makes them easier to differentiate them from lipids. (B) Short TE demonstrates peaks attributable to more metabolites, including lipids, glutamine and glutamate, and myo-inositol. [19].

Phase Correction

This correction consists of the multiplication of the complex spectrum by a complex phase factor equal to the initial phase of the FID. In particular, for frequency domain methods zero-phased spectra are required before quantification in order to obtain reliable metabolite estimates. On the other hand, for the time domain methods presented in this manuscript, the phase correction is not necessary because all processing and analysis is done considering the real and imaginary parts of the time domain signal.

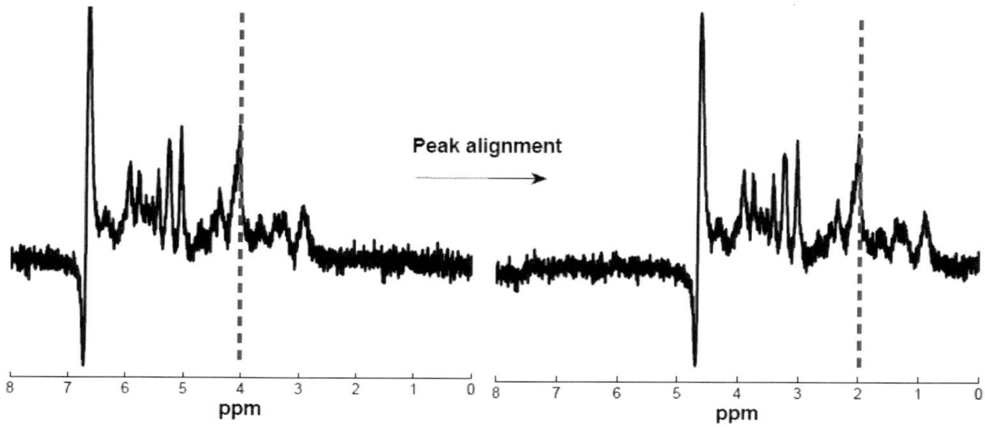

Figure 10. Real part of spectrum of an in vivo signal from mouse brain measured at 9.4 T in a Bruker scanner using a PRESS sequence and with acquisition parameters of TR=4 s, TE=12 ms, SW=4 KHz and 256 averages in a VOI of size 3×1.75×1.75 mm3. Normally, the carrier frequency is located at the water peak (i.e., 4.7 ppm), however, when this position is changed the metabolites of interest resonate at different locations. Nevertheless, the resonance frequency of other peaks in 1H MRS is known (e.g., NAA at 2.01 ppm) and can be used for peak alignment. [20].

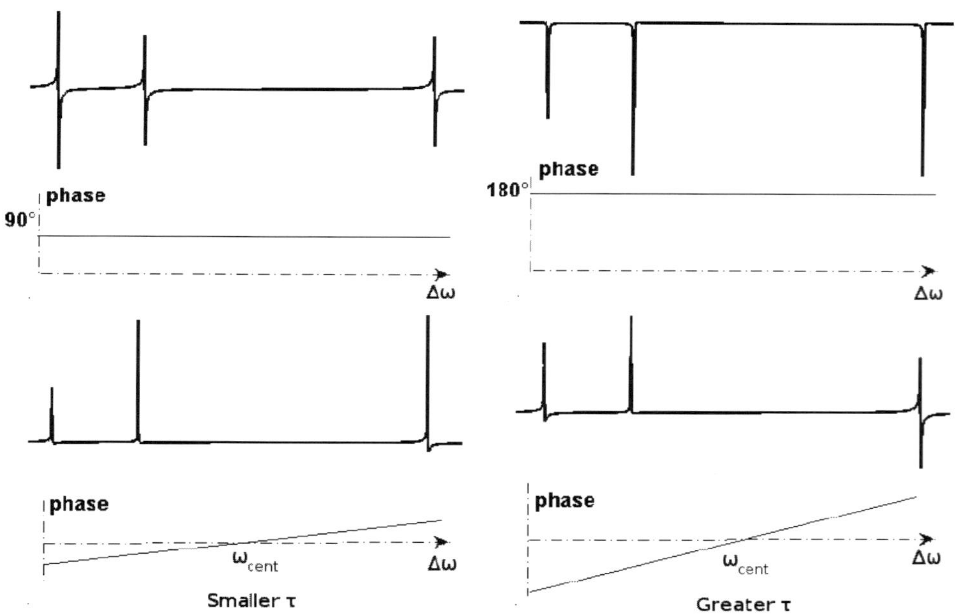

Figure 11. Top: zero order phase distortion influencing all peaks in the spectrum in the same way. Bottom: first order phase distortion influencing the peaks depending on the central frequency ωcent and the parameter τ which determines how steeply the first order phase changes with the frequenc [21].

Eddy Current Corrections

Eddy currents are caused by the scanner and cannot always be avoided. They are induced by the rapid switching in the magnet of the gradient coils and surrounding metal structures. These currents are a manifestation of Faraday's law of induction caused by the switching of the magnetic field gradients [14]. Klose [22] has developed a method to correct eddy currents (called ECC) by point-wise dividing the water suppressed signal by the phase term of the water unsuppressed signal.

Residual Water Suppression

In vivo ^{1}H-MRS signals contain a water resonance usually 10^3 to 10^4 larger than those of metabolites. Thus, to be able to identify metabolites and obtain a high SNR, an efficient suppression of the water peak is required. Water suppression techniques during acquisition provide a residual water signal at the level of most observable metabolites. Although these methods have a good performance, there is usually an incomplete water suppression and residual water remains in the signal. As a consequence, the metabolite peaks near the water resonance are altered and the tail of the residual water may affect the baseline of the signal. Therefore, numerical methods are required to suppress the unwanted peaks after acquisition.

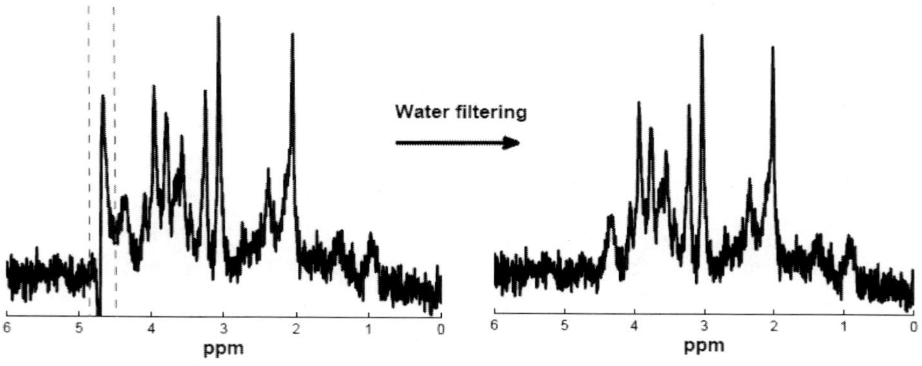

Figure 12. Real part of an in vivo spectrum from mouse brain measured at 9.4 T in a Bruker scanner using a PRESS sequence and with acquisition parameters of TR=4 s, TE=12 ms, SW=4 KHz and 256 averages in a VOI of size 3×1.75×1.75 mm3. Left: unfiltered signal. Right: water suppressed signal using the method HLSVD-PRO [23].

Baseline Correction/Estimation

Baseline estimation is very important in the quantification of in vivo MRS signals and can be the source of many systematic errors, therefore, an adequate correction/ estimation method is necessary. The baseline of in vivo MRS signals is characterized by a broad and non-flat unknown spectrum overlapping with the peaks of interest. Therefore, it can be identified with a fast decaying FID. In in vivo ^{1}H-MRS signals with long TE the baseline is not present, while short TE signals exhibit a significant baseline distortion.

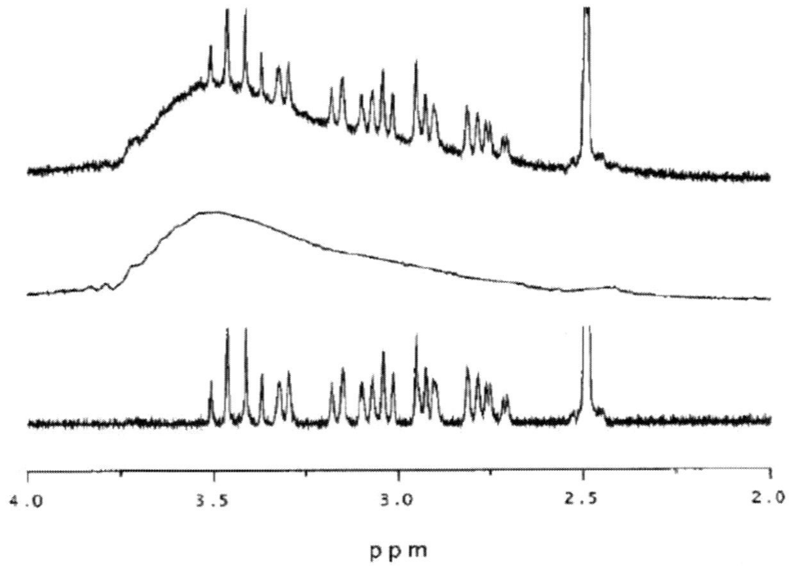

Figure 13. MR spectrum illustrating the baseline estimation method proposed by Golotvin and Williams [24]. Top: Original 400 MHz spectrum with a water hump. Middle: the baseline model consisting of fragments of the smoothed spectrum and regions interpolated by straight lines. Bottom: the spectrum after baseline subtraction.

Quantification

^{1}H-MRS signal can monitor the chemical environment if the well quantified. The quantification of MRS signal is the procedure to calculate the concentration of metabolite in the signal to obtain the chemical information of interest tissue. There are two approach in quantification of MRS signals: 1) Relative quantification: The results of the MRS are generally expressed as concentration ratios. Creatine peak or comparison with the healthy contralateral zone often serve as the reference values. 2) Absolute quantification: Measuring the true concentration of metabolites by MRS comes up against several technical difficulties: the peak area has to be determined accurately then converted into concentration after calibration.

Many methods have been developed to quantify the MRS signals accurate and also with more speeds. They can be discussed in three generation. The first generation was based on the peak picking. The concentration of each metabolite present based on the amplitude of each metabolite in its known frequency. This technique has lots of error and high sensitive to noise, baseline and any distortion.

The second generation was introduced based on the peak fitting. The model of peaks based on the Lorentzian, Gaussian or Voigt was used to minimize the estimation of signal. This technique was more robust in compare to the peak picking, but have poor performance in complex signal.

The third generation is introduced based on the metabolites' profile. All the information of the metabolite was generated and used in the processing procedure. This technique can quantify metabolites more accurate but it takes more time to process the signal. Currently many methods developed which employed the information of metabolites profile to quantify the MRS signal in less time and more accuracy.

The MRS quantification methods are implemented in time-domain or frequency-domain. In time domain methods, quantitation is carried out in the same domain as the domain where the signals are measured, giving more flexibility to the model function and allowing specific time-domain preprocessing.

In time domain methods, the object is to minimize the difference between the data and the model function, resulting in a typical nonlinear least-squares (NLLS) problem. The important feature of the interactive methods is they can use a basis set of metabolite profiles. VARPRO [25], the local optimization procedure based on Osborne's Levenberg-Marquardt algorithm [26], was the first widely used method for quantifying MRS data. It has been replaced later by AMARES [27], which proved to be better than VARPRO in terms of robustness and edibility. AMARES allows more prior knowledge and can also fit echo signals.

On the other hand, methods such as AQSES [28] or QUEST [25] make use of a metabolite basis set, which can be built up from simulated spectra (e.g., via programs based on quantum mechanics such as NMR-SCOPE [29]) or from in vitro spectra. The use of a metabolite basis set facilitates the disentangling of overlapping resonances when the corresponding metabolite profiles also contain at least one non-overlapping resonance.

Table 2. The comparison of three generation of quantification methods

Procedure	Preparation Time	Sensitivity to Baseline Imperfections	Ease of Use	Accuracy
Integration	+	-	+	*
Peak fitting with prior knowledge	+	-	+	**
Peak fitting with metabolite basis set	-	+	-	***

The frequency domain is naturally suited for frequency selective analysis with the advantage of decreasing the number of model parameters. Visual interpretation of the measured MRS signals and of the fitting results is best done in the frequency domain. The oldest and still widely used quantitation method in the frequency domain is based on the integration of the area under the peaks of interest [30].

Unfortunately, this method is not able to disentangle overlapping peaks and therefore to extract information from individual peaks or metabolite contributions. Residual baseline signals and low SNRs will also hamper good quantitation. Furthermore, an appropriate phasing is necessary when dealing with the real part of the frequency-domain MRS signal, which is far from trivial. Peak integration depends widely on the defined bounds.

Although frequency-domain fitting methods are equivalent to time-domain fitting methods from a theoretical point of view, a simple exact analytical expression of the discrete Fourier transform (DFT) of the model function is often not available for the Voigt and/or Gaussian lineshapes, even if numerical approximations exist.

Table 2. The comparison between some popular quantification methods

Methods	Prior-knowledge	Profile	Lineshape	Water	Baseline	Accuracy
VARPRO	✓					*
AMARES	✓			✓		**
AQSES		✓		✓	✓	**
QUEST		✓			✓	***
LCModel		✓	✓		✓	***

The model functions in the frequency domain are more complicated than in the time-domain and necessitate thereby more computation time. The frequency-domain methods which only use the real part of the spectrum in their model, such as LCModel [15], require a very good phasing to get the spectrum in its absorption mode. Quantification methods can be divided into three domains: time-domain methods, frequency-domain methods and methods which implement in time-frequency domain.

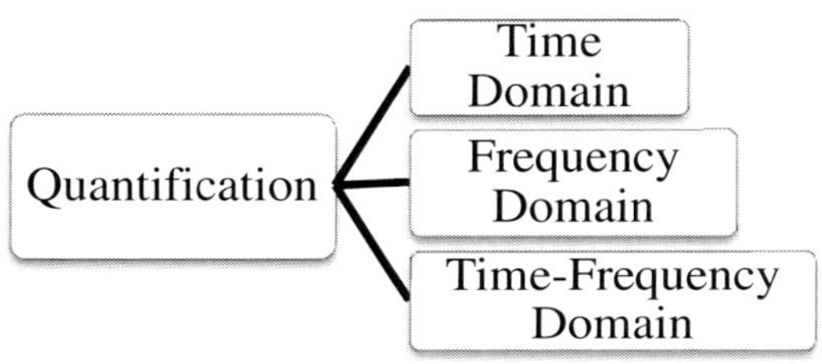

Time-domain and frequency-domain methods are usually divided into two main classes: black-box or non-interactive methods and methods based on iterative model function fitting or interactive methods. The time-domain methods can be expressed as the list below:

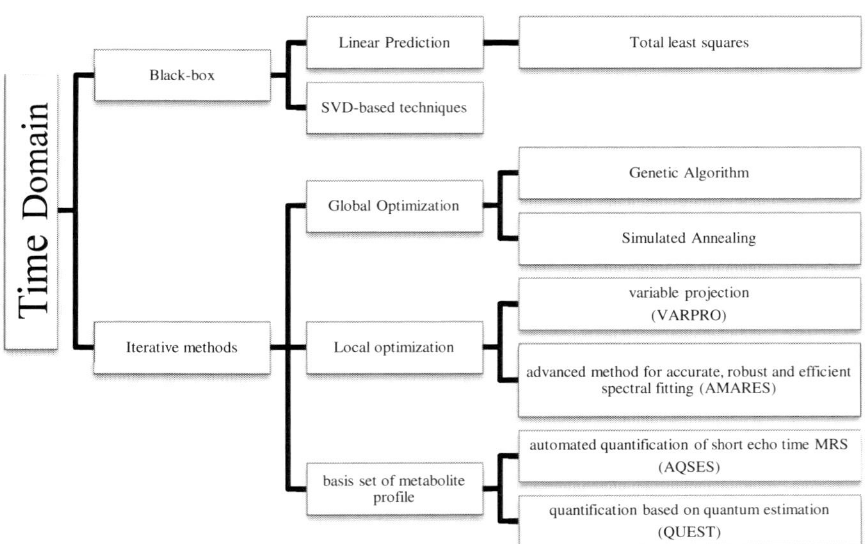

The frequency-domain is naturally suited for frequency selective analysis with the advantage of decreasing the number of model parameters. Visual interpretation of the measured MRS signals and of the fitting results is best done in the frequency domain. The frequency-domain methods can be expressed as the list below:

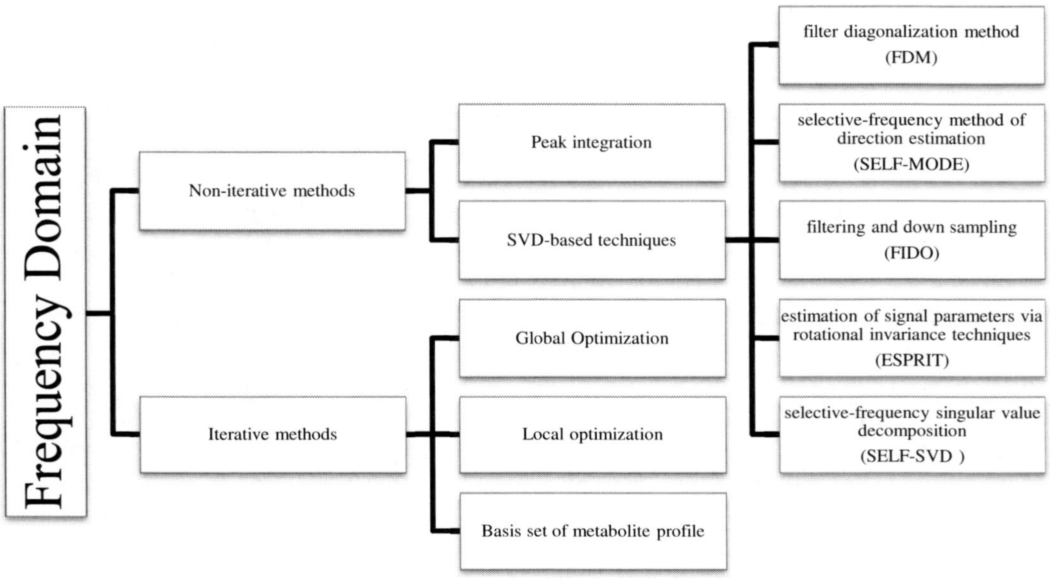

In order to utilize the advantages of time and frequency domain simultaneously, time-frequency domain method were implementted. The time-frequency domain methods can be expressed as the list below:

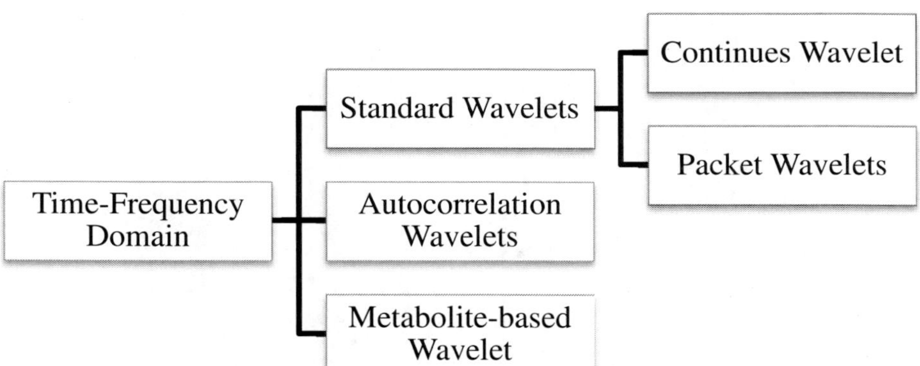

CLINICAL APPLICATION

Brain Tomur

The main application of ^{1}H-MRS is brain tumors. This technique is usually used as a complement to conventional MRI. Combined with conventional MRI, proton MR spectra may improve diagnosis and treatment of brain tumors. ^{1}H-MRS may help with differential diagnosis, histologic grading, degree of infiltration, tumor recurrence, and response to treatment mainly when radio necrosis develops and is indistinguishable from tumor by conventional MRI.

MRSI is usually preferable to SVS because of its spatial distribution. It allows the acquisition of a spectrum of a lesion and the adjacent tissues and also gives a better depiction of tumor heterogeneity. On the other hand, SVS is faster.

Elevation of Choline is seen in all neoplastic lesions. Choline peak may help with treatment response, diagnosis and progression of tumor. Its increase has been attributed to cellular membrane turnover which reflects cellular proliferation. Choline peak is usually higher in the center of a solid neoplastic mass and decreases peripherally. Cho signal is consistently low in necrotic areas.

Another ^{1}H-MRS feature seen in brain tumors is decrease NAA. This metabolite is a neuronal marker and its reduction denotes destruction and displacement of normal tissue. Absence of NAA in an intra-axial tumor generally implies an origin outside of the central nervous system (metastasis) or a highly malignant tumor that has destroyed all neurons in that location. Creatine signal, on the other hand, is slightly variable in brain tumors. It changes according to tumor type and grade. The typical ^{1}H-MRS spectrum for a brain tumor is one of high level of Choline, low NAA and minor changes in Creatine.

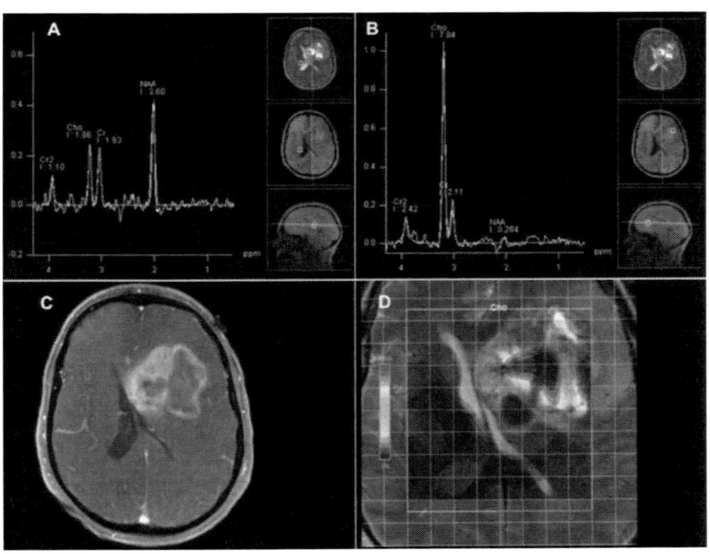

Figure14. Patient with glioblastoma with oligodendroglioma component, WHO grade IV of IV. (A) MRS spectrum from region of brain not affected by the tumor. (B) Spectrum from a voxel within the tumor, showing elevated choline. (C) T1 MR image and (D) color-rendered MRS image showing variations in the levels of choline within the tumor. [19].

Alzheimer's Disease

Pathologically, the condition shows predilection for the hippocampi and anteromedial temporal lobes. Susceptibility artifact encountered when imaging the medial temporal lobes makes reproducibility of high quality spectra difficult, thus convention is to place the voxel in the posterior cingulate gyrus, a spectroscopically homogeneous region [31]. Reduced NAA and raised myoinositol are consistently demonstrated on the spectra, even in the early stages. Moreover, it is thought that elevated myoinositol levels may provide the earliest imaging indicator of the disease as glial proliferation typically precedes significant neuronal loss or mitochondrial dysfunction.

MR spectroscopy is also proving useful in the diagnosis of mild cognitive impairment, a condition recognised as preceding the development of Alzheimer's disease [5]. What's more, researchers have reported that MR spectroscopy is capable of predicting which patients with mild cognitive impairment will go on to develop Alzheimer's disease. Significant decline in NAA/Cr ratio or rise in Cho/Cr ratio measured in the paratrigonal white matter may be detected before clinical onset of dementia.

Epilepsy

A major contribution of MRS to epilepsy is its ability to determine lateralization before surgical resection of the diseased brain region. ^{1}H-MRS identify abnormalities in high-energy metabolism, neuronal function and neurotransmitter levels, but information can only be obtained from restricted regions of the brain.

In temporal lobe epilepsy, ^{1}H-MRS is used to lateralize the seizure focus using a reduction in NAA/Cr, Ch as a guide. Similarly many dementias are characterized by unspecific reduction of NAA levels, associated with neuronal deficits [32].

Stroke

^{1}H-MRS studies in humans demonstrate that after acute cerebral infarction, lactate appears, while NAA and total Cr/PCr are reduced within the infarct compared to the contralateral hemisphere. Large variations in the initial concentrations of Cho have been observed in the region of infarction [33].

In stroke, these early observations have detected prolonged lactate elevation and loss of NAA in the affected region. One possible source of persistent lactate is production by ischemic but viable brain cells shifted toward anaerobic glycolysis. Such an "ischemic penumbra" is conceivably amenable to therapeutic interventions [33].

REFERENCES

[1] Hornak JP.; Rochester Institute of Technology, *Center for Imaging Science*; 1997.
[2] Hobbie RK, Roth BJ.; Intermediate physics for medicine and biology: Springer; 2007.
[3] Hornak JP.; *Molecules.* 1999;4(12):353-65.
[4] Ross B, Bluml S.; *The Anatomical Record.* 2001;265(2):54-84.
[5] Rosen Y, Lenkinski RE.; *Neurotherapeutics.* 2007;4(3):330-45.
[6] Vanhamme L, Sundin T, Hecke PV, Huffel SV; *NMR in Biomedicine.* 2001;14(4):233-46.
[7] Poullet J-B, Sima DM, Van Huffel S.; *Journal of Magnetic Resonance.* 2008;195(2):134-44.
[8] Delprat N, Escudie B, Guillemain P, Kronland-Martinet R, Tchamitchian P, Torrésani B.; *IEEE transactions on Information Theory.* 1992;38(2):644-64.
[9] Guillemain P, Kronland-Martinet R, *Martens B.;* 1989:38-60.
[10] Mainardi LT, Origgi D, Lucia P, Scotti G, Cerutti S.; *Medical engineering & physics.* 2002;24(3):201-8.
[11] Serrai H, Senhadji L, de Certaines JD, Coatrieux JL.; *Journal of Magnetic Resonance.* 1997;124(1):20-34.
[12] Suvichakorn A, Ratiney H, Bucur A, Cavassila S, Antoine J-P.; *Measurement Science and Technology.* 2009;20(10):104029.
[13] Becker ED.; Academic Press; 1999.
[14] de Graaf RA, Mason GF, Patel AB, Behar KL; *NMR in Biomedicine.* 2003;16(6-7):339-57.
[15] Provencher SW. ; *Magnetic Resonance in Medicine.* 1993;30(6):672-9.
[16] Negendank WG, Sauter R, Brown TR, Evelhoch JL, Falini A, Gotsis ED, et al.; *Journal of neurosurgery.* 1996;84(3):449-58.
[17] De Graaf RA.; John Wiley & Sons; 2008.
[18] Andrew ER. ; *Cambridge University Press*, 2009. 2009;1.
[19] Miller JC. Magnetic; *Massachusetts General Hospital Radiology Rounds.* 2012;10(7).
[20] GARCIA MIO.; 2011.
[21] Jiru F.; *European journal of radiology.* 2008;67(2):202-17.
[22] Klose U.; Magnetic *Resonance in Medicine.* 1990;14(1):26-30.
[23] Laudadio T, Mastronardi N, Vanhamme L, Van Hecke P, Van Huffel S.; *Journal of Magnetic Resonance.* 2002;157(2):292-7.
[24] Golotvin S, Williams A.; *Journal of Magnetic Resonance.* 2000;146(1):122-5.
[25] Ratiney H, Sdika M, Coenradie Y, Cavassila S, Ormondt Dv, Graveron-Demilly D. ; *NMR in Biomedicine.* 2005;18(1):1-13.
[26] Lootsma FA; Conference on Numerical Methods for Non-linear Optimization (1971: University of Dundee, Scotland); 1972: New York, Academic Press.
[27] Vanhamme L, van den Boogaart A, Van Huffel S. I; *Journal of Magnetic Resonance.* 1997;129(1):35-43.
[28] Poullet JB, Sima DM, Simonetti AW, De Neuter B, Vanhamme L, Lemmerling P, et al.; *NMR in Biomedicine.* 2007;20(5):493-504.
[29] Graverondemilly D, Diop A, Briguet A, Fenet B.; *Journal of Magnetic Resonance, Series A.* 1993;101(3):233-9.

[30] Meyer R, Fisher M, Nelson S, Brown T.; *NMR in Biomedicine*. 1988;1(3):131-5.
[31] Soares D, Law M.; *Clinical radiology*. 2009;64(1):12-21.
[32] Ranjeva J-P, Confort-Gouny S, Le Fur Y, Cozzone P.; *Child's Nervous System*. 2000;16(4):235-41.
[33] Saunders DE.; *British medical bulletin*. 2000;56(2):334-45.

In: Current Applications of Chemometrics
Editor: Mohammadreza Khanmohammadi

ISBN: 978-1-63463-117-4

Chapter 9

DATA HANDLING AND ANALYSIS IN METABOLOMICS

Marc Breit[1], Christian Baumgartner[1] and Klaus M. Weinberger[1,2*]

[1]Research Group for Clinical Bioinformatics, Institute of Electrical and Biomedical Engineering, University for Health Sciences, Medical Informatics and Technology (UMIT), Austria

[2]sAnalytiCo Ltd, Forsyth House, Cromac Square, Belfast, United Kingdom

ABSTRACT

Metabolomics, the systematic identification and quantitation of low molecular weight compounds, is, thanks to recent technological advances, currently making major inroads into clinical research and diagnostics. In this rapidly expanding field of analytical biochemistry, chemometric methods play a crucial role in the process of generating knowledge out of data, which are usually signals obtained by either mass spectrometry (MS) or nuclear magnetic resonance (NMR) spectroscopy. In the last decade, though, MS has established itself as the *de facto* standard for the analysis of small molecules, mainly because of its superior sensitivity and resolution.

Mathematical methods for data processing on the one side, and statistical and data mining approaches for data analysis on the other are the two main pillars of this entire process. Compared to the workflow in other functional genomics disciplines, the unique specialty of metabolomics lies in an additional level of scrutiny, namely a plausibility check of results from the perspective of biochemistry. Based on the existing body of knowledge and the detailed functional understanding of metabolic reactions, analytical results can be subjected to a thorough biochemical interpretation, usually in the context of annotated pathways, aiming at a verification of results and/or at generating new biological hypotheses.

In this chapter, emphasis is put on several steps along this workflow: data acquisition, algorithms for signal and data processing, statistical and data mining aspects for data analysis, and – eventually – on the interpretation of results in their biochemical context.

* Correspondence to: klaus.weinberger@sanalytico.com

List of Abbreviations

ANN	artificial neural networks
AV	associative voting
ATP	adenosine triphosphate
BALF	bronchoalveolar lavage fluid
CSF	cerebrospinal fluid
CO_2	carbon dioxide
CV	coefficient of variance
DBS	dried blood spots
DNA	deoxyribonucleic acid
EA	enrichment analysis
EBI	european bioinformatics institute
FATMO	fatty acid transport and mitochondrial oxidation
FDA	food and drug administration
GO	gene ontology
GSEA	gene set enrichment analysis
GWA	genome-wide association
HCA	hierarchical clustering analysis
H_2O	water
HMDB	human metabolome database
KEGG	kyoto encyclopedia of genes and genomes
kNN	k-nearest neighbor
LIMS	laboratory information management systems
LLOQ	lower limit of quantitation
MRM	multiple reaction monitoring
mRNA	messenger ribonucleic acid
MSEA	metabolite set enrichment analysis
MS	mass spectrometry
MSI	metabolomics standards initiative
ORA	overrepresentation analysis
NADH	nicotinamide adenine dinucleotide
NCBI	national center for biotechnology information
NH_3	ammonia
NMR	nuclear magnetic resonance
pBI	paired Biomarker Identifier
PCA	principal component analysis
PLS-DA	partial least square discriminant analysis
qPCR	quantitative polymerase chain reaction
RFM	random forest models
ROC	receiver operating characteristic
SEA	set enrichment analysis
SID	stable isotope dilution
SFR	stacked feature ranking
SMPDB	small molecule pathway database

SNP	single-nucleotide polymorphism
SOM	self-organizing maps
SOP	standard operating procedure
SSP	single sample profiling
SVM	support vector machines
uBI	unpaired biomarker identifier
Y2H	yeast-two-hybrid

INTRODUCTION

Metabolomics - History and Proof-of-Concept

When it became technically conceivable to obtain the complete sequence of the human genome in a reasonable timeframe, biologists and physicians alike were phrasing incredibly high expectations and predicting a quantum leap in evidence-based medicine and drug development [1,2]. Now, just a little more than a decade after the first relatively complete drafts have been released, there is a general agreement that the functional understanding of an organism and its (patho-)physiology requires much more than just deciphering its genetic material [3]. With a little exaggeration, one might actually argue that the most significant impact of the human genome project was that it did not meet these expectations straight away and, thus, promoted a wave of technology developments heralding an era of functional genomics with its comprehensive analytical platforms for messenger ribonucleic acid (mRNA) (transcriptomics), proteins and peptides (proteomics), and low molecular weight biochemicals (metabolomics).

Among these three disciplines, transcriptomics was the first to be covered by standardized commercial solutions. While the microarray-based technologies still offer insufficient precision and accuracy for truly quantitative workflows and require confirmation by less multiplexed approaches like quantitative polymerase chain reaction (qPCR), they still became an accepted standard for this level of expression profiling [4]. Interestingly, the very basic unsolved issues arising from the lack of biological understanding of these data sets did not seem to hinder the widespread use of transcriptomics (for a brief overview of strengths and weaknesses of the current omics platforms, see table 1).

Analogous developments for the most complex class of biomolecules, the proteome, led to a variety of competing platforms, none of which was able to establish a similarly accepted industry standard. One of the reasons may simply be its complexity and dynamic range which made it impossible to stick to the idea that one analytical platform could really generate a comprehensive data set – as one was accustomed to for both sorts of nucleic acids. The far more important shortcoming may have been the dramatic lack of reproducibility – both preanalytical and analytical – that was not appropriately considered when celebrating the first 'breakthrough' results, particularly in the field of clinical biomarker research in oncology [5-7].

The third of these disciplines, metabolomics, fundamentally differs from the other two as it is not gene-centric in annotation and, thus, cannot easily be linked to the data sets from transcriptomics or proteomics experiments. In addition, the basic qualitative workflow, i.e., the structural elucidation of metabolites, is not just another variation of sequencing a biopolymer (except for polymeric carbohydrates, of course) but can be a very tedious if not impossible task because mass spectrometry (MS) is sometimes not informative enough to unanimously identify a metabolite structure and nuclear magnetic resonance (NMR) spectroscopy very often not sensitive enough given the minute amounts of a particular biomolecule which can be available in biological samples. Still, metabolomics has gained a quite remarkable momentum, which is mainly due to two independent developments:

First, the sensitivity and robustness of triple quadrupole mass spectrometers were improved dramatically since these instruments were first conceived and developed in the late 1970s [8]. Also, experimental improvements such as scheduled multiple reaction monitoring (MRM) for hundreds of transitions (reviewed by Ebhardt [9]) and the use of stable isotope dilution (SID) for absolute quantitation [10] facilitated an approach to the metabolome which was – and still is – far from complete but extremely informative.

Second, and probably even more important, the detailed knowledge of metabolic pathways and their individual reactions allowed for a very straightforward biochemical interpretation once quantitative data became available. While the scientific communities of cell biologists and biochemists just begin to catalogue and map protein-protein-interactions (and do not even seriously try to predict functional relationships at the mRNA level), the vast majority of metabolic pathways is characterized and functionally understood in great detail (e.g., substrates and products linked to enzymatic reactions, molecular reaction mechanisms, equilibria, kinetics and energetics of these reactions, co-factors, etc.). It is this wealth of background knowledge that makes metabolomics the most promising area in functional genomics for future clinical applications [3,11-13].

To date, the most successful use case of mass spectrometry-based metabolomics is still the routine diagnostic application in neonatal screening [14]. Here, more than twenty so-called inborn errors of metabolism, i.e., monogenic disorders of amino acid metabolism and fatty acid transport and mitochondrial oxidation (FATMO), are detected with unmatched diagnostic performance based on highly standardized, multiplexed quantitative assays for just two classes of metabolites, amino acids and acylcarnitines. Yet, this application can only be seen as the starting point for a renaissance of classical biochemistry and a new era of multiparametric metabolic markers in clinical chemistry.

Application Areas

By now, metabolomics is utilized in a very wide range of applications, so it is virtually impossible to provide a comprehensive overview. Yet, a brief selection of examples should already indicate how versatile this analytical tool has become.

Fundamental insights into diabetology could be provided by mapping the genotype to the metabolic phenotype (now termed the ‘metabotype’) of individuals, through genome-wide association (GWA) studies combined with quantitative metabolomics experiments [15,16]. This combination has also led to breakthrough findings in agro-biotechnology, e.g., in population genetics research on livestock where major traits of different breeds of cattle could be linked to metabotypes in a strikingly plausible way [17].

Another very promising field of applications is in nutritional science, the so-called nutrimetabolomics [18]. Here, population-based studies have examined the consequences of nutritional habits [19] or individual food products or ingredients [20-22] on human metabolism. Metabolomics analyses are also particularly useful in characterizing microbial systems, for optimizing cell culture conditions, and in bioprocessing [23]. They have become instrumental in the detailed understanding of animal models such as transgenic mice [24] or mice exposed to radiation [25].

Table 1. Comparison of the strengths and weaknesses of the four main omics disciplines

Discipline / Main analytes	Completeness[1]	Qualitative analysis	Quantitative analysis	Functional understanding
Genomics / Deoxyribonucleic acid (DNA)	Very good; full genome sequence available	Very good; well-established and fully automated platforms for high-throughput sequencing	Only applicable for copy number variation or similar applications	Fair; GWA studies identify relevant variants (single-nucleotide polymorphisms (SNPs)) but often with limited odds ratio and explained variance; haplotype association studies still at an early stage
Transcriptomics / messenger ribonucleic acid (mRNA)	Good; virtually all genes covered on microarrays, some transcripts only accessible via sequencing	Good; Directly linked to genomic sequence information; also accessible via transcript sequencing	Fair; standardized platforms available; high-throughput results semi-quantitative at best; confirmation by qPCR required but more difficult to multiplex	Very poor; even if protein function is characterized, interpretation is further hindered by the lack of proportionality between mRNA and protein abundance
Proteomics / Peptides, proteins, post-translational modifications	Poor; only a minor percentage of the proteome accessible due to huge dynamic range and complexity, particularly poor coverage of post-translational modifications	Fair; In principle well-established workflow for identification (MS-based sequencing of peptides) but tedious due to complexity of proteome in biological matrices	Fair; standardized solutions for relative quantitation available, e.g., isobaric tags; absolute quantitation still very rarely achieved although stable isotope dilution works here as well	Poor; function of many individual proteins well described but protein interaction networks mainly based on yeast-two-hybrid (Y2H) data, so still very poor in information, particularly regarding protein complexes
Metabolomics / Metabolites (amino acids, carbohydrates, organic acids, lipids, etc.)	Fair; current platforms far from complete coverage but knowledge of rate-limiting steps in pathways allows for smart selection of key analytes	Poor; Identification / structural elucidation of metabolites can be very difficult (MS data not sufficiently informative / NMR requires too much material)	Very good; quantitation by stable isotope dilution can be multiplexed for hundreds of metabolites and reaches quality standards of clinical chemistry	Very good; most metabolic pathways understood in great detail (equilibria, mechanism, kinetics and energetics of reactions, co-factors, localization etc.) allowing for direct interpretation of results

[1]This column only refers to the human–omes.

The most promising field is definitely clinical research where multiparametric metabolic signatures have led to a paradigm shift in early diagnosis, staging and monitoring of chronic kidney disease [26-29], diabetes [13,30,31], or Alzheimer's disease [32,33].

Similarly groundbreaking findings which also demonstrate the potential for evidence-based drug development were presented in oncology [34-36], exercise physiology [37,38], or hypoxia-induced oxidative stress [39].

Fundamental Approaches

In metabolomics, depending on the definition, different fundamental approaches are distinguished, e.g., untargeted metabolomics, targeted metabolomics, metabolic fingerprinting, or *in situ* metabolomics [13,18,40,41]. Due to their widespread use, in this section, only the two approaches of untargeted and targeted metabolomics will be shortly outlined.

Untargeted metabolomics (also known as metabolic profiling) aims at providing a snapshot of the biochemical profile within a sample of interest, trying to measure as many compounds as possible at once, and thus ideally providing a holistic picture. This concept is basically unbiased, and the number of metabolites analyzed in one sample is primarily limited by analytical instrumentation and methods.

Targeted metabolomics, in contrast, aims at the quantitation of a selected subset of known compounds by determining absolute values of metabolite concentrations within a sample. This approach is to some extend biased through the pre-selection of metabolites (which is often hypothesis-driven), and necessary *a priori* knowledge for the identification and quantification of metabolites.

Basically in common to the presented approaches, and to some extent also defining the structure of this chapter, is the data handling and analysis workflow, which ranges from the acquisition of data, the processing of signals and data utilizing chemometric methods, and the analysis of data by means of statistical and data mining approaches, to the biochemical interpretation of results, mainly in the context of metabolic pathways.

Data Acquisition

Data Quality

In the process of acquiring biochemical information on a biological sample of interest, the quality of the analytical data is fundamentally decisive for any subsequent steps in the workflow of chemometric data examination (which, of course, is true for most areas of research). In general, the early and continuous consideration of regulatory directives and guidelines, if not anyhow required by the type of application (e.g., in clinical diagnostics), helps in establishing a foundation for generating reliable data. Crucial issues with respect to data quality lie in all steps of the workflow, ranging from pre-analytical sample handling (biobanking and sample preparation) and the actual chemical analytics (development and execution of analytical methods) to the area of bioinformatics (the engineering of software and applications for the integration, management and linkage of data). Basic aspects of quality assurance and quality control, which play essential roles in contributing to a high level of confidence in the obtained data, are also described within this section.

Sample Handling

The sample types most frequently analyzed in metabolomics experiments are plasma, serum, dried blood spots (DBS), and urine but, of course, other body fluids (e.g., cerebrospinal fluid (CSF), synovial fluid, sputum, or lavages such as bronchoalveolar lavage fluid (BALF)), all sorts of tissue homogenates and cell cultures samples (cells and supernatants) are also used. Standardized protocols are essential for sample collection, particularly in multi-centric studies. A key step is the freezing of samples; samples ought to be stored in a frozen state as soon as possible to minimize any further metabolic activity within the sample, especially to suppress and prevent oxidation but – as with most other recommendations – reproducibility is much more important than optimal conditions. The same is true for freeze/thaw cycles: it goes without saying that they should be kept to the absolute minimum but it is imperative that all the samples in one study are treated the same way.

To get from a complex biological sample to an extract containing the metabolites of interest and yet clean enough to be analyzed in state-of-the-art analytical instruments, common sample preparation steps are pipetting, chemical derivatization, extraction, filtration, and dilution [42]. An example for the efforts undertaken in optimizing sample preparation, in this case of extraction protocols, was recently discussed for the application in brain research [43]. The usage of laboratory automation systems (mainly robotic liquid handling systems) enables improvements in workflow standardization, pipetting accuracy, elimination of manual pipetting errors, and the reproducibility and traceability of results. An overview of the main platforms available for robotic liquid handling is given by Vogeser [42]. From a regulatory perspective, the 'General Principles of Software Validation; Final Guidance for Industry and FDA Staff' [44] is partially applicable to the engineering/programming of methods for automated sample preparation while the actual liquid handling part can be validated just like the analytical performance according to the 'Guidance for Industry - Bioanalytical Method Validation' [45].

Chemical Analytics

The primary analytical technologies utilized for the biochemical analysis of samples in metabolomics are mass spectrometry and NMR spectroscopy [40,46]. When suitably sensitive and robust instruments became available, integrated technology platforms were developed [11] and typically built around mass spectrometers as their core analytical devices [47-50]. Due to the strengths and weaknesses of the different technologies and the combinations thereof (e.g., sensitivity and quantitation capabilities of triple quadrupole instruments vs. resolution and mass accuracy of time-of-flight systems), the appropriate selection of instruments needs to be decided depending on the specific application.

In addition to these technological issues, the development of very well characterized bioanalytical methods and protocols is crucial for the generation of reliable data in any metabolomics experiment. Examples for bioanalytical method development are described in specific detail for the quantitation of prostaglandins and fatty acid metabolites [51] or for the general analysis of metabolites from blood plasma [52]. The definition, documentation, and validation of bioanalytical methods is recommended to be performed in consideration of the 'Guidance for Preparing Standard Operating Procedures (SOPs)' [53], and the 'Guidance for Industry - Bioanalytical Method Validation' [45].

Bioinformatics

From a bioinformatics perspective, key steps during the process of data acquisition are the basic handling of generated signals and raw data, the integration and management of proprietary information from different sources, and the interconnection of all this with publicly annotated knowledge. The handling of generated signals and raw data may be directly performed utilizing the software applications provided by the vendors of mass spectrometry or other analytical systems.

The integration and management of proprietary information (e.g., of samples, measurements results, instruments or metabolite details) may be executed utilizing laboratory information management systems (LIMS), which support the overall metabolomics workflow in a comprehensive manner, e.g., MetDAT [54] or SetupX [55]. In an idealized scenario, generated data might be directly linked to existing publicly annotated data, stored in online databases, as for example in the Human Metabolome Database (HMDB) [56] or MetLin [57]. A comprehensive review on metabolomics-related databases is given by Wishart[58]. Coming back to the quality aspects mentioned in the beginning of this section, the regulatory guideline 'General Principles of Software Validation; Final Guidance for Industry and FDA Staff' [44] should be taken into account when engineering bioinformatics software, metabolomics applications, or chemometric algorithms. When the acquisition of data is completed, and data are stored and managed in an appropriate way, they may be subject to the next step of basic processing.

Data Processing

Data Characteristics

In metabolomics experiments and studies, data on chemical compounds obtained through mass spectrometry analytics are essentially composed of the measurements of ion intensities, their mass domain (actually the mass to charge ratio, m/z), and their time domain (the retention time in case of chromatographic or electrophoretic separation). By nature, metabolites differ in their biological and physical properties, such as their polarity, solubility, abundance, or ionization behavior.

Due to this fact and the diversity of technical parameters of mass spectrometry platforms (e.g., chromatographic performance, ionization efficiency, resolution of mass analyzers, or sensitivity and accuracy of ion detectors), the definition and usage of a common standard for the processing of raw data is still pending.

With respect to both, metabolic profiling and targeted metabolomics, the chemometric methods discussed in this section are the processing of crude signals, the detection and extraction of features, the normalization of data, and the quantitation of metabolite concentrations, with the earlier steps being more relevant in metabolic profiling, and the later ones being key in targeted metabolomics.

Signal Processing

Signals obtained by mass spectrometry are subject to noise related to the analytical technique used (electronic noise), to interferences caused by the complexity of the sample itself (chemical noise), and to potential errors in the preparation of samples (e.g., variability in the derivatization reactions or potential cross-contaminations). The correction of the baseline and procedures for the reduction of noise are, therefore, essential steps in the processing of metabolomics signals. Typical approaches for modeling the baseline shape are the estimation of a local average or minimum intensity [59-61], low order polynomial Savitzky-Golay filters [62] or polynomial models [63]. For reducing the noise in mass spectral data, common approaches are decomposition methods, like the wavelet transformation [60,64] or smoothing filters, such as the Gaussian filter or moving average/median filters [65,66]. When performing the noise reduction procedures, one must be aware of the potential bias, which may be introduced [67,68]. For deriving high quality data, a sensitive usage of baseline shape estimation and smoothing techniques is important, in keeping relevant signals, and at the same time reducing background noise sufficiently.

Feature Extraction

Utilizing data compression, features can be extracted from the raw signals or peaks, which are supposed to represent chemical compounds [69]. One approach relies on the assorting of MS data on the mass domain (m/z points) into bins of a certain width [62,70]. Another methodology is known as peak picking, the detection of peaks by monitoring specific ions or transitions. Peak picking approaches are described in literature as methods for pattern classification [71], extracting derivatives [68,72], refining heuristics [73], peak shape model approximations [74], or the application of wavelets [60,64]. Especially in large experiments, MS-based data may also be subject to systemic drifts; on the mass domain this may be corrected through calibration or appropriate mass binning, on the time domain by the usage of non-parametric alignment methods, such as dynamic programming. Through the deconvolution of isotopes, a profile can be corrected from isotopic artifacts, and simplified to unique metabolites; at the same time, the accuracy of quantitative data is markedly improved by adding up the peak intensities of the major isotopes belonging to each metabolite [75].

Normalization

In metabolite measurements, data are prone to various sources of hooded inter-sample variance, such as differences in the concentration or homogeneity of samples, the degradation of samples over time, or drifts of the analytical system. Normalization here refers to the data adjustment described in this section, prior to further statistical treatment [61,76]. One approach for inter-sample normalization is based on the estimation of a scaling factor, as by the summation of measurements [77], the mapping of intensity distribution [77,78], the sub-selection of peaks/signals [62,79], or by the introduction of class information [61]. Another approach performs normalization of all measured concentrations in a sample based on some other biological properties [79], e.g., by creatinine or the urine volume in urinary substances, or by the cell count, dry weight, protein content or deoxyribonucleic acid (DNA) levels in cell culture or tissue samples. The relevance of this latter form of data normalization is widely discussed and questioned, since it often leads to confusion or misleading results, for example when analyzing urinary samples obtained from patients with impaired kidney function. As an accepted gold standard is still missing, the technique of normalization needs to be decided on a case-by-case basis.

Quantitation

Quantitation of metabolite concentrations is, in principle, performed by referring to calibration curves. Common techniques are the quantitation by reference to internal standards (standards are added prior to sample extraction), to external standards (standards are added after sample extraction), or by standard addition (where samples are incrementally spiked with standards). In general, two basic concepts are differentiated, relative quantitation and absolute quantitation [50,80]. Relative quantitation is typically applied in untargeted metabolomics, aims at the determination of relative metabolite abundances, and is performed e.g., by normalizing the signal intensity of a compound referring to a certain metabolite (similar to the principle of a house-keeping gene in expression profiling) or – more often – by simply comparing peak intensities between two samples or cohorts. Absolute quantitation is usually only feasible in targeted metabolomics, aims at the measurement of absolute values of metabolite concentrations in SI units (mol/l, mol/g tissue etc.), and is achieved by utilizing non-naturally-occurring or (preferably) isotopically labeled but structurally identical internal standards for each metabolite or for groups of metabolites with very similar physicochemical properties. For adequately handling the chemical diversity of the sample matrix, methodologies using selected sets of multiple internal standards have been developed, for example being specific to compound classes [51,81] or to retention time regions [82,83]. From the perspective of data quality, all steps referring to calibration and quantification of metabolite concentrations essentially need to be performed in accordance with given regulatory guidelines, e.g., the 'Guidance for Industry - Bioanalytical Method Validation' [45]. Based on the results of the quantitation, a measurement matrix containing the absolute metabolite concentrations can be created, which may then be the input for the comprehensive data analysis.

DATA ANALYSIS

Overview

For the analysis of metabolomics data, univariate, multivariate and machine learning methods are employed, in both unsupervised and supervised manners, depending on the specific biological question. Computational methods for analyzing and interpreting metabolomics data are discussed in great detail in the existing literature, for example by Li [46] covering the main application areas of biomarker discovery, data clustering and visualization, classification and prediction, and methods for the identification/structural elucidation of metabolites. Here, both unsupervised methods (without prior knowledge on the sample identity) and supervised methods (using *a priori* information on the samples) are described in depth. The most widely used unsupervised methods comprise principal component analysis (PCA), self-organizing maps (SOM), and clustering techniques such as hierarchical clustering analysis (HCA), k-means, and fuzzy k-means. Also, the application of supervised methods is very common, e.g., k-nearest neighbor (kNN) classification, partial least square discriminant analysis (PLS-DA), artificial neural networks (ANN), and support vector machines (SVM). Common pitfalls occurring in the analysis of data from metabolomics experiments were compiled and discussed by Broadhurst [84]. For a general overview of data analysis workflows in metabolomics, please refer to more specialized literature [36,61,82,85]. In this chapter, a selection of topics with special relevance for metabolomics data analysis is discussed, such as preparation, variance, and correlation of data, as well as approaches for the search for biomarkers or biomarker signatures.

Preparation

Regarding the preparation of metabolomics data, the transformation of data as well as the handling of missing values play essential roles. Due to their biological nature, data in metabolomics experiments generally tend to have a large inter-parameter dynamic range, a fairly small intra-parameter dynamic range, and a non-normal distribution of concentration values, which makes appropriate data transformations necessary. For transforming MS data in metabolomics (as well as in other application areas), log transformations or similar approaches have been proposed [67,86,87], ensuring that multiplicative errors are converted into additive errors, and variance can be stabilized.

Missing values (non-detects) are a consequence of the utilized instrument set-up and analytical protocols (defining e.g., detection limits and quantification thresholds), molecular sample interactions, and the dynamics of metabolic pathways. In case of a large number of missing values in a certain feature, it is definitely recommended to discard those [61,82] unless the qualitative information (metabolite detectable/non-detectable) proves to have relevant, e.g., diagnostic, meaning. Basic approaches, using pre-specified values (like zero, or the minimum value in a dataset observed), means, or medians, for imputing missing data points, are known for having serious limitations and inducing a bias, e.g., by reducing the variance of the dataset in an unjustified manner [88]. Instead, it is recommended to use multivariate methods, as described by Stacklies [89], for estimating missing values. In general, though, the gain in predictive abilities through artificial data point estimation is very limited, and the entire process can be primarily seen as a means of preparing data for the smoother application of multivariate statistical tools.

Variance

Variance in metabolomics studies is influenced by the naturally occurring biological variance of metabolites, as well as an unavoidable variance resulting from the analytical technology used. Biological variability of metabolites can be surprisingly low if the underlying homeostatic control is strict (e.g., glucose and most amino acids in peripheral blood) but can also reach a coefficient of variance (CV, the ratio of standard deviation to average abundance) of up to 50% or even more, particularly in less well regulated sample types such as urine [90,91]. This fact must be taken into account when planning the experimental design of a study and estimating the appropriate cohort sizes and numbers of samples and time points. The technical/analytical variance of metabolite data on the other hand generally ranges between a minimum CV level of around 2 - 5%, which can be achieved for small numbers of specifically optimized analytes, and CV values above 20% for semi-quantitative parameters, depending on the compound classes analyzed, the effort put into optimizing the assay, and the availability of an authentic internal standard. The relevant FDA 'Guidance for Industry – Bioanalytical Method Validation' [45] demands a CV of less than 15% throughout the dynamic range of a validated quantitative assay while allowing up to 20% at the lower limit of quantitation (LLOQ).

Correlation

In the course of data analysis, also the interdependence of abundances between different metabolites should be turned to account. Correlation between features may result from specific properties of the analytical platform, as for example, from the same molecule, several ionization and derivatization products may be obtained [92].

It can also reflect specific chemical characteristics of certain metabolites, such as obtaining multiple signals related to the isotopic pattern of the same parent ion. Most important for data interpretation is, of course, correlation because of biological mechanisms. As a result of efforts to link the correlation of metabolites to metabolite reaction networks, various authors differentiated somewhat artificially between positive correlation, negative correlation, and correlation dependent on disease states [93-95].

In this context, extensive investigations by means of network-based approaches were performed, e.g., trying to specify structural properties of biochemical networks by the usage of topological network descriptors [96].

In this kind of network analysis, correlation-based distance metrics are often developed based on the proximity of metabolites, which is used for the visualization of networks, by clustering or graph-based representations.

However, the possibility to make implications on biochemical functions based on these correlation-based distance metrics is limited, and should only be seen as a means of generating hypotheses for experimental confirmation.

Biomarker Identification

In recent years, the quest for metabolic biomarkers for clinical applications has gained increasing attention.

A multitude of approaches for selection and ranking of biomarker candidates was developed, based primarily on supervised methods, as described in the beginning of this section. A comprehensive overview of these biomarker discovery strategies is given by Baumgartner [97]. For independent samples, the most common approaches are random forest models (RFM) [98], associative voting (AV) [36], stacked feature ranking (SFR) [99] and the unpaired biomarker identifier (uBI) [100]. For dependent samples, typical methods are paired null hypothesis testing [101] and the paired Biomarker Identifier (pBI) [100]. A validation of putatively identified biomarker candidates should, of course, always be based on a replication of study results in independent cohorts but this process can be streamlined by first scrutinizing the data in a thorough biochemical plausibility check, which may already eliminate the majority of false-positive hits.

Biochemical Interpretation

Knowledge Annotation

Thanks to the detailed level on which the main pathways of metabolism and their molecular reactions have been characterized by generations of biochemists, an elaborate body of knowledge is nowadays available in public databases. A structured review of metabolomics-related data sources, distinguishing chemical bioinformatics databases, metabolic pathways databases, metabolomics databases, pharmaceutical product databases, and toxic substance databases is given by Wishart [58].

Since the focus of this section is on the functional interpretation of biochemical interactions in the context of metabolic pathways, only pathway databases will be subject to further discussion. Popular examples of a growing number of these are BioCyc [102], the Human Metabolome Database (HMDB) [56], Reactome [103], or the Small Molecule Pathway Database (SMPDB) [104] but the entire field was pioneered by the Kyoto Encyclopedia of Genes and Genomes (KEGG) [105] which – despite its name – did not follow the gene-centric approach of the National Center for Biotechnology Information (NCBI; www.ncbi.nlm.nih.gov) or the European Bioinformatics Institute (EBI; www.ebi.ac.uk) but rather identified the pathway as the central biological entity around which they tailored their repository.

While there are countless – typically more intuitive than systematic – strategies for the biochemical interpretation of metabolomics data, the following paragraphs will only describe three basic concepts which can be seen as standard tools for getting a first overview and generate hypotheses for further elaboration.

These three methods are metabolite set enrichment, shell-wise exploration of metabolic reactions, and metabolic route finding.

Metabolite Set Enrichment

Enrichment analysis (EA) is a general concept that has been used in transcriptomics and proteomics experiments for quite some time to boost the statistical power of studies and deduce some functional insight. In gene-centrically annotated classes of biomolecules (nucleic acids and proteins), the grouping of analytes is usually based on the gene ontology (GO) but, of course, this does not work for metabolites. Still, a strategy analogous to a gene set enrichment analysis (GSEA) may also be helpful in the functional interpretation of metabolomics data. According to Chagoyen [106], two basic approaches are differentiated, overrepresentation analysis (ORA), and set enrichment analysis (SEA), with SEA using quantitative information (e.g., metabolite concentrations) for analysis. A further specialized concept, single sample profiling (SSP), which compares concentrations of metabolites of one sample with literature reference values, is described by Xia [107]. In the concise case of a metabolite set enrichment analysis, the process can be described as the mapping of a given set of metabolites, statistically identified as significantly different between two biological states or cohorts, onto a set of metabolic pathways, resulting in a ranking of these pathways according to the number of altered metabolites or some derived statistical value, e.g., a cumulative p-value [3].

Selected software tools providing functionality for EA in metabolomics, are MBRole (Metabolites Biological Rule) [108], MSEA (Metabolite Set Enrichment Analysis) [107], or MetaboAnalyst [109]. To reduce the risk of false positive hits, instead of using a generic reference pathway, these analyses should always be based on species-specific pathway lists or reactomes.

Shell-Wise Exploration

One of the most intuitively helpful ways of biochemical interpretation is a shell-wise exploration of metabolic reactions around metabolites of interest, e.g., those top-ranked by statistical significance (corrected p-value) or diagnostic performance (area under the curve in a receiver operating characteristic (ROC) analysis) [3].

One particular metabolite is selected as the seed node, and a shell of all reactions synthesizing or degrading this metabolite are displayed. This leads to a set of enzymes, whose activities could directly be responsible for the altered concentration of the metabolite of interest, and an extended set of metabolites of secondary interest. In a next step, each of these secondary metabolites can again be used as a new seed node (compare Figure 1) and so on.

This way of looking at the role of single enzymatic reactions has long been established in neonatal screening for monogenic disorders [14] but was recently confirmed to be equally relevant for multifactorial phenotypes/diseases in a series of large-scale studies [110,111]. Since the ratio of product and substrate concentrations of each reaction can be taken as a rather immediate measure of enzymatic activity, these ratios are far more informative than the individual concentration data and less prone to confounding factors such as nutritional uptake. Using the ratios as additional traits in GWA studies identified more hits in or near coding sequences and improved the statistical significance by many orders of magnitude [15,112].

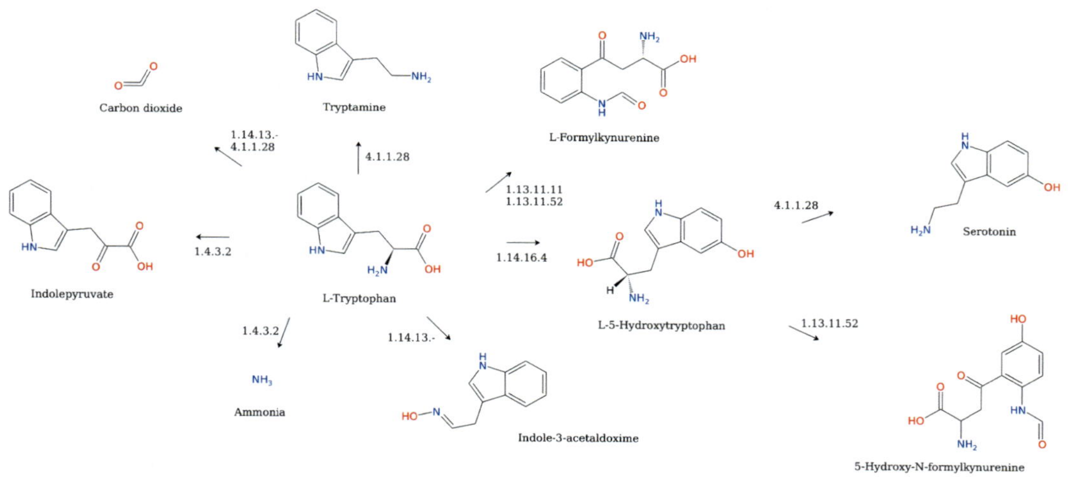

Figure 1. Systematic sketch of the analysis of tryptophan metabolism through shell-wise exploration. As the first seed node, L-tryptophan is chosen, leading to a set of eight either synthesizing of degrading enzymatic reactions. As the secondary seed node, 5-hydroxy-L-tryptophan is used for further expansion (not all reactions shown for this second step).

Route Finding

When working on the biochemical interpretation of metabolomics results, one should keep in mind that the basic concept of metabolic pathways as confined entities, often linear or circular in shape is very old and partially outdated. Of course, these pathways are part of a larger and densely woven network but the ‘old-fashioned’ textbook representation is deeply routed in most biochemists and keeps them from seeing connections across the artificial boundaries of these pathways. To overcome these limitations, allow for generating new ‘cross-border’ hypotheses, and treat the entire set of metabolic pathways as one integrated source of knowledge, the mundane concept of route finding as it is used in car navigation systems can be applied.

Route finding algorithms provide the functionality to find and depict all possible routes between two metabolites, also across artificial pathway boundaries. These algorithms can be used to identify the shortest route, routes up to a maximum length, edge-disjoint paths (routes not sharing a certain enzyme), or node-disjoint paths (routes not sharing a certain metabolite) between two metabolites.

Some route finding tools apply atom-mapping rules to take reaction mechanisms into account [113] or use weighted graphs [114]. As one rather simple example (chosen not to create a too complicated graph), MetaRoute [115], a web-based tool for route finding based on the information in KEGG [105], was used to find all routes between serine and pyruvate in H. sapiens (see Figure 2). It is important to note at this point, that common co-factors and small inorganic molecules (e.g., adenosine triphosphate (ATP), nicotinamide adenine dinucleotide (NADH), carbon dioxide (CO_2), water (H_2O), ammonia (NH_3), etc.) need to be excluded to avoid overly complex results graphs and to prevent too many false positive hits.

Hypothesis Generation

All three concepts, metabolite set enrichment, shell-wise exploration, and route finding, can be used for deriving deeper insights from multivariate metabolic data sets, and as a starting point for the generation of hypotheses inspiring new experiments and studies for verification or validation of results. Nevertheless, results need to undergo additional plausibility checks. Major factors influencing or potentially skewing results are redundancies in metabolism itself (groups of compounds are often metabolized by one enzyme or several enzymes catalyze the same reactions), disturbances of signals of endogenous metabolites by drug compounds, or analytical or statistical artifacts in general.

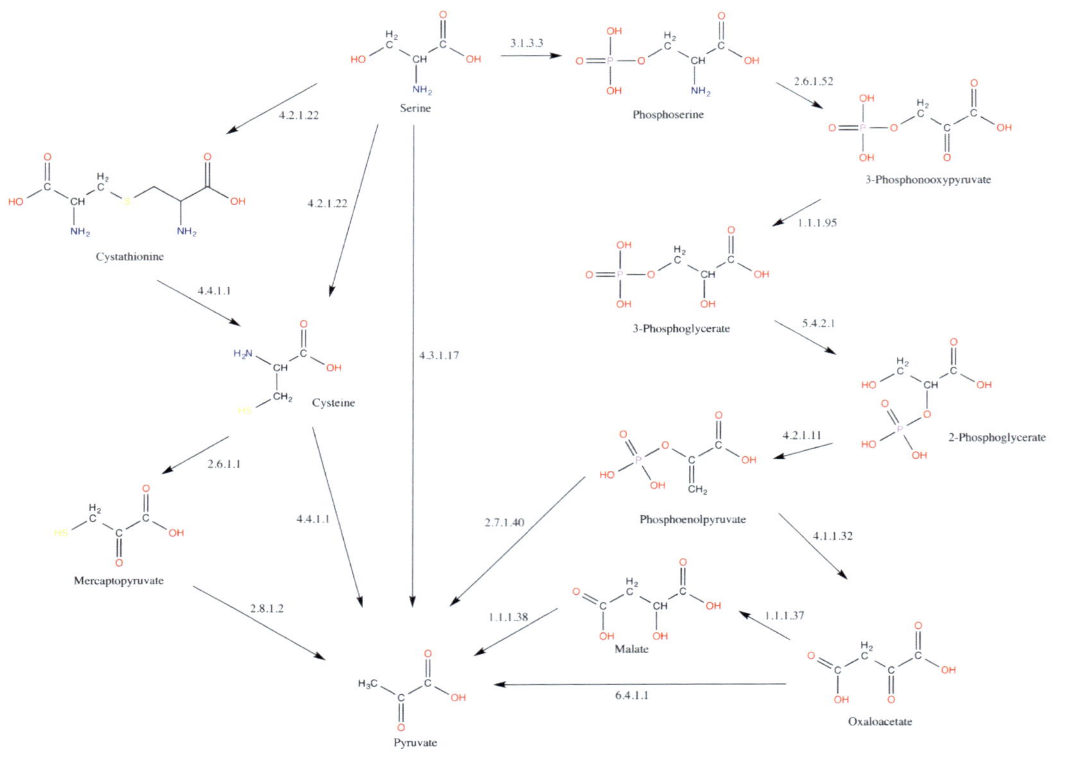

Figure 2. Route finding across metabolic pathways. Source is serine, target is pyruvate, species is Homo sapiens, calculated using MetaRoute [115], results redrawn for better graphical quality.

To discuss just one example: when deducting hypotheses from metabolite set enrichment analyses, one needs to keep in mind that ontologies for metabolomics analyses are basically existing (e.g., the concept of pathways as it is used in KEGG), and GO-derived semantic similarity measures can also be applied [116] but this still does not warrant a balanced assessment. Slightly exaggerated, one might state that one reference pathway, e.g., 'glycerophospholipid metabolism', covers or directly influences at least one third of the entire human metabolism while other reference pathways, say: 'lysine degradation', are still complex but basically only reflect the catabolism of one selected amino acid. To use both *pari passu* in an EA can hardly reflect their biological importance.

Regarding the generation of hypotheses based on the shell-wise exploration of metabolism, a basic discussion is ongoing with respect to the usage of ratios mentioned above. On the one hand, it can certainly be argued that ratios are artificial parameters, uncommon and too much driven by preformed hypotheses. Moreover, if a certain information content is present in a given data set, this information should be found through appropriate mathematical analyses (which it eventually could). On the other hand, it may just as well be argued that these ratios are *in praxi* improving the statistical performance, and that it is exactly the purpose of the accumulated biological knowledge to support or direct the data analysis and exclude biochemically unlikely or impossible findings.

Finally, route finding approaches offer a unique opportunity to generate completely new hypotheses, which are surpassing the possibilities based on 'pure', conventional pathway analysis. This approach could additionally be extended and supported by the application of further network-oriented tools, e.g., novel ways of network inference or more intuitive means of network visualization [117-119].

CONCLUSION

Metabolomics is often described as the youngest omics discipline although a detailed analysis of metabolism by organic chemists actually predates nucleic acid or protein technologies by roughly a century. It offers, however, the possibility to gain major additional insights into basic biochemistry, the pathophysiology of diseases, the mode-of-action of drugs, or novel metabolic biomarkers for clinical diagnostics. When high-quality data have been generated, the combination of bioinformatic and chemometric methods for the processing of data, as well as mathematical, statistical and data mining approaches, build the foundation for extracting knowledge out of multivariate data sets. In metabolomics, an additional opportunity lies in the biochemical interpretation of results. Thanks to the detailed functional understanding of metabolic pathways, plausibility of results can be examined much more thoroughly within the respective biochemical context. Despite the multitude of existing algorithms, concepts, tools and more, widely accepted gold standards or real expert systems for most of these steps are still missing, and the actual interpretation of results is often viewed as a niche for the dying race of experienced biochemists.

In addition to the topics covered in this chapter, several promising trends and initiatives can be observed. From the technological perspective, the manufacturers of MS instruments are, partly in collaboration with vendors of liquid handling robotics systems, working towards fully integrated platforms, e.g., clinical black-box MS analyzers, for the usage in routine diagnostics. In parallel, many interesting and promising initiatives are on their way in the area of microfluidics solutions, although most of these are still in an experimental and research-focused phase. Moreover, a multitude of standardization initiatives exists, e.g., mzML for defining formats for the exchange of data [120], or the definition of a common ontology, e.g., by the metabolomics standards initiative (MSI) [121]. Despite the plethora of these programs and initiatives, a summary is undertaken by Enot [3].

Also in the area of modeling and simulation of biochemical networks, really promising research is performed, as for example the modeling of cell cultures [122], or network simulations of inborn errors of metabolism [123]. One must admit that all of these approaches are basically still of a theoretical nature and a successful transfer to clinical applications has not been demonstrated but they may be one of the most exciting areas for relevant innovation.

Furthermore, the integration of data from different omics experiments is on its way, e.g., the combination of proteomics and metabolomics data for the identification of putative biomarkers [124] but one should be very cautious not to expect the solution of all problems from a broader data basis since these combinations are not necessarily improving the diagnostic performance [29]. Still, the ultimate goal should be the discovery of sets of multi-parametric marker candidates which, after successful independent validation by means of clinical studies, can act as novel biomarkers in routine clinical diagnostics.

REFERENCES

[1] Hood LE, Hunkapiller MW, Smith LM. *Genomics*. 1987;1(3):201-12.

[2] Hood L. *JAMA*. 1988;259(12):1837-44.

[3] Enot DP, Haas B, Weinberger KM. *Mol. Biol*. 2011;719:351-75.

[4] Ishii M, Hashimoto S, Tsutsumi S, Wada Y, Matsushima K, Kodama T, Aburatani H. *Genomics*. 2000;68(2):136-43.

[5] Petricoin EF, Ornstein DK, Paweletz CP, Ardekani A, Hackett PS, Hitt BA, Velassco A, Trucco C, Wiegand L, Wood K, Simone CB, Levine PJ, Linehan WM, Emmert-Buck MR, Steinberg SM, Kohn EC, Liotta LA. *J. Natl. Cancer Inst*. 2002;94(20):1576-8.

[6] Wulfkuhle JD, Liotta LA, Petricoin EF. *Nat. Rev. Cancer*. 2003;3(4):267-75.

[7] Petricoin EF, Ornstein DK, Liotta LA. *Urol. Oncol*. 2004;22(4):322-8.

[8] Yost RA, Enke CG. *Anal. Chem*. 1979;51(12):1251-64.

[9] Ebhardt HA. *Methods Mol. Biol*. 2014;1072:209-22.

[10] Claeys M, Muscettola G, Markey SP. *Biomed. Mass. Spectrom*. 1976;3(3):110-6.

[11] Weinberger KM, Graber A. *Screening Trends in Drug Discovery*. 2005;(6):42-45.

[12] Weinberger KM, Ramsay SL, Graber A. *Biosystems Solutions*. 2005;36-37.

[13] Weinberger KM. *Ther Umsch*. 2008;65(9):487-91. [Article in German]

[14] Röschinger W, Olgemöller B, Fingerhut R, Liebl B, Roscher AA. *Eur. J. Pediatr*. 2003;162Suppl 1:S67-76.

[15] Gieger C, Geistlinger L, Altmaier E, Hrabé de Angelis M, Kronenberg F, Meitinger T, Mewes HW, Wichmann HE, Weinberger KM, Adamski J, Illig T, Suhre K. *PLoS Genet*. 2008;4(11):e1000282.

[16] Shin SY, Fauman EB, Petersen AK, Krumsiek J, Santos R, Huang J, Arnold M, Erte I, Forgetta V, Yang TP, Walter K, Menni C, Chen L, Vasquez L, Valdes AM, Hyde CL, Wang V, Ziemek D, Roberts P, Xi L, Grundberg E; Multiple Tissue Human Expression Resource (MuTHER) Consortium, Waldenberger M, Richards JB, Mohney RP, Milburn MV, John SL, Trimmer J, Theis FJ, Overington JP, Suhre K, Brosnan MJ, Gieger C, Kastenmüller G, Spector TD, Soranzo N. *Nat. Genet*. 2014;46(6):543-50.

[17] Weikard R, Altmaier E, Suhre K, Weinberger KM, Hammon HM, Albrecht E, Setoguchi K, Takasuga A, Kühn C.. *Physiol Genomics*. 2010;42A(2):79-88.

[18] Astarita G, Langridge J. *J. Nutrigenet. Nutrigenomics*. 2013;6(4-5):181-200.

[19] Altmaier E, Kastenmüller G, Römisch-Margl W, Thorand B, Weinberger KM, Illig T, Adamski J, Döring A, Suhre K. *Eur. J. Epidemiol.* 2011 Feb;26(2):145-56.

[20] Wang-Sattler R, Yu Y, Mittelstrass K, Lattka E, Altmaier E, Gieger C, Ladwig KH, Dahmen N, Weinberger KM, Hao P, Liu L, Li Y, Wichmann HE, Adamski J, Suhre K, Illig T. *PLoS One*. 2008;3(12):e3863.

[21] Altmaier E, Kastenmüller G, Römisch-Margl W, Thorand B, Weinberger KM, Adamski J, Illig T, Döring A, Suhre K. *Mol. Nutr. Food Res*. 2009;53(11):1357-65.

[22] Jaremek M, Yu Z, Mangino M, Mittelstrass K, Prehn C, Singmann P, Xu T, Dahmen N, Weinberger KM, Suhre K, Peters A, Döring A, Hauner H, Adamski J, Illig T, Spector TD, Wang-Sattler R. *Transl Psychiatry*. 2013;3:e276.

[23] Sonntag D, Scandurra FM, Friedrich T, Urban M, Weinberger KM. *BMC Proc.* 2011;5Suppl 8:P27.

[24] Astarita G, McKenzie JH, Wang B, Strassburg K, Doneanu A, Johnson J, Baker A, Hankemeier T, Murphy J, Vreeken RJ, Langridge J, Kang JX. *PLoS One*. 2014;9(4):e96221.

[25] Laiakis EC, Hyduke DR, Fornace AJ. *Radiat. Res*. 2012;177(2):187-99.

[26] Lundin U, Modre-Osprian R, Weinberger KM. Targeted metabolomics for clinical biomarker discovery in multifactorial diseases. In: Ikehara K, editor. Advances in the Study of Genetic Disorders. Croatia: InTech; 2011.

[27] Argilés Á, Siwy J, Duranton F, Gayrard N, Dakna M, Lundin U, Osaba L, Delles C, Mourad G, Weinberger KM, Mischak H. *PLoS One.* 2013;8(5):e62837.

[28] Duranton F, Lundin U, Gayrard N, Mischak H, Aparicio M, Mourad G, Daurès JP, Weinberger KM, Argilés A. *Clin. J. Am. Soc. Nephrol.* 2014;9(1):37-45.

[29] Nkuipou-Kenfack E, Duranton F, Gayrard N, Argilés A, Lundin U, Weinberger KM, Dakna M, Delles C, Mullen W, Husi H, Klein J, Koeck T, Zürbig P, Mischak H. *PLoS One*. 2014;9(5):e96955.

[30] Altmaier E, Ramsay S L, Graber A, Mewes H W, Weinberger K M, Suhre K. *Endocrinology* 2008;(149):3478-3489.

[31] Suhre K, Meisinger C, Döring A, Altmaier E, Belcredi P, Gieger C, Chang D, Milburn MV, Gall WE, Weinberger KM, Mewes HW, Hrabé de Angelis M, Wichmann HE, Kronenberg F, Adamski J, Illig T. *PLoS One.* 2010;5(11):e13953.

[32] Astarita G, Jung KM, Vasilevko V, Dipatrizio NV, Martin SK, Cribbs DH, Head E, Cotman CW, Piomelli D. *PLoS One*. 2011;6(10):e24777.

[33] Mapstone M, Cheema AK, Fiandaca MS, Zhong X, Mhyre TR, MacArthur LH, Hall WJ, Fisher SG, Peterson DR, Haley JM, Nazar MD, Rich SA, Berlau DJ, Peltz CB, Tan MT, Kawas CH, Federoff HJ. *Nat. Med.* 2014;20(4):415-8.

[34] Mazurek S, Eigenbrodt E. *Anticancer Res*. 2003;23(2A):1149-54.

[35] Christofk HR, Vander Heiden MG, Harris MH, Ramanathan A, Gerszten RE, Wei R, Fleming MD, Schreiber SL, Cantley LC. *Nature.* 2008;452(7184):230-3.

[36] Osl M, Dreiseitl S, Pfeifer B, Weinberger K, Klocker H, Bartsch G, Schäfer G, Tilg B, Graber A, Baumgartner C. *Bioinformatics*. 2008;24(24):2908-14.

[37] Netzer M, Weinberger KM, Handler M, Seger M, Fang X, Kugler KG, Graber A, Baumgartner C. *J. Clin. Bioinforma*. 2011;1(1):34.

[38] Netzer M, Kugler KG, Müller LA, Weinberger KM, Graber A, Baumgartner C, Dehmer M. *J. Theor. Biol*. 2012;310:216-22.

[39] Pichler Hefti J, Sonntag D, Hefti U, Risch L, Schoch OD, Turk AJ, Hess T, Bloch KE, Maggiorini M, Merz TM, Weinberger KM, Huber AR. *High Alt. Med. Biol.* 2013;14(3):273-9.
[40] Shulaev V. *Brief Bioinform.* 2006;7(2):128-39.
[41] Patti GJ, Yanes O, Siuzdak G. *Nat. Rev. Mol. Cell. Biol.* 2012;13(4):263-9.
[42] Vogeser M, Kirchhoff F. *Clin. Biochem.* 2011;44(1):4-13.
[43] Urban M, Enot DP, Dallmann G, Körner L, Forcher V, Enoh P, Koal T, Keller M, Deigner HP. *Anal Biochem.* 2010;406(2):124-131.
[44] Food and Drug Administration (FDA). General Principles of Software Validation; Final Guidance for Industry and FDA Staff. Rockville: *Center for Biologics Evaluation and Research* (CBER). 2002.
[45] Food and Drug Administration (FDA).Guidance for Industry - Bioanalytical Method Validation. Rockville: *Center for Drug Evaluation and Research* (CDER). 2001.
[46] Li F, Wang J, Nie L, and Zhang W. Computational Methods to Interpret and Integrate Metabolomic Data. In: Roessner U, editor. Metabolomics. Croatia: Intech; 2012;99-130.
[47] Dettmer K, Aronov PA, Hammock BD. *Mass Spectrom Rev.* 2007;26(1):51-78.
[48] Lu W, Bennett BD, Rabinowitz JD. J Chromatogr B Analyt. *Technol. Biomed. Life Sci.* 2008;871(2):236-42.
[49] Grebe SK, Singh RJ. *Clin. Biochem. Rev.* 2011;32(1):5-31.
[50] Lei Z, Huhman DV, Sumner LW. *J Biol Chem.* 2011;286(29):25435–42.
[51] Unterwurzacher I, Koal T, Bonn GK, Weinberger KM, Ramsay SL. *Clin. Chem. Lab.* Med. 2008;46(11):1589-97.
[52] Roberts LD, Souza AL, Gerszten RE, Clish CB. *Curr. Protoc. Mol. Biol.* 2012;30(2):1-24.
[53] Environmental Protection Agency (EPA).Guidance for Preparing Standard Operating Procedures (SOPs). EPA QA/G-6. Washington: *Office of Environmental Information.* 2007.
[54] Biswas A, Mynampati KC, Umashankar S, Reuben S, Parab G, Rao R, Kannan VS, Swarup S. *Bioinformatics.* 2010;26(20):2639-40.
[55] Scholz M, Fiehn O. Pac. *Symp. Biocomput.* 2007:169-80.
[56] Wishart DS, Jewison T, Guo AC, Wilson M, Knox C, Liu Y, Djoumbou Y, Mandal R, Aziat F, Dong E, Bouatra S, Sinelnikov I, Arndt D, Xia J, Liu P, Yallou F, Bjorndahl T, Perez-Pineiro R, Eisner R, Allen F, Neveu V, Greiner R, Scalbert A. *Nucleic Acids Res.* 2013;41(Database issue):D801-7.
[57] Zhu ZJ, Schultz AW, Wang J, Johnson CH, Yannone SM, Patti GJ, Siuzdak G. *Nat. Protoc.* 2013;8(3):451-60.
[58] Wishart DS. Chapter 3: Small molecules and disease. *PLoS Comput. Biol.* 2012;8(12):e1002805.
[59] Haimi P, Uphoff A, Hermansson M, Somerharju P. *Anal. Chem.* 2006;78(24):8324-31.
[60] Karpievitch YV, Hill EG, Smolka AJ, Morris JS, Coombes KR, Baggerly KA, Almeida JS. *Bioinformatics.* 2007;23(2):264-5.
[61] Enot DP, Lin W, Beckmann M, Parker D, Overy DP, Draper *J. Nat. Protoc.* 2008;3(3):446-70.
[62] Wang W, Zhou H, Lin H, Roy S, Shaler TA, Hill LR, Norton S, Kumar P, Anderle M, Becker CH. *Anal. Chem.* 2003;75(18):4818-26.

[63] Bylund D. Chemometric tools for enhanced performance in liquid chromatography-mass spectrometry. Comprehensive Summaries of Uppsala Dissertations from the Faculty of Science and Technology. Uppsala University. 2001.

[64] Zhao Q, Stoyanova R, Du S, Sajda P, Brown TR. *Bioinformatics*. 2006;22(20):2562-4.

[65] Zhu W, Wang X, Ma Y, Rao M, Glimm J, Kovach JS. *Proc. Natl. Acad. Sci. USA*. 2003;100(25):14666-71.

[66] Jonsson P, Johansson AI, Gullberg J, Trygg J, A J, Grung B, Marklund S, Sjöström M, Antti H, Moritz T. *Anal. Chem*. 2005;77(17):5635-42.

[67] Listgarten J, Emili A. Mol Cell Proteomics. 2005;4(4):419-34.

[68] Fredriksson MJ, Petersson P, Axelsson BO, Bylund D. *J. Sep. Sci.* 2009;32(22):3906-3918.

[69] Katajamaa M, Oresic M. *J. Chromatogr A*. 2007;1158(1-2):318-28.

[70] Beckmann M, Parker D, Enot DP, Duval E, Draper *J. Nat Protoc*. 2008;3(3):486-504.

[71] Tibshirani R, Hastie T, Narasimhan B, Soltys S, Shi G, Koong A, Le QT. *Bioinformatics*. 2004;20(17):3034-44.

[72] Vivó-Truyols G, Torres-Lapasió JR, van Nederkassel AM, Vander Heyden Y, Massart DL. *J. Chromatogr A*. 2005;1096(1-2):133-45.

[73] Morris JS, Coombes KR, Koomen J, Baggerly KA, Kobayashi R. *Bioinformatics*. 2005;21(9):1764-75.

[74] Tan CS, Ploner A, Quandt A, Lehtiö J, Pawitan Y. *Bioinformatics*. 2006;22(12):1515-23.

[75] Eibl G, Bernardo K, Koal T, Ramsay SL, Weinberger KM, Graber A. *Rapid Commun. Mass Spectrom*. 2008;22(14):2248-52.

[76] Goodacre R, Vaidyanathan S, Dunn WB, Harrigan G,Kell DB. *Trends Biotechnol*. 2004;22:245–52.

[77] Torgrip RJO, Aberg KM, Alm E, Schuppe-Koistinen I, Lindberg J. *Metabolomics*. 2008;4:114–21.

[78] Dieterle F, Ross A, Schlotterbeck G, Senn H. *Anal. Chem*. 2006;78:4281–90.

[79] Warrack BM, Hnatyshyn S, Ott KH, Reily MD, Sanders M, Zhang H, Drexler DM. J. Chromatogr B Analyt. *Technol. Biomed. Life Sci*. 2009;877:547–52.

[80] Bueschl C, Krska R, Kluger B, Schuhmacher R. *Anal. Bioanal. Chem*. 2013;405(1):27-33.

[81] Hermansson M, Uphoff A, Käkelä R, Somerharju P. *Anal. Chem*. 2005;77(7):2166-75.

[82] Bijlsma S, Bobeldijk I, Verheij ER, Ramaker R, Kochhar S, Macdonald IA, van Ommen B, Smilde AK. *Anal. Chem*. 2006;78(2):567-74.

[83] Sysi-Aho M, Katajamaa M, Yetukuri L, Oresic M. *BMC Bioinformatics*. 2007;8:93.

[84] Broadhurst DI, Kell DB. *Metabolomics*. 2006;2(4):171-196.

[85] Brown M, Dunn WB, Ellis DI, Goodacre R, Handl J, Knowles JD, O'Hagan S, Spasic I, Kell DB. *Metabolomics*. 2005;1(1):39-59

[86] Purohit PV, Rocke DM, Viant MR, Woodruff DL. *OMICS*. 2004;8(2):118-30.

[87] Lu C, King RD. *Bioinformatics*. 2009;25(16):2020-7.

[88] Jain RB, Caudill SP, Wang RY, Monsell E. *Anal Chem*. 2008;80(4):1124-32.

[89] Stacklies W, Redestig H, Scholz M, Walther D, Selbig J. *Bioinformatics*. 2007;23(9):1164-7.

[90] Crews B, Wikoff WR, Patti GJ, Woo HK, Kalisiak E, Heideker J, Siuzdak G. *Anal. Chem*. 2009;81(20):8538-44.

[91] Parsons HM, Ekman DR, Collette TW, Viant MR. *Analyst.* 2009;134(3):478-85.

[92] Draper J, Enot DP, Parker D, Beckmann M, Snowdon S, Lin W, Zubair H. BMC *Bioinformatics.* 2009;10:227.

[93] Camacho D, de la Fuente A, Mendes P. *Metabolomics.* 2005;1(1):53-63.

[94] Mendes P, Camacho D, de la Fuente A. *Biochem. Soc. Trans.* 2005;33(Pt 6):1427-9.

[95] Steuer R. *Brief Bioinform.* 2006;7(2):151-8.

[96] Mueller LAJ, Kugler KG, Netzer M, Graber A, Dehmer M. *Biol. Direct.* 2011;6:53.

[97] Baumgartner C, Osl M, Netzer M, Baumgartner D. *J Clin Bioinforma.* 2011;1(1):2.

[98] Enot DP, Beckmann M, Overy D, Draper J. *Proc Natl. Acad. Sci. USA.* 2006;103(40):14865-70.

[99] Netzer M, Millonig G, Osl M, Pfeifer B, Praun S, Villinger J, Vogel W, Baumgartner C. *Bioinformatics.* 2009;25(7):941-7.

[100] Baumgartner C, Lewis GD, Netzer M, Pfeifer B, Gerszten RE. *Bioinformatics.* 2010;26(14):1745-51.

[101] Lehmann EL, Romano JP. Testing statistical hypotheses. 3 edition. New York. NY: Springer Verlag. 2005.

[102] Karp PD, Ouzounis CA, Moore-Kochlacs C, Goldovsky L, Kaipa P, Ahrén D, Tsoka S, Darzentas N, Kunin V, López-Bigas N. *Nucleic Acids Res.* 2005;33(19):6083-9.

[103] Joshi-Tope G, Gillespie M, Vastrik I, D'Eustachio P, Schmidt E, de Bono B, Jassal B, Gopinath GR, Wu GR, Matthews L, Lewis S, Birney E, Stein L. *Nucleic Acids Res.* 2005;33(Database issue):D428-32.

[104] Jewison T, Su Y, Disfany FM, Liang Y, Knox C, Maciejewski A, Poelzer J, Huynh J, Zhou Y, Arndt D, Djoumbou Y, Liu Y, Deng L, Guo AC, Han B, Pon A, Wilson M, Rafatnia S, Liu P, Wishart DS. *Nucleic Acids Res.* 2014;42(Database issue):D478-84.

[105] Kanehisa M, Goto S. *Nucleic Acids Res.* 2001;28(1):27-30.

[106] Chagoyen M, Pazos F. *Brief Bioinform.* 2013;14(6):737-44.

[107] Xia J, Wishart DS. Nucleic *Acids Res.* 2010;38(Web Server issue):W71-7.

[108] Chagoyen M, Pazos F. *Bioinformatics.* 2011;27(5):730-1.

[109] Xia J, Mandal R, Sinelnikov IV, Broadhurst D, Wishart DS. *Nucleic Acids Res.* 2012;40(Web Server issue):W127-33.

[110] Suhre K, Shin SY, Petersen AK, Mohney RP, Meredith D, Wägele B, Altmaier E; CARDIoGRAM, Deloukas P, Erdmann J, Grundberg E, Hammond CJ, de Angelis MH, Kastenmüller G, Köttgen A, Kronenberg F, Mangino M, Meisinger C, Meitinger T, Mewes HW, Milburn MV, Prehn C, Raffler J, Ried JS, Römisch-Margl W, Samani NJ, Small KS, Wichmann HE, Zhai G, Illig T, Spector TD, Adamski J, Soranzo N, Gieger C. *Nature.* 2011;477(7362):54-60.

[111] Suhre K, Gieger C. *Nat. Rev. Genet.* 2012;13(11):759-69.

[112] Illig T, Gieger C, Zhai G, Römisch-Margl W, Wang-Sattler R, Prehn C, Altmaier E, Kastenmüller G, Kato BS, Mewes HW, Meitinger T, Hrabé de Angelis M, Kronenberg F, Soranzo N, Wichmann HE, Spector TD, Adamski J, Suhre K. *Nature Genetics.* 2010;42:137–141.

[113] Arita M. *Genome Res.* 2003;13(11):2455-66.

[114] Croes D, Couche F, Wodak SJ, van Helden *J. Nucleic Acids Res.* 2005;33(Web Server issue):W326-30.

[115] Blum T, Kohlbacher O. *Bioinformatics.* 2008;24(18):2108-9.

[116] Guo X, Shriver CD, Hu H, Liebman MN. *AMIA Annu. Symp. Proc.* 2005:972.

[117] Emmert-Streib F. *PLoS One*. 2013;8(2):e56461.
[118] Emmert-Streib F, Zhang SD, Hamilton P. *J. Transl. Med.* 2014;12:26.
[119] Triapthi S, Dehmer M, Emmert-Streib F. *Bioinformatics*. 2014. pii: btu384. [Epub ahead of print]
[120] Martens L, Chambers M, Sturm M, Kessner D, Levander F, Shofstahl J, Tang WH, Römpp A, Neumann S, Pizarro AD, Montecchi-Palazzi L, Tasman N, Coleman M, Reisinger F, Souda P, Hermjakob H, Binz PA, Deutsch EW. *Mol. Cell. Proteomics*. 2011 Jan;10(1):R110.000133.
[121] Sansone SA, Fan T, Goodacre R, Griffin JL, Hardy NW, Kaddurah-Daouk R, Kristal BS, Lindon J, Mendes P, Morrison N, Nikolau B, Robertson D, Sumner LW, Taylor C, van der Werf M, van Ommen B, Fiehn O. *Nat. Biotechnol.* 2007;25(8):846-8.
[122] Quek LE, Dietmair S, Krömer JO, Nielsen LK. *Metab. Eng*. 2010;12(2):161-71.
[123] Modre-Osprian R, Osprian I, Tilg B, Schreier G, Weinberger KM, Graber A. *BMC Syst. Biol.* 2009;3:2.
[124] Blanchet L, Smolinska A, Attali A, Stoop MP, Ampt KA, van Aken H, Suidgeest E, Tuinstra T, Wijmenga SS, Luider T, Buydens LM. *BMC Bioinformatics*. 2011;12:254.

In: Current Applications of Chemometrics
Editor: Mohammadreza Khanmohammadi

ISBN: 978-1-63463-117-4

Chapter 10

ON SOME EMERGING CONCEPTS IN THE QSAR PARADIGM

Rahul Balasaheb Aher, Pravin Ambure and Kunal Roy*
Drug Theoretics and Cheminformatics Laboratory,
Department of Pharmaceutical Technology,
Jadavpur University, Kolkata, India

ABSTRACT

Quantitative structure-activity relationship (QSAR) is a well-established predictive tool in various fields of the chemical research. It captures structural information in the numerical form from a set of molecules of some level of similarity as descriptors, and correlates this to the response property/activity/toxicity using statistical tools for the predictive model development. Since it utilizes some experimental information of the chemicals as the observed response for model development, it has found application in different fields for the new chemical development in a rational way. Due to its widespread application and existing shortcomings in the current methodologies, novel tools are being continuously developed. This chapter is written in order to get the readers acquainted with some recent selected findings in the QSAR concepts and techniques emerged from 2006 onwards.

We have organized these emerging concepts in QSAR methodologies under the broad categories of 'novel descriptors', 'emerging concepts in chemometric techniques' and 'development in validation and applicability domain of QSAR models'. The discussed novel descriptors will include topological maximum cross correlation (TMACC) descriptor, hybrid ultrafast shape descriptor, substructure-pair descriptor (SPAD), ligand-receptor interaction fingerprint (LiRIf), extended topochemical atom (ETA) descriptors etc. The methodology section was divided into novel variable selection and novel QSAR model development subtopics. The novel method for optimal selection of variables includes gravitational search algorithm (GSA), random replacement method, combination of modified particle swarm optimization (MPSO) and partial least squares (PLS), advanced replacement method (RM) and enhanced replacement methods (ERM). The novel QSAR model development methods comprise different multi-task support

* Corresponding Author: kunalroy_in@yahoo.com, kroy@pharma.jdvu.ac.in

vector regression (SVR) algorithms, combination of R-group signatures and the SVM algorithm, multi-stage adaptive regression, Comparative Occupancy Analysis (CoOAn) etc. The section of new developments in validation and applicability of QSAR will include different expression of predictive squared correlation coefficient (Q^2), r_m^2 and ${}^cR_p^2$ metrics, Concordance Correlation Coefficient (*CCC*) etc. for a stricter test of validation of regression-based QSAR models, and finally a novel *k*-nearest neighbor's method and model disturbance index (MDI) based method to assess applicability domain of a QSAR model. The information provided in this chapter on the recent developments in the QSAR field would be beneficial to the QSAR researchers and chemoinformaticians.

INTRODUCTION

Quantitative structure–activity relationship (QSAR) is one of the major chemoinformatic techniques, which have found its application and gained popularity in different fields of research. These areas include medicinal, agricultural and environmental (toxicity prediction), along with the emerging area of nanoparticle toxicity prediction (nano-QSAR). One of the main reasons for gaining the applicability of QSAR over the other computational method in varied areas is due the fact that it utilizes the experimental data for deriving the mathematical models, and these models could be utilized for the reliable prediction prior to the actual experimentation.

QSAR modeling deals with the development of a quantitative relationship between an experimental activity/property and structural features of a molecule. The approach depends on being able to represent the structure of a molecule in quantitative terms (descriptors) and then to develop a relationship between the quantitative values representing a structure and corresponding experimental activity/property value. "*The molecular descriptor is the final result of a logic and mathematical procedure which transforms chemical information encoded within a symbolic representation of a molecule into a useful number or the result of some standardized experiment.*" [1]

The QSAR technology has gone through profound changes in the last decade. There are various areas in the QSAR technique where these changes have occurred. These areas comprise of novel descriptors, novel methodologies, and novel validation parameters including methods defining applicability domain of QSAR models. These changes were essential to resolve some existing shortcomings of QSAR techniques. In this chapter, we have reviewed and discussed the advances in QSAR field embracing each mentioned area separately.

NOVEL DESCRIPTORS

Till date a wide range of descriptors have been reported and these can be computed using various available software tools [2-4]. Based on the information that can be extracted, descriptors can be classified into 0D (molecular formula information like molecular weight, number of atoms, atom types, sum of atomic properties etc.), 1D (global molecular properties like pK_a, solubility, logP, functional groups etc.), 2D (structural patterns like connectivity indices, wiener index), 3D (steric and electrostatic field etc.).

Developing a Novel Descriptor

Several QSAR scientists are always interested in developing new descriptors rather than just using the existing ones, especially while working on unique datasets, e.g., toxicity of nano- materials [5]. The basic requirements to develop a new descriptor are mentioned here:

a) *Invariance property*: It signifies the ability of an algorithm for their calculation to provide a descriptor value that is independent of the specific characteristics of the molecular representation, such as atom numbering, molecular conformations, spatial reference frame, etc.
b) *Degeneracy*: It implies the ability of a descriptor to avoid equal values for different molecules. So, the descriptors can be classified into, no degeneracy at all, low, intermediate, or high degeneracy categories. For instance, the number of atoms in a molecule and the molecular weights are high degeneracy descriptors, while, generally 3D-descriptors show low or no degeneracy at all.
c) *An unambiguous algorithm*: A molecular descriptor must be defined by a computable mathematical expression whose terms have to be non-ambiguous and clearly obtainable from the molecular structure.
d) *An acceptable numerical range*: The values of molecular descriptors must be in a suitable numerical range, avoiding singular points and values as 10^{13} or 10^{-9}.

The following general characteristics are desired for a descriptor as suggested by Randic [6]:

1. Simple
2. A structural interpreter
3. Shows a good correlation with at least one property
4. Preferably discriminate among isomers
5. Possible to generalize to “higher” descriptors
6. Not based on experimental properties
7. Is not trivially related to other descriptors
8. Is possible to construct efficiently
9. Should use familiar structural concepts
10. Change gradually with gradual change in structures

In this section, we will discuss some novel descriptors, which emerged since 2007.

Topological MAximum Cross-Correlation (TMACC) Descriptors

Autocorrelation descriptors are one of the important descriptors in 2D-QSAR. Melville et al. [7] investigated these descriptors, while focusing on whether the grid independent descriptors (GRIND) approach can be used in 2D-QSAR and the result of this approach is TMACC descriptors.

Melville et al. replaced the force field interactions measured on a grid with atomic physicochemical values, and the Euclidean 3D distance is replaced with a topological distance based on the shortest bond distance between atoms.

The first step of calculating the TMACC descriptor is to assign a set of numerical values to each atom, consisting of interested physicochemical properties. In their study, four atomic properties were used, e.g., Gasteiger partial charges to represent electrostatics, Crippen-Wildman molar refractivity parameters to represent steric properties and polarizability, Crippen-Wildman log*P* parameters representing hydrophobicity, and the recently introduced log*S* parameters representing solubility. While employing the log*S* descriptors, Melville et al. did some minor modifications to the SMARTS strings, in order to get the correct frequencies of atom type.

To take into account the different scales used by each set of atomic parameters, each contribution was scaled by the largest absolute value, so that the positive and negative values took maximum values of +1 and -1, respectively.

For the Gasteiger partial charges, they took maximum values for positive and negative charges from the "fragment like" subset of the ZINC database, consisting of 49,134 molecules, carrying out the calculation with Open Babel 2.0.0.

For all data sets, non-polar hydrogen atoms were treated implicitly, and their atomic value was added to the value of the heavy atom to which it was bonded. Polar hydrogen atoms were treated explicitly, like any other atom.

The standard equation (eq. 1) for calculating an autocorrelation descriptor, x_{ac}, is

$$x_{ac}(p,d) = \sum p_i p_j \quad \ldots \tag{1}$$

where, p is a property and d is a topological distance between atoms i and j, the shortest number of bonds between atoms. The sum is computed over all atom pairs that are separated by the distance, d.

The TMACC descriptor extends this equation in three ways. First, each atomic property that can take positive and negative values, each one is considered as a separate property; for example, partial charge was separated into a positive and negative charge property. Second, cross-correlation, as well as autocorrelation values were checked to allow the property type for atom i to differ from that calculated for atom j.

For instance, positive charge-positive charge interactions, positive charge-negative charge interaction and positive charge-negative log*S* interactions are considered. Third, like the GRIND descriptor, the maximum value calculated for any given distance was kept. The TMACC equation is therefore represented as eq. 2

$$x_{tmacc}(p,q,d) = \max(p_i q_j, q_i p_j) \quad \ldots \tag{2}$$

In Eq. (2), q is the property of the second atom. There are two terms to consider for each atom pair, because when $p \neq q$, $q_j p_i \neq p_j q_i$.

Comparison to an implementation of HQSAR demonstrates that the TMACC performs competitively, and, the latter is marginally superior. Adding different atom types can also easily extend the TMACC descriptor. The weakness of the TMACC descriptor is that it is currently insensitive to chirality. Also, the use of physicochemical cross-correlations means that identification of properties responsible for activity is difficult.

Hybrid Ultrafast Shape Descriptor

MACCS key descriptors are binary in nature, which encode the presence or absence of sub-structural fragments. Such descriptors have been reported to perform well in the domain of similarity-based virtual screening. In contrast, it has been observed that 3D methods as well as existing 2D methods do not perform well in terms of the number of actives retrieved during screening. Most 3D methods have performed less well since 3D descriptors have to deal with translational and rotational variance besides a large number of conformations. It was reported that one of the key features in discriminating active from inactive molecules is the molecular shape. However, the procedure of calculating molecular shape is a challenging task, and also time consuming. A recent shape descriptor proposed by Ballester and Richards [8] called Ultrafast Shape Recognition (USR) has shown to avoid the alignment problem, and is up to 1500 times faster to calculate than other current methodologies. This shape descriptor makes the assumption that a molecule's shape can be uniquely defined by the relative position of its atoms and that 3D shape can be characterized by 1D distribution.

Cannon et al. [9] have introduced a new hybrid descriptor which is a combination of the MACCS key 166 bit packed descriptor (binary in nature and encodes topological information) and Ultrafast Shape Recognition (USR) descriptor (based on 12 floating point numbers) that combines both 2D and 3D information. The hybrid descriptor can improve the virtual screening performance as suggested by Baber et al. [10]. In the USR descriptor, 4 additional floating-point numbers were added, which incorporates information concerning fourth moment of the inter-atomic distance distributions making the hybrid descriptor of total 182 components. Further, the descriptor efficiency was assessed using the Random Forest classifier to conduct a virtual screen and rank molecules taken from the WADA 2005 dataset and the National Cancer Institute (NCI) database based on their probability of being active. The hybrid descriptor's performance was assessed against the USR descriptor (with three moments), the USR descriptor with four and five moments (UF4, UF5) and the MACCS key descriptor on an external validation set and was found to be superior in terms of performance.

Topological Descriptors for Analyzing Biological Networks

Graph-theoretical methods are utilized to derive descriptors like topological descriptors, and are proven as powerful tools to perform biological network analysis. However, the most of the developed descriptors does not have the ability to take vertex- and edge-labels into account. Vertices in such biological networks represent proteins, transcription factors or metabolites etc., which are connected by edges representing interactions, concentrations or reactions, respectively. This feature is important to characterize biological networks like protein-protein interaction networks, transcriptional regulatory networks, and metabolic networks more meaningfully instead of only considering pure topological information.

Dehmer et al. [11] emphasized on analyzing a special type of biological networks, namely bio-chemical structures. In this study, first, they developed some novel information-theoretic descriptors having the ability to include vertex- and edge-labels when measuring the information content of a chemical structure and then investigate some useful properties thereof. Second, they applied the mentioned measures combined with other well-known descriptors to supervised machine learning methods for predicting Ames mutagenicity. Moreover, the influence of reported topological descriptors was investigated on the prediction performance of the underlying graph classification problem.

This study demonstrated that the application of entropic measures to molecules representing graphs is useful to characterize such structures significantly. It was observed that if one extends the measures for determining the structural information content of unlabeled graphs to labeled graphs, the uniqueness of the resulting indices is higher. Further development of such methods might be valuable and fruitful for solving problems within biological network analysis.

Substructure-Pair Descriptor (SPAD)

QSAR analysis is helpful for designing bioactive peptides and a descriptor for capturing various properties of peptides is essential for this. The atom pair holographic (APH) code has been designed for the description of peptides, and it represents peptides as the combination of 36 types of key atoms and their intermediate binding between two key atoms. Osoda and Miyano [12] have developed a novel descriptor, substructure-pair descriptor (SPAD), which captures different characteristics of peptides from APH. It was proposed that the combination of APH and SPAD might lead to better QSAR for peptides with many types of amino acid inductions. The substructure pair descriptor (SPAD) represents peptides as the combination of 49 types of key substructures and the sequence of amino acid residues between two substructures. The size of the key substructures is larger and the length of the sequence is longer than traditional descriptors. APH captures internal characters of an amino acid induction, while SPAD captures the relationship between two amino acid inductions. A comparison of scope of each descriptor shows that SPAD captures different properties from APH and has greater descriptive power than APH. For instance, similarity searches on C5a inhibitor data set and kinase inhibitor data set showed that order of inhibitors become three times higher by representing peptides with SPAD, respectively. Hence, it was proposed that SPAD [12] is a novel and powerful descriptor for studying SAR of various types of peptides.

Ligand–Receptor Interactions Fingerprint (LiRIf)

Rabal and Oyarzabal [13] described the development and implementation of a simple and robust method for representing biologically relevant chemical space (BRCS), independently of any reference space, and analyzing chemical structures accordingly. This led to development of a novel descriptor, ligand–receptor interaction fingerprint (LiRIf) that converts structural information into a one-dimensional string accounting for the plausible ligand–receptor interactions (LRI) as well as for topological information. Exploiting ligand–receptor interactions as a descriptor enables the clustering, profiling, and comparison of libraries of compounds from a computational and medicinal chemistry view. The process for describing and clustering fragments follows four main sequential stages: (1) assigning atom types, (2) analysis and characterization of the direct chemical environment of the cleavage site or attachment point, (3) exploration of the remainder of the fragment after excision, and (4) generation of a fingerprint (LiRIf) encoding all extracted features that account for ligand–receptor interactions.

The proposed methodology assists the following tasks, (i) definition of the BRCS, (ii) data organization from a LRI perspective, (iii) data visualization leading to an easy interpretation of trends in structure–activity relationships (SAR) by medicinal chemists, (iv) data analysis to enable the comparison in a reference independent space as well as the profiling and identification of relevant chemical features, and (v) data mining to help search for structures that contain key interactions or specific features.

Extended Topochemical Atom (ETA) Indices

Extended topochemical atom (ETA) indices developed by Roy et al. [14, 15], have been extensively applied for toxicity and ecotoxicity modeling in the field of QSAR. The extended topochemical atom (ETA) indices were introduced as an extension of topochemically arrived unique (TAU) parameters developed in the late 1980s [16]. The ETA scheme includes various basic parameters such as α (related to size or bulk), ε (related to the electronegativity of atoms) and β (related to electronic contribution). Some of the basic ETA indices have also been incorporated in the popular software package DRAGON (version 6) [17]. Recently, Roy et al. have derived additional novel indices and evaluated for modeling a range of fundamental physicochemical properties [14], including octanol–water partition coefficient (*log P*), water solubility (*ln S*) and molar refractivity (R_m). Additionally, it has been attempted to model different aromatic substituent constants, including hydrophobic substituent constant (π), molar refractivity (*MR*), Hammett electronic constants (σ_m and σ_p), using these newly derived topological parameters, together with first generation ETA parameters.

The second-generation ETA indices are summarized in Table 1 and these are well described in the literature [14].

Table 1. Second generation indices under the ETA Scheme

Sl. No.	Definition	Significance
1	$\Delta\alpha_A = \left\langle \frac{\sum\alpha - \left[\sum\alpha\right]_R}{N_V} \right\rangle$	A measure of count of non-hydrogen heteroatoms [N_V stands for total number of atoms excluding hydrogen's]
2	$\Delta\alpha_B = \left\langle \frac{\left[\sum\alpha\right]_R - \sum\alpha}{N_V} \right\rangle$	A measure of count of hydrogen bond acceptor atoms and/or polar surface area

Table 1. (Continued)

Sl. No.	Definition	Significance
3	$\varepsilon_1 = \frac{\sum \varepsilon}{N}$	A measure of electronegative atom count [N stands for total number of atoms including hydrogen's]
4	$\varepsilon_2 = \frac{\sum \varepsilon_{EH}}{N_V}$	A measure of electronegative atom count [EH stands for excluding hydrogen's]
5	$\varepsilon_3 = \frac{\left[\sum \varepsilon\right]_R}{N_R}$	[R stands for reference alkane]
6	$\varepsilon_4 = \frac{\left[\sum \varepsilon\right]_{SS}}{N_{SS}}$	[SS stands for saturated carbon skeleton]
7	$\varepsilon_5 = \frac{\sum \varepsilon_{EH} + \sum \varepsilon_{XH}}{N_V + N_{XH}}$	[XH stands for those hydrogen's which are connected to a heteroatom]
8	$\Delta\varepsilon_A = \varepsilon_1 - \varepsilon_3$	A measure of contribution of unsaturation and electronegative atom count
9	$\Delta\varepsilon_B = \varepsilon_1 - \varepsilon_4$	A measure of contribution of unsaturation
10	$\Delta\varepsilon_C = \varepsilon_3 - \varepsilon_4$	A measure of contribution of electronegativity
11	$\Delta\varepsilon_D = \varepsilon_2 - \varepsilon_5$	A measure of contribution of hydrogen bond donor atoms
12	$\psi = \frac{\alpha}{\varepsilon}$	A measure of hydrogen bonding propensity of the atoms
13	$\psi_1 = \frac{\sum \alpha}{\left[\sum \varepsilon\right]_{EH}} = \frac{\sum \alpha / N_V}{\varepsilon_2}$	A measure of hydrogen bonding propensity of the molecules and/or polar surface area
14	$\Delta\psi_A = \langle 0.714 - \psi_1 \rangle$	A measure of hydrogen bonding propensity of the molecules
15	$\Delta\psi_B = \langle \psi_1 - 0.714 \rangle$	A measure of hydrogen bonding propensity of the molecules
16	$\Delta\beta = \sum \beta_{ns} - \sum \beta_s$	A relative measure of relative unsaturation content
17	$\Delta\beta' = \frac{\Delta\beta}{N_V}$	A measure of relative unsaturation content
18	$\sum \beta_{ns(\delta)}$	A measure of lone electrons entering into resonance
19	$\sum \beta'_{ns(\delta)} = \frac{\sum \beta_{ns(\delta)}}{N_V}$	A measure of lone electrons entering into resonance

Superaugmented Eccentric Distance Sum Connectivity Indices

Four highly discriminating fourth-generation topological indices (TIs), termed as superaugmented eccentric distance sum connectivity indices, as well as their topochemical versions (denoted by ${}^{SED}\xi^{c}_{C1}$, ${}^{SED}\xi^{c}_{C2}$, ${}^{SED}\xi^{c}_{C3}$ and ${}^{SED}\xi^{c}_{C4}$), have been conceptualized in the study carried out by Gupta et al. [18]. The values of these indices for all possible structures with three, four, and five vertices containing one heteroatom were computed using an *in-house* computer program.

Superaugmented eccentric distance sum connectivity indices, ${}^{SED}\xi^{c}_{N}$, proposed in this study are defined as the inverse of the summation of quotients of the product of adjacent vertex degrees and the product of the squared distance sum and eccentricity of the concerned vertex for all vertices in a hydrogen-suppressed molecular graph. It can be expressed as follows (eq.3):

$$ {}^{SED}\xi^{C}_{N} = \left[\sum_{i=0}^{n} \frac{M_i}{E_i^N * S_i^2} \right]^{-1} \quad \ldots \qquad (3)$$

where, M_i is the product of degrees of all the vertices (v_j), adjacent to vertex i and can be easily obtained by multiplying all the non-zero row elements in augmentative adjacency matrix, E_i is the eccentricity, S_i is the distance sum of vertex i, and n is the number of vertices in the graph, and the N is equal to 1, 2, 3, 4 for superaugmented eccentric distance sum connectivity indices -1, -2, -3, -4, respectively.

Similarly, the topochemical version of superaugmented eccentric distance sum connectivity indices can be defined as the inverse of the summation of quotients of the product of adjacent vertex chemical degrees and the product of the squared chemical distance sum and chemical eccentricity of the concerned vertex for all vertices in a hydrogen-suppressed molecular graph.

It can be expressed as follows (eq.4):

$$ {}^{SED}\xi^{C}_{CN} = \left[\sum_{i=0}^{n} \frac{M_{ic}}{E_{iC}^N * S_{iC}^2} \right]^{-1} \quad \ldots \qquad (4)$$

where, M_{ic} is the product of chemical degrees of all the vertices (v_j), adjacent to vertex i and can be easily obtained by multiplying all the non-zero row elements in additive chemical adjacency matrix, E_{ic} is the chemical eccentricity, S_{ic} is the chemical distance sum of vertex i, and n is the number of vertices in the graph, and the N is equal to 1, 2, 3, 4 for superaugmented eccentric distance sum connectivity topochemical indices -1, -2, -3, -4, respectively (denoted by ${}^{SED}\xi^{c}_{C1}$, ${}^{SED}\xi^{c}_{C2}$, ${}^{SED}\xi^{c}_{C3}$ and ${}^{SED}\xi^{c}_{C4}$).

The superaugmented eccentric distance sum connectivity topochemical indices exhibited exceptionally high discriminating power, low degeneracy, and high sensitivity toward both the presence and the relative position of heteroatom(s) for all possible structures with five vertices containing at least one heteroatom.

Fast Frequent Subgraph Mining (FFSM)

Khashan et al. [19] presented a novel approach to generate fragment-based molecular descriptors. The molecules are represented by labeled undirected chemical graph. Fast frequent subgraph mining (FFSM) is used to find chemical fragments (sub-graphs) that occur in at least a subset of all molecules in a dataset. The collection of frequent sub- graphs (FSG) forms dataset-specific descriptors whose values for each molecule are defined by the number of times each frequent fragment occurs in this molecule.

Fragment based descriptors present an important class of molecular descriptors that have been used in both similarity searching and QSAR modeling. A common feature to all present methods is that chemical fragments are identified a priori; thus frequently the total number of such descriptors generated for a molecular dataset is exceedingly large and/or fragment descriptors are generic.

Khashan et al. have used a novel approach to generate fragment descriptor where unique chemical fragments are identified based only on the dataset under study. For any dataset, FFSM was employed to search for common FSG that are found in at least a subset of molecules in the dataset. Once these FSG are identified, the frequency of each substructure in each molecule in the dataset is calculated; thus each common FSG serves as a chemical descriptor type and the frequency becomes a descriptor value. The detailed description of the FSG mining approach is given in the literature [19].

Unlike holistic molecular descriptors that characterize properties of the entire molecule, fragment descriptors offer the advantage of model interpretation in terms of significant chemical functional groups that influence compounds' biological activity. Such explanation especially with respect to the differences between active and inactive molecules might provide useful guidance to medicinal chemists with respect to rational design of new biologically active chemical entities.

Emerging Concepts in Chemometric Techniques

We have categorized the novel concepts in QSAR methodology under two subheadings: *novel variable selection methods and novel QSAR modeling approaches.*

Novel Variable Selection Method

Behnam et al., 2013 [20] introduced the gravitational search algorithm (GSA) for the selection of novel features, and applied for the anticancer potency modeling of a set of imidazo[4,5-b] pyridine derivatives. The GSA method is a swarm-based optimization algorithm, which mimics the law of gravity and the motion and interactions of masses, to select the descriptors for QSAR modeling. They have also compared the performance of GSA with the genetic algorithm (GA) and found that GSA finds the global optimum faster than GA and also exhibits higher convergence rate. Thus, GSA method has certain advantages over the GA, which is a more established heuristic search algorithm.

Bahl et al., 2012 [21] developed the novel variable selection method (random replacement method) and applied for the multiple linear regression analysis on two model data sets. They have also extended the application of random replacement method (RRM) for the selection of basis functions in the spline regression, and this approach is named as random function approximation (RFA). The comparison of RFA model with that of multivariate adaptive regression splines and genetic function approximation showed that, there are improvement in the quality of the models based on the coefficient of determination, R^2 and Q^2_{Loo}.

Khajeh et al., 2012 [22] reported the novel and efficient variable (descriptor) selection strategy for selecting and building the optimal linear model for predicting the entropy of formation by using a structurally wide variety of organic compounds in the QSPR analysis. This method is a combination of "modified particle swarm optimization" (MPSO) and "partial least squares" (PLS) methods, and can also be used as an alternative variable selection technique in QSAR/QSPR studies.

Mercader et al., 2011 [23] developed the advanced replacement method (RM) and enhanced replacement method (ERM) method for the selection of optimal set of descriptors from a larger pool of variables, for the development of QSAR and QSPR models. They have proposed the three different alternatives for the initial steps involved in the RM and the ERM analysis, out of which one new alternative have shown the superior results in the selection of the optimum set of descriptors from the greater pool than the genetic algorithm. Both RM and ERM First Step Modification (RMfsm and ERMfsm) methods are more efficient than the older algorithms in terms of providing better statistical values and used lower computational demand.

Novel QSAR Modeling Approaches

Lingling et al., 2013 [24] reported a two step modeling approach to study the selectivity and activity of histone deacetylase inhibitors (HDACIs). They have performed the novel two-step hierarchical QSAR modeling work on the HDAC inhibitors, in which initially the HDACIs were assigned to a class (1 or 2) based on their activity using binary QSAR models and then their activity values were predicted using class-specific continuous QSAR models. As per this approach, the selectivity of HDAC inhibitor can be determined initially, followed by the next prediction of the IC_{50} values from the continuous models.

Rosenbaum et al., 2013 [25] presented the two different multi-task support vector regression (SVR) algorithms and their application on the multi-target QSAR models, in order to design a lead molecule in a multi-target drug design process. These two algorithms are top-down domain adaption multi-task (TDMT) SVR and graph-regularized multi-task (GRMT) SVR. This multi-task learning is a valuable approach for inferring the multi-target QSAR models for lead identification and optimization, and its application would be more beneficial if the knowledge can be transferred from a similar task with a lot of in-domain knowledge to a task with little in-domain knowledge.

Hongming et al., 2013 [26] developed the Free-Wilson like local QSAR models by combining R-group signatures and the SVM algorithm. They have used the R-group signatures as descriptors to build the nonlinear SVM models. The free Wilson QSAR analysis has the limitation of not predicting the activity of the compounds, which are outside the scope of developed model as defined by the training set R-groups. This methodology can make the predictions of compounds containing R- groups, which are not present in the training set also. Thus, this novel method overcomes the flaws associated with the Free-Wilson methodology, without sacrificing the interpretability of the developed model.

Wei et al., 2013 [27] utilized the novel regression method of multi-stage adaptive regression (MAR) to develop the QSAR model for predicting the activity of iNOS inhibitors. This method is based on the multiple regression analysis, polynomial regression and adaptive algorism. The obtained results revealed that the MAR model has a better predictive ability and more reliability as compared to the best multiple linear regression (BMLR) model.

Verma et al., 2012 [28] reported the simple and easily interpretable 3D-QSAR formalism based on the comparison of local occupancies of fragments of an aligned set of molecules in a 3D-grid space namely, Comparative Occupancy Analysis (CoOAn). This formalism extracts the crucial position-specific molecular features and correlates them quantitatively with their biological endpoints. The steps in formalism includes superimposition of the molecules, their hypothetical breakdown into constitutive fragments, construction of a 3D-grid around the aligned set of molecules, determination of the occupancies of different fragments/groups in the grid cells, and final correlation with the biological activities using a chemometric method.

Akyuz et al., 2012 [29] combined the electron conformational (EC) and genetic algorithm (GA) method (4D-QSAR) for identifying the pharmacophoric features and predicting the anti-HIV activity. The purpose of EC-GA method is not only to develop a correlation between the molecular descriptors and activity, but also to describe the pharmacophoric group using the conformational flexibility of the HEPT derivatives. In EC-GA method, which involves conformational and alignment freedom, the heavily populated conformers at the room temperature are taken into consideration by using the Boltzmann weighting, for pharmacophore identification and bioactivity predictions of the compounds.

Pissurlenkar et al., 2011 [30] introduced the formalism of ensemble QSAR (*e*QSAR). In this approach, the biological activity is correlated with the descriptors for a set of low energy conformers, rather the descriptor calculated for the single lowest energy conformation. This formalism is based on the assumption that chemical behavior in complex biological systems is context-dependent, and therefore a molecule can exist and interact in a variety of conformations. By using this approach, it is also possible to determine whether the structural changes have a favorable or unfavorable effect on the binding affinity.

Liu et al., 2011 [31] developed the multi-target QSAR model by using HIV-1 and HCV datasets jointly. They have applied the accelerated gradient descent algorithm of multi-task learning (MTL) framework for the development of QSAR model. This novel MTL-based method is more efficient in terms of both convergence speed and learning accuracy. Such type of methodology is useful in the design of inhibitors, which are acting on the multiple targets.

Shih et al., 2011 [32] proposed the first combination approach for combining the 3D-pharmacophore model, CoMFA, and CoMSIA models for predicting the inhibitory activity against B-Raf kinase, and have used the same training set for the model development. Initially the 3D-pharmacophore models were generated, and were used to align the diverse inhibitors structures for generating CoMFA, and CoMSIA models. This approach could be used to screen inhibitor database, optimize inhibitors and to identify novel inhibitors with significant potency.

Goulon et al., 2011 [33] discussed the application of a new machine learning approach called Graph Machines-based QSAR approach (descriptor independent QSAR method), for the prediction of the adsorption enthalpies of alkanes on zeolites. In this approach, the molecules are considered as the structured data and are represented by the graph. This method is different from the classical QSAR method in which the molecules are described by using vectors composed of descriptors. This novel approach avoids the use of computation and selection of descriptors, which is often a major issue in the QSAR application.

Hao et al., 2011 [34] developed the novel genetic algorithm-support vector machine (GA-SVM) hyphenated approach for the development of a QSAR model by using a series of $P2Y_{12}$ (members of the G-protein coupled receptor family) antagonists. The GA-SVM approach showed superior results in terms of their performance when compared with the other approaches like GA combined with partial least squares (G-PLS), random forest (GA-RF), and Gaussian process (GA-GP).

Manoharan et al., 2010 [35] proposed the concept of fragment-based QSAR approach (FB-QSAR). This hybrid approach incorporates both the essential elements of classical Free-Wilson model and Fujita-Ban model. They have validated the methodology of FB-QSAR on a dataset consisting of 52 Hydroxy ethylamine (HEA) inhibitors as anti-Alzheimer agents. Such novel approach offers a weighing advantage of fragment contribution for the enhanced activity.

Dong et al., 2010 [36] have presented a formalism of structure-based multimode QSAR (SBMM QSAR) to highlight the issue of encoding of structural information of the binding site into the QSAR models. In this formalism, the structure of target protein is used to characterize the small molecule ligand, and the issues of conformations and multiple poses of ligand are systematically treated with the modified Lukacova-Balaz scheme. This approach is mostly applicable to the QSAR analysis, when the receptor target protein structure is available.

Zhou et al., 2010 [37] have presented the methodology of combining particle swarm optimization algorithm (PSO) and genetic algorithm (GA) to select simultaneously the features subset and to optimize the kernel parameters of the support vector machine (SVM). The logic of combining both the method is to utilize full advantage of both the methods in optimizing features subset and kernel parameters. This method was evaluated by predicting the activities of four peptide datasets and protein structural class of one of the dataset.

Ajmani et al., 2009 [38] introduced the concept of Group-Based QSAR (G-QSAR) in which the fragment-based descriptor is correlated with the response variable. There are also other methods, which uses the information from the fragments, for the calculation of descriptors such as Free-Wilson approach, H-QSAR, 2-D and topological QSAR. But the methodology of G-QSAR differs from the other methods in two aspects. Firstly, the fragmentation of molecules is done with a set of predefined rules before the calculation of corresponding fragment-based descriptors. Secondly, the G-QSAR method also takes into the consideration of cross/interaction terms as a descriptors to account for the fragment interactions in the QSAR model, whereas the other methods does not considers the contribution of fragment interaction. This method provides a similar or better predictive QSAR model along the hints of site of improvement/modification.

Martins et al. 2009 [39] has presented a novel 4D-QSAR formalism known as LQTA-QSAR. This approach makes the use of molecular dynamics (MD) trajectories and topology information. It calculates the intermolecular interaction energies at each grid point by considering probes and all aligned conformations, which are resulting from MD simulations, and these interaction energies are considered as the descriptors and used for the QSAR analysis. This paradigm is a user-friendly computational method, which involves less calculation consumption time.

Mazanetz et al., 2009 [40] developed the new 3D-QSAR method based on the Active Site Pressurization (ASP) method, which is novel molecular dynamics methodology. This method helps in taking the effect of target flexibility in the computational studies of molecular recognition and binding. The structure-based drug design (SBDD) differs from ASP in which the SBDD often do not take into consideration of protein flexibility, but relies on the static crystal structure, whereas the ASP considers the protein flexibility.

Cui et al., 2009 [41] developed a QSAR model using the novel adaptive weighted least square support vector machine (AWLS-SVM) method. This method combines the outlier detection approach and the adaptive weight value for the training set sample, and was developed in order to eliminate the influence of unavoidable outliers in the training set. This method takes the advantage of robust 3σ principle and adaptive weight, to detect the marked outliers and eliminates the effect of un-marked outliers on the model performance, and develops a model with good precision.

Wendt et al., 2008 [42] proposed a novel procedure for the 3D-QSAR analyses namely Quantitative Series Enrichment Analysis (QSEA) based on the topomer technologies. For the conventional 3D-QSAR analysis, the compounds can be included in the analysis only when if it is possible to align the 3D conformations of their structures onto the 3D model, which may be either a pharmacophore or a receptor binding site. It is due to the fact that a 3D-QSAR analysis is 3D-alignment dependent, which is totally contrary with the 2D-QSAR analysis (alignment-independent QSAR).So, even if the activity data is available for the 3D-QSAR analyses, the model building may fail due to failure in the alignment process and due to the compromised SAR information. The proposed novel tool helps to overcome the limitation of 3D-QSAR, as in the new method there is no requirement of alignment either on the pharmacophore or the receptor-binding site.

Novel Validation Metrics

Among all the steps involved in QSAR model development, the validation of the models plays the key role for its subsequent application on new set of chemicals. A new European legislation on chemicals—REACH (Registration, Evaluation, Authorization and restriction of Chemicals) came into force in 2007, which deals with risk assessment of chemicals for their safe use, thus contributing to the human health and environment. This law allows and encourages the use of QSAR model predictions when the experimental data are not sufficiently available or as supplementary information, provided validity of the model is justified.

According to these OECD principles [43], a QSAR model should have: (1) a defined end point; (2) an unambiguous algorithm; (3) a defined domain of applicability; (4) appropriate measures for goodness-of-fit, robustness and predictivity, and (5) a mechanistic interpretation, if possible. Herein, we are going to discuss some recent developments in the validation metrics.

Different Expressions of Predicted Variance (Q^2)

Whenever a QSAR model is built, it is necessary to check the ability of the model to predict activity/property of new compounds, which are not used in the model. Validation of models by means of new compounds whose data is not involved in the model development process is generally referred to as external validation. R^2_{pred} (Q^2_{F1}) is one of the useful parameters utilized in external validation, which has been subjected to a lot of modifications in recent years.

The three different expressions for calculating the external Q^2 as discussed by Shi et al. [44], Schüürmann et al. [45] and Consonni et al. [46] are noted here.

The original version of R^2_{pred} (Q^2_{F1}) (Shi et al. [44]) is given by the following equation:

$$Q^2_{F1} = 1 - \frac{\sum (Y_{(obs(test)} - Y_{pred(test)})^2}{\sum (Y_{obs(test)} - \bar{Y})^2} \quad \ldots \quad (5)$$

where, $Y_{obs(test)}$ and $Y_{pred(test)}$ are the observed and predicted response values respectively for test set and $\bar{Y}$ is the mean observed response value of training set. The model is said to be acceptable if the value of Q^2_{F1} is greater than 0.5. This parameter takes values in a standardized range (i.e., less than or equal to 1) thus allowing easy comparison of different QSAR models in terms of the performance of fitting and predictive abilities.

Schüürmann et al., [45] via a mathematical proof demonstrated that Q^2_{F1} yields a systematic overestimation of the prediction capability that is triggered by the difference between the training and test set activity means. They proposed another expression for calculation of Q^2 based on prediction of test set compounds denoted by Q^2_{F2} as given in eq.6. They suggested that Q^2_{F2} provides a more reliable estimate of the external predictive ability and recommended that the OECD guidelines should be revised.

$$Q^2_{F2} = 1 - \frac{\sum (Y_{(obs(test)} - Y_{pred(test)})^2}{\sum (Y_{obs(test)} - \bar{Y}_{test})^2} \quad \ldots \quad (6)$$

Here, $\bar{Y}_{test}$ refers to the mean observed response of the test set compounds. Q^2_{F2} differs from Q^2_{F1} only for the mean used in the denominator.

Consonni et al. [46] further discussed these parameters and pointed that unlike Q^2_{F2}, Q^2_{F1} computes the external predictive ability by including the information from the training set defined in terms of $\bar{Y}$. In the expression of Q^2_{F2}, no information about the reference model is accounted, since $\bar{Y}_{test}$ only encodes information derived from the test/external set, and this information changes continuously on the basis of the compounds present in the test set. It was suggested that to assess the predictability, predictions for all the test compounds should be evaluated independently of test set composition, which can be arbitrary or dependent on the size and distribution of the new data. Further they proposed another parameter (Q^2_{F3}) for validation of a QSAR model. This parameter is defined according to the following (eq.7):

$$Q^2_{F3} = 1 - \frac{\left[\sum (Y_{(obs(test)} - Y_{pred(test)})^2\right] / n_{ext}}{\left[\sum (Y_{obs(train)} - \bar{Y}_{train})^2\right] / n_{tr}} \quad \ldots \tag{7}$$

In Eq. 7, n_{tr} refers to the number of compounds in the training set. Here the summation in the numerator deals with the external test set compounds while that in the denominator deals with the training set compounds. Considering that the number of test and training compounds are usually different, divisions by n_{ext} and n_{tr} make the two values comparable. However, the value of Q^2_{F3}, despite measuring the model predictive ability, is sensitive to training set data selection and tends to penalize models fitted to a very homogeneous dataset even if predictions are close to the truth.

Comparing the nature of Q^2 values calculated using three different functions, Consonni et al. concluded that Q^2_{F1} and Q^2_{F2} are based on the sum of squares of the external/test set referring to the training set mean and test set mean, respectively; Q^2_{F3} is instead based on the mean squares of the training set in order to be independent of the distribution of test compounds.

Q^2_{F1} and Q^2_{F2} were found to suffer from some drawbacks when the external test compounds are not uniformly distributed over the range of the training set. Q^2_{F3} appeared independent of the external compounds property distribution.

The r^2_m Metrics

Roy and coworkers [47, 48] have proposed a novel metric r^2_m as an additional validation parameter. This metric is calculated based on the correlations between observed and predicted values with (r^2) and without (r^2_0) intercept for the least squares regression lines as shown in the following equation (eq. 8):

$$r^2_m = r^2 \times \left(1 - \sqrt{r^2 - r^2_0}\right) \ldots \tag{8}$$

The metric r^2_m does not consider the differences between individual responses and the training set mean and thus avoid overestimation of the quality of predictions due to a wide response range. Initially, the r^2_m metric was used for the external validation using a test set, but later it was used also for the internal validation employing LOO-predicted values. Similarly, based on the predicted response values of both the training and test sets, values of $r^2_{m(overall)}$ could be calculated [48].

For the calculation of the r_m^2 metrics, one can plot the observed response values in the *y*-axis and predicted values in the *x*–axis. However, the opposite may also be done, which will lead to a different value of the r_m^2 metric (r'^2_m) unless the predictions are perfect, i.e., when there is no intercept in the least squares regression line correlating observed and predicted values. This is because of the fact that the correlation between the observed (*y*) and predicted (*x*) values is same to that between the predicted (*y*) and observed (*x*) values in presence of an intercept of the corresponding least squares regression lines. But, this is not true when the intercept is set to zero [48]. The r'^2_m metrics is expressed by the following (eq. 9):

$$r'^2_m = r^2 \times \left(1 - \sqrt{r^2 - r'^2_0}\right) \ldots \tag{9}$$

Squared correlation coefficient values between the observed and predicted values of the test set compounds with intercept (r^2) and without intercept (r_0^2), were calculated for calculation of r_m^2 metrics. Change of the axes gives the r'^2_0 value. When the observed values of the test set compounds (y-axis) are plotted against the predicted values of the compounds (x-axis) setting intercept to zero, the slope of the fitted line gives the value of *k* and interchanging the axes gives the value of *k*′. The underlying formulas are as follows for calculation of r_0^2, r'^2_0, *k* and *k*′.

$$r_0^2 = 1 - \frac{\sum (Y_{(obs} - k \times Y_{pred})^2}{\sum (Y_{obs} - \bar{Y}_{obs})^2} \ldots \tag{10}$$

$$r'^2_0 = 1 - \frac{\sum (Y_{pred} - k' \times Y_{obs})^2}{\sum (Y_{pred} - \bar{Y}_{pred})^2} \ldots \tag{11}$$

$$k = \frac{\sum (Y_{obs} \times Y_{pred})}{\sum (Y_{pred})^2} \ldots \tag{12}$$

$$k = \frac{\sum (Y_{obs} \times Y_{pred})}{\sum (Y_{obs})^2} \ldots \tag{13}$$

In cases where the Y-range is not very wide, either of the r_m^2 and r'^2_m metrics may penalize heavily the quality of predictions. However, the average of the r_m^2 and r'^2_m values of test compounds ($\overline{r^2_{m(test)}}$) appears to better reflect the quality of predictions than the original r_m^2 metrics [47]. In general, the difference between r_m^2 and r'^2_m values i.e., $\Delta r^2_{m(test)}$ should be low for good models. It has been shown that the value of $\Delta r^2_{m(test)}$ should preferably be lower than 0.2 provided that the value of $\overline{r^2_{m(test)}}$ is more than 0.5 [48]. Similarly, $\overline{r^2_{m(LOO)}}$ and $\Delta r^2_{m(LOO)}$ parameters can be used for the training set, and $\overline{r^2_{m(overall)}}$ and $\Delta r^2_{m(overall)}$ can be used for the overall set i.e., training plus test set [48].

The ${}^cR_p^2$ Metric

Y-Randomization is performed in order to ensure the robustness of the developed QSAR model. In this technique, the activity/property values of the training set compounds are randomly shuffled keeping the descriptor matrix unchanged and new random models are built based on the shuffled activity values. For an acceptable QSAR model, the average correlation coefficient (R_r) of randomized models should be less than the correlation coefficient (R) of non-randomized model. The extent of difference in the values of the mean squared correlation coefficients of the randomized (R_r^2) and that of the non-randomized (R^2) models is reflected in the value of R_p^2 [47] parameter. This parameter penalizes the model R^2 for a small difference between the values of the squared correlation coefficients of the non- randomized (R^2) and the randomized (R_r^2) models as per the following equation (eq. 14):

$$R_p^2 = R^2 \times \sqrt{R^2 - R_r^2} \quad \ldots \qquad (14)$$

The threshold value of R_p^2 is 0.5 and a QSAR model exceeding this stipulated value might be considered to be robust and not the outcome of mere chance only. However in an ideal case, the average value of R^2 for the randomized models should be zero, i.e., R_r^2 should be zero. Thus, in such a case the value of R_p^2 should be equal to the value of R^2 for the developed QSAR model. Thus, the corrected formula of R_p^2 (${}^cR_p^2$) [49] is given as (eq. 15)

$${}^cR_p^2 = R \times \sqrt{R^2 - R_r^2} \quad \ldots \qquad (15)$$

Concordance Correlation Coefficient (*CCC*)

Chirico et al. [50] proposed an external validation parameter *CCC* denoted as $\hat{\rho}_c$ (eq. 16), which is given below.

$$\hat{\rho}_c = \frac{2\sum_{i=1}^{n}(x_i - \bar{x})(y_i - \bar{y})}{\sum_{i=1}^{n}(x_i - \bar{x})^2 + \sum_{i=1}^{n}(y_i - \bar{y})^2 + n(\bar{x} - \bar{y})^2} \quad \ldots \quad (16)$$

where, x and y correspond to the abscissa and ordinate values of the graph plotting the experimental data values versus those values calculated using the model or vice versa, n is the number of compounds, and $\bar{x}$ and $\bar{y}$ correspond to the averages of abscissa and ordinate values, respectively.

CCC measures both precision (i.e., how far the observations are from the fitting line) and accuracy (i.e., how far the regression line deviates from the concordance line), thus any divergence of the regression line from the concordance line gives a value of *CCC* smaller than 1. The key point is that this result is obtained even if the Pearson's correlation coefficient equals 1. The greater simplicity and its independence of the axes disposition are the advantages over Golbraikh and Tropsha method and no training set information is involved, so it can be considered a true external validation measure. Studying on huge number of simulated dataset it was proposed that the *CCC* as a complementary or alternative measure for a QSAR model to be externally predictive.

The $r^2_{m(rank)}$ Metric

Different validation metrics like Q^2_{Fn} and r^2_m etc. represent the measures for predictive quality of a QSAR model. However, none of them provides any information regarding the rank-order predictions for the test set. Thus, Roy et al. [51] have introduced a new variant of the r^2_m metrics to incorporate the concept of ranking order predictions while calculating the common validation metrics originally using the Pearson's correlation coefficient-based algorithm. The ability of this new metric to perform the rank-order prediction was determined based on its application in judging the quality of predictions of regression — based QSAR/QSPR models for four different data sets.

Unlike other r^2_m metrics, the $r^2_{m(rank)}$ metric is calculated based on the correlation of the ranks obtained for the observed and the predicted response data. In this case, the observed and predicted response data of the molecules are ranked and the (Pearson's) correlation coefficients of the corresponding ranks are determined with ($r^2_{(rank)}$) and without intercept ($r^2_{0(rank)}$). The values of $r^2_{(rank)}$ and $r^2_{0(rank)}$ thus calculated based on the rank-order are used to determine the value of the $r^2_{m(rank)}$ metric.

The values of $r^2_{(rank)}$ and $r^2_{0(rank)}$ differ from each other based on the difference in ranking of the two variables. An ideal ranking where the observed and the predicted response data perfectly match with each other yields zero difference between the two values for each molecule and the $r^2_{m(rank)}$ metric thus calculated attains a value of unity. An increase in difference between the two rank-orders for different molecules is marked by a decrease in the value of this metric with the proposed threshold value for acceptance being 0.5.

Different validation metrics calculated in each case were compared for their ability to reflect the rank-order predictions based on their correlation with the conventional Spearman's rank correlation coefficient [52]. Based on the results of the sum of ranking differences analysis performed using the Spearman's rank correlation coefficient as the reference, it was observed that the $r^2_{m(rank)}$ metric exhibited the least difference in ranking from that of the reference metric. Thus, the close correlation of the $r^2_{m(rank)}$ metric with the Spearman's rank correlation coefficient inferred that the new metric could aptly perform the rank-order prediction for the test data set and can be utilized as an additional validation tool, besides the conventional metrics, for assessing the acceptability and predictive ability of a QSAR/QSPR model.

APPLICABILITY DOMAIN (AD)

The AD of a (Q)SAR is the physico-chemical, structural, or biological space, knowledge or information on which the training set of the model has been developed, and for which it is applicable to make predictions for the new compounds [53]. During QSAR model development, the applicability domain of the molecules plays a crucial role for estimating the uncertainty in the prediction of a particular molecule's property based on how similar it is to the molecules used to build the model (i.e., training set molecules).

There are four major classical approaches for defining the interpolation regions in a multivariate space: Range-based methods (descriptor ranges, principal components ranges etc.), Geometric methods, Distance based methods (Mahalanobis distance, Euclidean distance, City block distance, Leverage approach, Hotelling T^2 test) and Probability Density Distribution based methods. These approaches differ from one another in the algorithm used to characterize the AD within the descriptor space, for the reliable prediction of the model. The threshold for all kinds of distance methods and Hotelling T^2 is the largest distance between the training set data points and the center of the training data set. There are some new approaches for determining AD reported recently and these are discussed below [54].

A cluster-based approach to find applicability domain has been proposed by Stanforth et al. [55]. This method involves application of an intelligent version of the *k*-means clustering algorithm for modeling the training set, as a collection of clusters in the descriptor space, while the test compounds of each individual cluster are assigned a fuzzy membership from which an overall distance may be calculated.

Sahigara et al., 2012 [54] compared the results of all the different AD approaches mentioned above by using some selected datasets. They have concluded that different AD approaches has their own strengths and limitations and it depends on the model builder to select the most appropriate approach.

Sahigara et al., 2013 [56] also proposed the novel descriptor-based AD method, which uses the *k*-Nearest Neighbours (*k*NN) principle to define the AD of the QSAR model. The workflow of this approach includes three stages. The first stage of the approach involves a pattern defining of thresholds corresponding to the various training samples, which are later used for deriving the decision rule. The second stage involves deciding the criteria for the retainment of test set sample within the AD. The third stage takes into account the different model statistics and error terms for checking the reliability in the derived results. The features that characterize the proposed AD involve adaptability to local density of samples, kernel density estimators (KDE), low sensitivity to the smoothing parameter and versatility to implement various distances measures.

Recently, Yan et al., 2014 [57] proposed a model disturbance index (MDI) to define an applicability domain for QSAR modeling.

Conclusion

The chapter has highlighted various recent changes in the QSAR concepts emerged in terms of novel descriptors, novel methodologies and novel validation metrics including applicability domain of models. Several advances in QSAR concepts proposed in the last few years illustrate the great curiosity of the scientific society in this theoretical approach. The improvements in the current QSAR techniques are essential to recover the shortcomings of the classical QSAR techniques.

This overview will help the interested readers to understand the progress in QSAR techniques in the recent decade and the need of such advances in near future to make the QSAR technique more fruitful by eliminating all types of possible errors/limitations associated.

References

[1] Todeschini, R.; Consonni, V. Handbook of molecular descriptors: John Wiley Sons, 2008.

[2] Yap, C. W. PaDEL descriptor: An open source software to calculate molecular descriptors and fingerprints. *J. Comput. Chem*, 2011, *32*(7), 1466-1474.

[3] Guha, R. Chemistry Development Kit (CDK) descriptor calculator GUI (v 0.46). 2006.

[4] Cerius2 Version 4.10. *Accelrys, Inc, San Diego, CA*, 2005.

[5] Puzyn, T.; Gajewicz, A.; Leszczynska, D.; Leszczynski, J. Nanomaterials- the Next Great Challenge for Qsar Modelers. *Recent Advances in QSAR Studies*: Springer:383-409.

[6] Randic, M. Molecular bonding profiles. *J. Math. Chem.*, 1996, *19*(3), 375-392.

[7] Melville, J. L.; Hirst, J. D. *J. Chem. Inf. Model*, 2007, *47*(2), 626-634.

[8] Ballester, P. J.; Richards, W. G. *J. Comput. Chem.*, 2007, *28*(10), 1711-1723.

[9] Cannon, E. O.; Nigsch, F.; Mitchell, J. B. O. *Chem. Cent. J.*, 2008, *2*(1), 1-9.

[10] Baber, J. C.; Shirley, W. A.; Gao, Y.; Feher, M. *J. Chem. Inf. Model*, 2006, *46*(1), 277-288.

[11] Dehmer, M. M.; Barbarini, N. N.; Varmuza, K. K.; Graber, A. A. *BMC Struct. Biol.*, 2010, *10*(1), 18.

[12] Osoda, T.; Miyano, S. *J. Cheminform.*, 2011, *3*(1), 1-9.

[13] Rabal, O.; Oyarzabal, J. *J. Chem. Inf. Model*, 2012, *52*(5), 1086-1102.

[14] Roy, K.; Das, R. N. *SAR QSAR Environ. Res.*, 2011, *22*(5-6), 451-472.
[15] Roy, K.; Ghosh, G.. *J. Chem. Inf. Comput. Sc.i*, 2004, *44*, 559-567.
[16] Pal, D. K.; Sengupta, C.; De, A. U. *Indian J. Chem., Sect B: Org. Chem. Incl. Med. Chem.*, 1988, *27*(8), 734-739.
[17] Todeschini, R.; Consonni, V.; Mauri, A.; Pavan, M. DRAGON-Software for the calculation of molecular descriptors. *version 6*, 2004.
[18] Gupta, M.; Gupta, S.; Dureja, H.; Madan, A. K. *Chem. Biol. Drug Des.*, 2012, *79*(1), 38-52.
[19] Raed, K.; Weifan, Z.; Alexander, T. *Mol. Inf*, 2014, *33*(3), 201-215.
[20] Mohseni Bababdani, B.; Mousavi, M. *Chemometr. Intell. Lab*, 2013, *122*, 1-11.
[21] Bahl, J.; Ramamurthi, N.; Gunturi, S. B. *J. Cheminform*, 2012, *26*(3-4), 85-94.
[22] Khajeh, A.; Modarress, H.; Zeinoddini-Meymand, H. *J. Cheminform*, 2012, *26*(11-12), 598-603.
[23] Mercader, A. G.; Duchowicz, P. R.; Fernandez, F. M.; Castro, E. A. *J. Chem. Inf. Model*, 2011, *51*(7), 1575-1581.
[24] Zhao, L.; Xiang, Y.; Song, J.; Zhang, Z. *Bioorg Med Chem Lett*, 2013, *23*(4), 929-933.
[25] Rosenbaum, L.; Darr, A.; Bauer, M. R.; Boeckler, F. M.; Zell *J. Cheminform*, 2013, *5*(1), 33.
[26] Chen, H.; Carlsson, L.; Eriksson, M.; Varkonyi, P.; Norinder, U.; Nilsson, I. *J. Chem. Inf. Model*, 2013, *53*(6), 1324-1336.
[27] Long, W.; Xiang, J.; Wu, H.; Hu, W.; Zhang, X.; Jin, J., et al. *Chemometr. Intell. Lab*, 2013, *128*, 83-88.
[28] Verma, J.; Malde, A.; Khedkar, S.; Coutinho, E. *Mol. Inf.*, 2012, *31*(6-7), 431-442.
[29] Akyuz, L.; Saripinar, E.; Kaya, E.; Yanmaz, E. *SAR QSAR Environ. Res*, 2012, *23*(5-6), 409-433.
[30] Pissurlenkar, R. R. S.; Khedkar, V. M.; Iyer, R. P.; Coutinho, E. C. *J. Comput. Chem.*, 2011, *32*(10), 2204-2218.
[31] Liu, Q.; Zhou, H.; Liu, L.; Chen, X.; Zhu, R.; Cao, Z. *BMC Bioinformatics*, 2011, *12*(1), 294.
[32] Shih, K.-C.; Lin, C.-Y.; Zhou, J.; Chi, H.-C.; Chen, T.-S.; Wang, C.-C., et al. *J. Chem. Inf. Model.*, 2010, *51*(2), 398-407.
[33] Goulon, A.; Faraj, A.; Pirngruber, G.; Jacquin, M.; Porcheron, F.; Leflaive, P., et al. *Catal Today*, 2011, *159*(1), 74-83.
[34] Hao, M.; Li, Y.; Wang, Y.; Zhang, S. *Anal. Chim. Acta*, 2011, *690*(1), 53-63.
[35] Manoharan, P.; Vijayan, R. S. K.; Ghoshal, N. *J. Comput. Aided. Mol. Des.*, 2010, *24*(10), 843-864.
[36] Dong, X.; Ebalunode, J. O.; Cho, S. J.; Zheng, W. *J. Chem. Inf. Model*, 2010, *50*(2), 240-250.
[37] Zhou, X.; Li, Z.; Dai, Z.; Zou, X. *J. Mol. Graph Model*, 2010, *29*(2), 188-196.
[38] Ajmani, S.; Jadhav, K.; Kulkarni, S. A. *QSAR Comb Sci*, 2009, *28*(1), 36-51.
[39] Martins, J. P. A.; Barbosa, E. G.; Pasqualoto, K. F. M.; Ferreira, M. r. M. C. *J. Chem. Inf. Model.*, 2009, *49*(6), 1428-1436.
[40] Mazanetz, M. P.; Withers, I. M.; Laughton, C. A.; Fischer, P. M. *QSAR Comb. Sci.*, 2009, *28*(8), 878-884.
[41] Cui, W.; Yan, X. *Chemometr. Intell. Lab.*, 2009, *98*(2), 130-135.
[42] Wendt, B.; Cramer, R. D. *J. Comput. Aided. Mol. Des.*, 2008, *22*(8), 541-551.

[43] Organization for Economic Co-operation and Development. Guidance document on the validation of (quantitative) structure-activity relationship ((Q)SAR) models. OECD Series on Testing and Assessment 69. OECD Document ENV/JM/MONO. 2007, 55-65.
[44] Shi, L. M.; Fang, H.; Tong, W.; Wu, J.; Perkins, R.; Blair, R. M., et al. *J. Chem. Inf. Comput. Sci.*, 2001, *41*(1), 186-195.
[45] Schuurmann, G.; Ebert, R.-U.; Chen, J.; Wang, B.; Kuhne, R. *J. Chem. Inf. Model.*, 2008, *48*(11), 2140-2145.
[46] Consonni, V.; Ballabio, D.; Todeschini, R. *J. Chem. Inf. Model*, 2009, *49*(7), 1669-1678.
[47] Pratim Roy, P.; Paul, S.; Mitra, I.; Roy, K. *Molecules*, 2009, *14*(5), 1660-1701.
[48] Ojha, P. K.; Mitra, I.; Das, R. N.; Roy, K. *Chemometr. Intell. Lab.*, 2011, *107*(1), 194-205.
[49] Mitra, I.; Saha, A.; Roy, K. *Molecular Simulation*, 2010, *36*(13), 1067-1079.
[50] Chirico, N.; Gramatica, P. *J. Chem. Inf. Model*, 2011, *51*(9), 2320-2335.
[51] Roy, K.; Mitra, I.; Ojha, P. K.; Kar, S.; Das, R. N.; Kabir, H. *Chemometr. Intell. Lab.*, 2012, *118*, 200-210.
[52] Caruso, J. C.; Cliff, N. *Educ. Psychol. Meas.*, 1997, *57*(4), 637-654.
[53] Jaworska, J.; Nikolova-Jeliazkova, N.; Aldenberg, T. *ATLA-NOTTINGHAM-*, 2005, *33*(5), 445.
[54] Sahigara, F.; Mansouri, K.; Ballabio, D.; Mauri, A.; Consonni, V.; Todeschini, R. *Molecules*, 2012, *17*(5), 4791-4810.
[55] Stanforth, R. W.; Kolossov, E.; Mirkin, B. *QSAR Comb. Sci.*, 2007, *26*(7), 837-844.
[56] Sahigara, F.; Ballabio, D.; Todeschini, R.; Consonni, V. *J. Cheminform.*, 2013, *5*, 27.
[57] Yan, J.; Zhu, W.; Kong, B.; Lu, H.; Yun, Y.; Huang, H.; Liang, Y. *Mol. Inform.*, 2014, *33*(8), 503-513.

In: Current Applications of Chemometrics
Editor: Mohammadreza Khanmohammadi

ISBN: 978-1-63463-117-4

Chapter 11

CHEMOMETRIC TREATMENT OF SERUM SPECTRA FOR OBTAINING INFORMATION ABOUT THE CLINICAL PARAMETERS

D. Perez-Guaita*[1], *J. Ventura-Gayete*[2], *C. Pérez-Rambla*[2], *M. Sancho-Andreu*[2], *S. Garrigues*[1] *and M. de la Guardia*[1]*

[1]Department of Analytical Chemistry, University of Valencia, Burjassot, Valencia, Spain

[2]University Hospital Doctor Peset Aleixandre, Valencia, Spain

ABSTRACT

The great amount of clinical analysis carried out daily in the hospitals justifies the attempts to develop fast screening tools in order to avoid the confirmation analysis of samples far from pathological values. Infrared (IR) is a promising technique on the clinical field that has been used for the determination of clinical parameters in sera. However, serum and related fluids are complex matrices which produce overlapped spectra and chemometric treatment is a necessary and critical step in order to extract as much as possible information. This chapter focuses on the extraction of the information about the concentration levels of clinical compounds in serum samples through chemometric treatment of IR spectra. To do it we have made a revision of the regression algorithms used for predicting the concentration of clinical parameters and evaluated the diagnostic capability of attenuated total reflectance (ATR) Fourier transform (FT)-IR measurements of serum samples for the detection of samples with normal values of target analytes using discrimination methods, such as linear discriminant analysis and partial least squares discriminant analysis.

Keywords: Clinical Diagnosis, Chemometrics, LDA, PLS-DA, Screening, Serum, Vibrational Spectroscopy

---

* Corresponding Author: miguel.delaguardia@uv.es

INTRODUCTION

The determination of different clinical parameters, such as cholesterol, creatinine, glucose, High Density Lipoproteins (HDL), Low Density Lipoproteins (LDL), urea, albumin, globulin and immunoglobulin is essential for clinical diagnosis of several illnesses and the monitoring of medical treatments. Those parameters are also good indicators of the general state of the patient, assessing the correct function of vital organs, such as liver and kidney [1] Because of that, a great amount of these determinations are carried out every day at clinical laboratories through enzymatic methods using high-sophisticated auto-analyzers. Unfortunately, these complex determinations imply the use of costly instrumentation and specific reagents, making the analysis expensive and avoiding their portability to the point-of-care testing.

So, a lot of studies have been made in order to propose alternatives to the commonly used enzymatic methods or to develop efficient screening tools. Among them one of the most promising techniques is vibrational spectroscopy, which provides a great amount of molecular information about the samples from a rapid and direct spectra acquisition [2, 3]. Infrared spectroscopy is nowadays widely used on the clinical field [4] and several methodologies have been proposed for the diagnosis [5] or determination of clinical parameters on biofluids [6]. Infrared (IR) provides several advantages over the classic clinical methods, being i) a versatile technique that can extract information from solid [7], liquid [8] and gas samples [9], covering all kind of samples including noninvasive measurements of thumbs [10] or microscopic imaging of tissues [11], ii) an excellent tool for the point-of-care analysis with compactable and portable instrumentation and iii) a cost-effective alternative to the reagent consuming enzymatic methods.

Unfortunately, regarding the analysis of biochemical parameters of serum, IR technique lacks from sensitivity and selectivity. Because of that, besides the classical acquisition of the serum spectrum by attenuated total reflectance (ATR) or transmission measurements of dry films, several research groups around the globe are currently introducing technical improvements on IR analysis of serum; including the use of quantum cascade lasers [12], the introduction of microfluidics [13] for preprocessing the samples or the measurements of ATR of dry-films from organic extracts [14]. The selectivity problem is normally faced with chemometrics, i.e., the extraction of the information of the composition of the serum through multivariate calibration methods [3]. Figure 1 shows an overview of the problem of serum analysis, evidencing the complex composition of serum samples and the resultant overlapped bands of its spectra in the mid infrared (MIR) and near (NIR) infrared range. As it can be seen, serum is a heterogeneous matrix mainly composed by water, which is normally subtracted by drying the sample or using a water blank as a background. From the non-aqueous part of the serum, the main compounds are proteins, whose bands dominate the spectra. Other compounds also present on the IR spectra provided overlapped bands on the 900-1300 cm^{-1} region, corresponding to the ν(C-O) and ν(P-O) stretching vibrations from saccharides and phospholipids respectively or the bands between 2800 and 3100 cm^{-1}, correspond to the ν(C-H) vibrations of mainly lipids.

As we have stressed before, the use of multivariate modelling is mandatory for the quantification of the clinical parameters from their overlapped bands on the IR spectra. In this chapter we have focused on the chemometric methods used for building those models, including a revision of the mainstream use of PLS regression methods and a new introduction of classification methods for the screening of samples with healthy parameter levels.

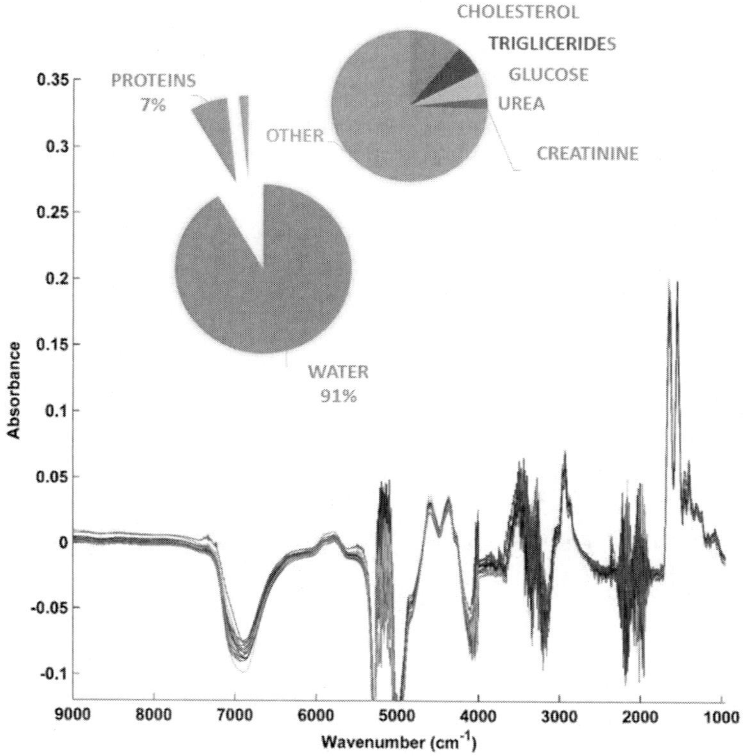

Figure 1. IR spectra and main approximate composition of sera samples. Note: Mid infrared (1000-4000 cm^{-1}) spectra were obtained using an attenuated total reflectance accessory, and near infrared spectra (4000-9000 cm^{-1}) were obtained using transmission mode.

Chemometric Modelling of Sera Samples

Chemometric modelling is normally performed through powerful software packages commercially available (e.g., *Matlab* from *Mathworks*, *Xlstat* from *Addinsoft* or *The Unscrambler* from *Camo*). Three steps are normally required, from the spectra acquisition until the end of the analytical procedure:

i) **Preprocessing**. The first step is the preprocessing of data in order to adequate them for the modelling, with the aim of removing externals sources of variation such as baseline shifts, pathlength changes, noise and other spectral artefacts related to the experimental procedure (atmospheric contributions, scattering or ATR correction). The pre-treatment of data can also select the appropriate variables or enhance the differences among spectra by the use of derivatives or mean centering of signals.

ii) **Modelling.** This is the main process and intends to build a mathematical relationship between the spectra matrix and the reference data. The reference data can be a concentration vector (regression) or a class vector (classification). Most of the algorithms available need the optimization of several parameters as the spectral region selected or other specific features of each method (e.g., number of latent variables (LV) in PLS). The selection of those conditions is critical and it is normally performed using internal parameters as the root mean square error of cross validation (RMSECV) and the information concerning the bands of the analyte and other interferences.

iii) **Validation**. Models need to be validated with an external set of samples (i.e., samples not included nor in the calibration of the model nor in the selection of the parameters of the model), especially when dealing with supervised methods, in order to obtain feasible conclusions about the relationship between spectral data and the composition of samples. The error measured on the validation is normally computed as the root mean square error of prediction (RMSEP) established from the regression methods or from the number of misclassified samples (NMC) in the case of sample classification, giving an estimation of the accuracy of the proposed methodology.

Regression Methods

In the same way than IR spectroscopy is used for the multicomponent determination on a wide range of samples [15], regression methods can be employed to extract as much as possible information from the IR spectra of serum thus making possible the simultaneous prediction of the concentration of some of the commonly required clinical parameters. Table 1 summarizes some of the works published for the prediction of clinical parameters from the vibrational spectra of serum and related fluids (whole blood and plasma). As it can be seen, spectra from both, MIR and NIR ranges, acquired from different kind of measurement modes can be treated using multivariate regression algorithms for the evaluation of several clinical parameters through the use of regression algorithms.

Partial Least Squares

PLS regressions establish a relationship between the predictive X matrix of spectra, and the predicted y-vector of parameter concentrations using least squares algorithms. In short, those algorithms extract a set of LV, which explain the sources of variation in the spectra correlated to the concentrations [33, 34]. In other words, the spectra are represented in the space of variables in order to reveal new directions that are linear combinations of the old predictive variables (wavenumbers), which present the best correlation to the concentration vector.

Considering the prediction of a concentration vector of n samples from their spectra containing j variables, the relationship between the concentration vector y (n x 1) and the spectra matrix X (n x j) can be described as indicated in equation 1:

$$y = Xb^T + e \qquad \text{(Equation 1)}$$

where b (1 x j) is the regression vector calculated for a specific number of LVs and e (n x 1), is the vector of residuals. PLS can be described as a powerful predictive version of principal component regression (PCR), because latent variable selection is performed according to the covariance matrix between spectral data and the investigated parameters.

The main drawback of PLS technique is the necessity of a robust calibration data set composed of representative spectra of the type of samples under analysis. This fact causes that the calibration procedure can be potentially time consuming and cost-extensive due to the need of several reference values.

Even though, Table 1 evidences that PLS is the most popular algorithm employed for the determination of clinical parameters on serum, plasma and whole blood. Normally the regions for the modelling of each analyte are selected taking into account the absorbance bands of the analyte [20] or by combining different regions and selecting those which offer the best root mean square error of prediction [24]. However, some studies use complex algorithms for the selection of the spectral ranges, based for example on the loading vectors [29]. The chemometric selection of the spectral ranges is specially performed on the NIR range, where the bands are strongly overlapped and it is difficult to assign specific spectral signatures to each considered compound. An example of those sophisticated methods for variable selection is the searching combination moving window partial least squares regression, which according to Kang et al. [30], "can select the optimized combination spectral region for each blood component successfully within the complex blood matrix over the low concentration range."

Locally Weighted (LW)-Partial Least Squares

In order to overcome a lack of linearity of the relationship between signals and analyte concentrations and to facilitate the selection of proper calibration sets, LW-PLS can be seen as a suitable approach. Local regression approximations are based on the use of specific calibration equations for each sample to be analyzed, using small calibration sets tailored to the unknown sample from a large library of samples.

We have evidenced the advantages and drawbacks of the application of LW-PLS to the direct determination of clinical compounds in human serum from the ATR-FTIR spectra [27].

Briefly, the LW-PLS approach consisted in four steps: (i) development of a principal component analysis (PCA) model at F components, (ii) computation of the Euclidean distance between the query and the calibration samples in the scores space, (iii) selection of the N nearest neighbors to the query and (iv) calculation of a local PLS regression employing the selected samples and LV latent variables.

Table 1. Published literature about determination of biochemical parameters in human serum plasma and blood based on the chemometric treatment of IR spectra

Calibration method	Matrix	Analyte	Sample Pre-processing	Measurement Mode	Ref.
PLS	Serum	HDL LDL	MIR	Transmission-Dry films	[16]
	Blood	GLU	MIR	Transmission	[17]
	Serum	TP, HSA, CHOL, HDL, LDL, TRY, GLU, URE	MIR	Transmission-Dry films	[6]
	Serum	GLU	MIR	ATR/Transmission Dry films	[18]
	Serum	CRE	MIR	Transmission-Dry films	[19]
	Whole blood	TP, ALB, CHOL, TRI GLU, URE	MIR	ATR	[20]
	Plasma	CRE	MIR	Transmission-Dry films	[13]
	Blood	GLU	MIR	Transmission	[21]
	Serum	HSA, IG, GLB, AGC	MIR	ATR	[22]
	Whole blood	HEM	MIR	ATR	
	Serum	TRI, GLU, LAC	MIR	Transmission-QCL	[12]
	Plasma	TP, HAS, CHOL,TRI, GLU, LAC	MIR	Transmission-QCL	[23]
	Serum	HSA, CHOL, HDL, LDL, GLU, URE	MIR	ATR	[24]
	Serum	HDL, LDL, CHOL, TRI	MIR	ATR-Dry films	[14]
	Serum	LDL	NIR	Transmission	[25]
	Serum	TP, HAS, TRI, CHOL, GLU, URE, HDL, LDL, VLDL	NIR	Transmission	[26]
LW-PLS	Serum	TP, TRI, GLU, URE	MIR	ATR	[27]
Iterative Spectra Subtraction	Plasma	16 different kinds of plasma proteins, GLU, LAC, URE	MIR	Transmission-Dry films	[28]
PLS Selection region based on loading vector	Serum	TP, HSA, GLB, CHOL, TRI, GLU,	MIR	Transmission	[29]
SCMWPLS	Serum	CHOL, GLU, URE	NIR	Transmission	[30]
PLS, MLR, ANN	Plasma	TRI	NIR	Transmission	[31]
SCMWPLS	Serum	ALB, IG	NIR	Transmission	[32]

Note: ALB, albumin; ANN, artificial neural network; CHOL, cholesterol; CRE, creatinine; GLB, globulin; GLU, glucose; HDL, high density lipoprotein; HEM, hemoglobin; IG, Immunoglobulin; ICR, independent component regression; Ig, immunoglobulin; LAC, lactate; LDL, low density lipoprotein; MIR, mid infrared; MLR, multiple linear regression; NIR, near infrared; PCR, principal component regression; PLS, partial least squares; SCMWPLS, searching moving window partial least squares; TRI, triglycerides; TP, total protein; UAC, uric acid; URE, urea.

The prediction of triglycerides, urea, total protein and glucose was investigated trough the modelling of the spectral data using PLS and LW-PLS model calculation and comparison of their respective prediction capabilities. Results evidenced that LW-PLS modelling improved between 5 and 30 % the prediction obtained using PLS.

Other Algorithms

Using an iterative algorithm developed by Petitbois et al. [28], spectra of dry films have been employed for the determination of numerous clinical parameters in plasma. The algorithm established a direct correlation of the spectra bands with concentration of the target analytes, selecting the analyte most correlated to one of the IR bands. The concentration of this analyte was determined by a simple univariate regression on this band and then a calculation of the contribution of this component to the spectra was established. This contribution was subtracted to the sample spectrum, and the remaining spectra was used for establishing the correlation for the determination of the next components considered one by one.

In addition, in a recent study Al-Mbaideen and Benaissa [25] have used independent component regression (ICR) for improving the results obtained for serum analysis by NIR-PLS. On the other hand, multiple linear regression (MLR) or artificial neural network (ANN) [31] are examples of other regression algorithms that can be used for the treatment of serum infrared spectra.

A direct comparison of the errors obtained by using different algorithms is difficult to perform because the calibration and validation sets employed on each study are different. In general only the main compounds are predicted with comparable accuracy to the tolerance of the clinical hospitals, and for example errors lower than 10% were found for PLS -ATR - FTIR prediction of glucose, urea, cholesterol, triglycerides, total proteins and albumin in whole blood [20]. However, the studies of our group have evidenced that using a large amount of samples coming from different kind of patients to build the calibration models, the quantification of minor compounds is dramatically affected by the high heterogeneity of samples and band-overlapping [24], thus being necessary to increase the number and variability of samples employed for calibration in order to minimize the prediction error.

Classification Methods

The general aim of the aforementioned works was to evaluate infrared spectroscopy as an alternative technique to the commercially available enzymatic methods by performing prediction models in order to obtain the concentration of as many as possible analytes in an as accurately as possible way. However, in our knowledge the use of ATR-FTIR as a screening tool in order to determine which serum or blood samples contains the parameters under routine study inside the so called normal values has not been evaluated in the scientific literature and it could be a key question for a clever use of vanguard fast analytical tools as IR-based ones. However, the high molecular information about the samples provided by the IR spectra has allowed that linear discriminant analysis (LDA) and partial least squares-discriminant analysis (PLS -DA) based on FTIR spectra can be used for the discrimination of healthy and pathological samples according to different illness as cancer and inflammation in gastric tissues [35, 36] or β-thalassemia in hemolisates [37]. In addition, it has been evidenced by the group of Khanmohammadi [38] that LDA of FTIR-ATR spectra of blood can discriminate between cancerous and healthy samples based on the bio-molecular changes in blood composition.

On using the appropriate chemometric tool it could be possible the evaluation of the capability of IR serum spectra for the classification of sera samples according to the "healthy" or "unhealthy" levels of their clinical parameters. The final goal is to provide a tool for avoiding the need of confirmation analysis of samples far from pathological values, which are the majority of those obtained in the case of primary care laboratories in our societies.

A method for the screening and diagnosis of healthy levels based on ATR -FTIR spectra of serum samples should take into account three important parameters: (i) the percentage of samples with normal values of the considered parameter classified as normal (true negatives), (ii) the percentage of samples classified as normal with values of the parameter out of the normal values (false negatives) and (iii) the concentration values of these false negatives. The first parameter informs about the effectiveness of the method to avoid the need of the confirmation analysis of samples with normal values and the other two parameters evaluate the associated risks of the method in sample screening. Since the evaluation of a sample as "healthy" is associated to parameter values between fixed concentration normal levels, it is expected that the classification of a continuous parameter in classes with arbitrary cut-offs can create grey zones of diagnosis in the proximities of the limit considered (see Figure 2) which could be the reason for false positive or false negative results. Thus, the benefits of the use of classification methods based on IR signals in a screening or diagnosis are strongly dependent on this grey zone. The narrower the grey zone was, the greater the amount of true negatives and the less the amount and the risk of false negatives.

Classification Algorithms Used

Two different classification methods, LDA and PLSDA have been used for the treatment of IR serum data. A brief explanation of the classification methods employed and the details of the samples used on the evaluation of serum parameters are detailed below. LDA is a traditional way of doing discriminant analysis introduced by R. Fisher [39]. It is similar to principal component analysis and it is also based on a decomposition of whole data in new directions, but instead of searching those that best describe the data, it searches for the vectors in the underlying space that best discriminate among classes. Therefore, given a number of independent variables which describe the observations, LDA creates a linear combination of the variables which yields the largest mean differences between the desired classes [40]. The general aim of the method is to maximize the between-class variance while minimizing the within-class variance. For the LDA study, routine samples obtained from clinical laboratories were used and reference data were obtained using an Abbott architect c16000 auto-analyzer (Libetrtyville, IL, U.S.A) based commercial methods described in the literature [24]. Samples were chosen in order to cover a large interval of different values of the considered analytes. 1500 samples were analyzed, including 750 samples from primary care patients, 550 samples from hospital patients and 100 samples from pre-dialysis patients. The range of concentration parameters for the samples under study were 1-4.7 g/dL for albumin, 45-545 mg/dL for cholesterol, 0.4-16.54 mg/dL for creatinine, 32-490 mg/dL for glucose, 5-119 mg/dL of HDL, 14-354 mg/dL for LDL, and 9-249 mg/dL for urea.

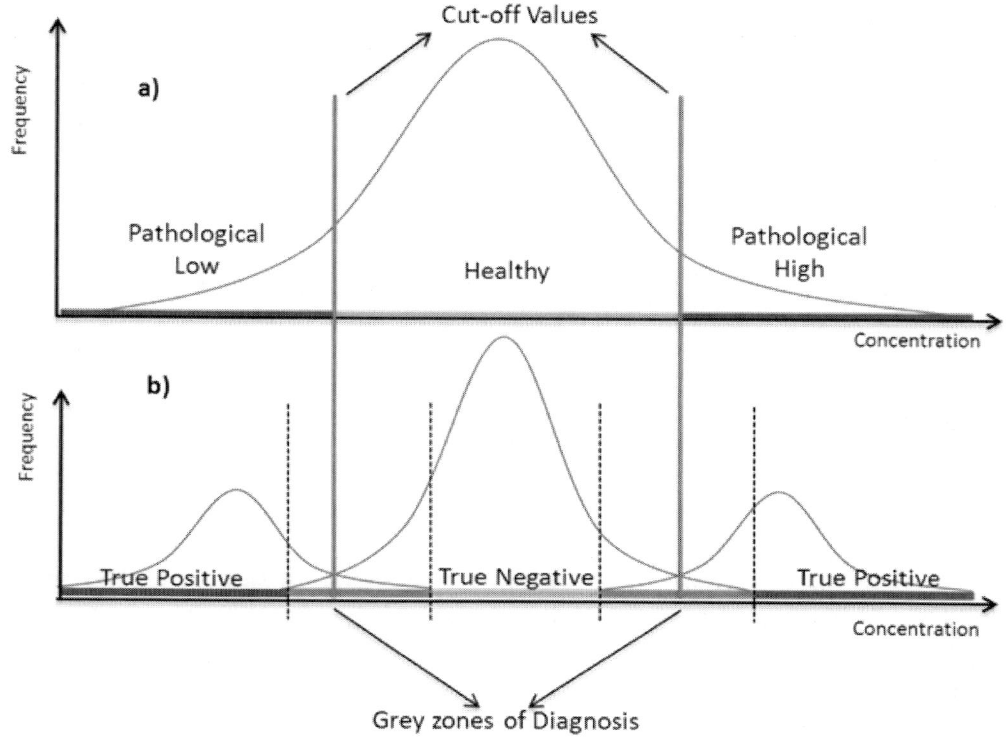

Figure 2. a) A priori distribution of sera parameter values according to the concentration of target analyte and b) reclassification expected taking into account the viability of pathological level samples.

PLS-DA is a variant of the aforementioned PLS algorithm used when the Y vector is categorical, replaced by the set of dummy variables describing the categories, expressing the class membership of the objects under study. Normally, a number is assigned to the y value of each class (Normally +1 and -1) and then a classical PLS regression is performed. For the classification of external samples, their spectra are introduced on the model, predicting an y value and calculating the probability to be in a particular class as a function of this y value. Similar to the PLS regression, latent variables are built to find a compromise between describing the variables spectra (X) and the prediction of the classes (Y). This technique is tailored to deal with a much larger number of predictors than observations and with multicollineality [41]. For PLS-DA modeling of protein concentrations in sera, 320 serum samples provided by the Protein Analysis Department of the Doctor Peset hospital, were considered [22]. Reference data were obtained using a Paragon capillary electrophoresis equipment from Beckman-Coulter Inc. (Brea, CA, USA). Samples were chosen randomly and covered a wide range of concentrations; 1.90-4.89 g/dL for albumin, 0.36-2.79 g/dL for immunoglobulin, 2.03-4.83 g/dL for total globulin.

Discriminant Capability of the ATR-FTIR Spectrum

It was observed that serum spectra differ as a function of the class where the sample was classified according to the concentration of target analytes. This effect was evidenced in the average spectra obtained for each class (see Figures 3 and 4).

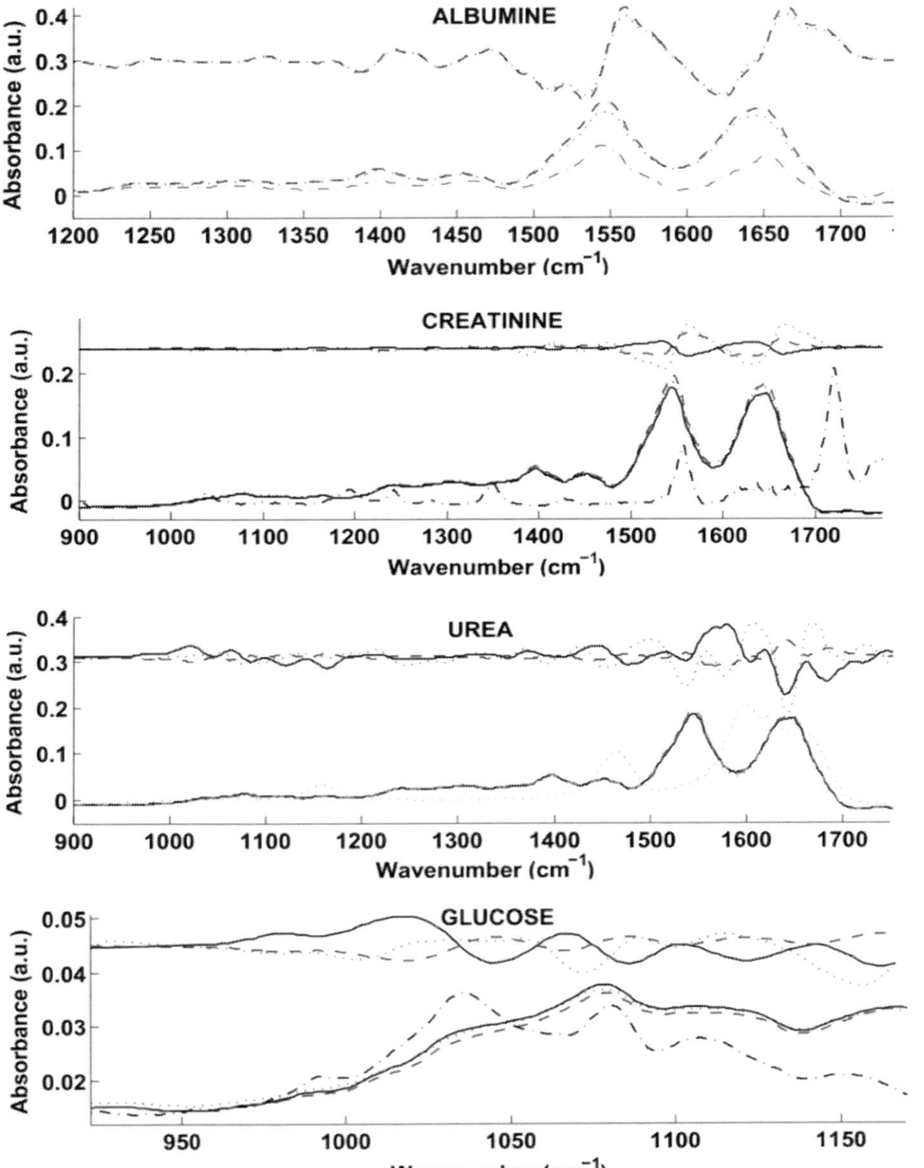

Figure 3. Mean spectra for samples with abnormally low values of the analyte (dotted line), abnormally high values of the analyte (solid line) or inside the normality values (dashed line) before (bottom) and after (top) preprocessing. Dash-dotted lines represent spectra of standards of 3 g/dL of an albumin suspension, 1 g/dL of glucose, 3 g/dL of urea and 0.3 g/dL of creatinine. Only the regions employed for model building are shown.

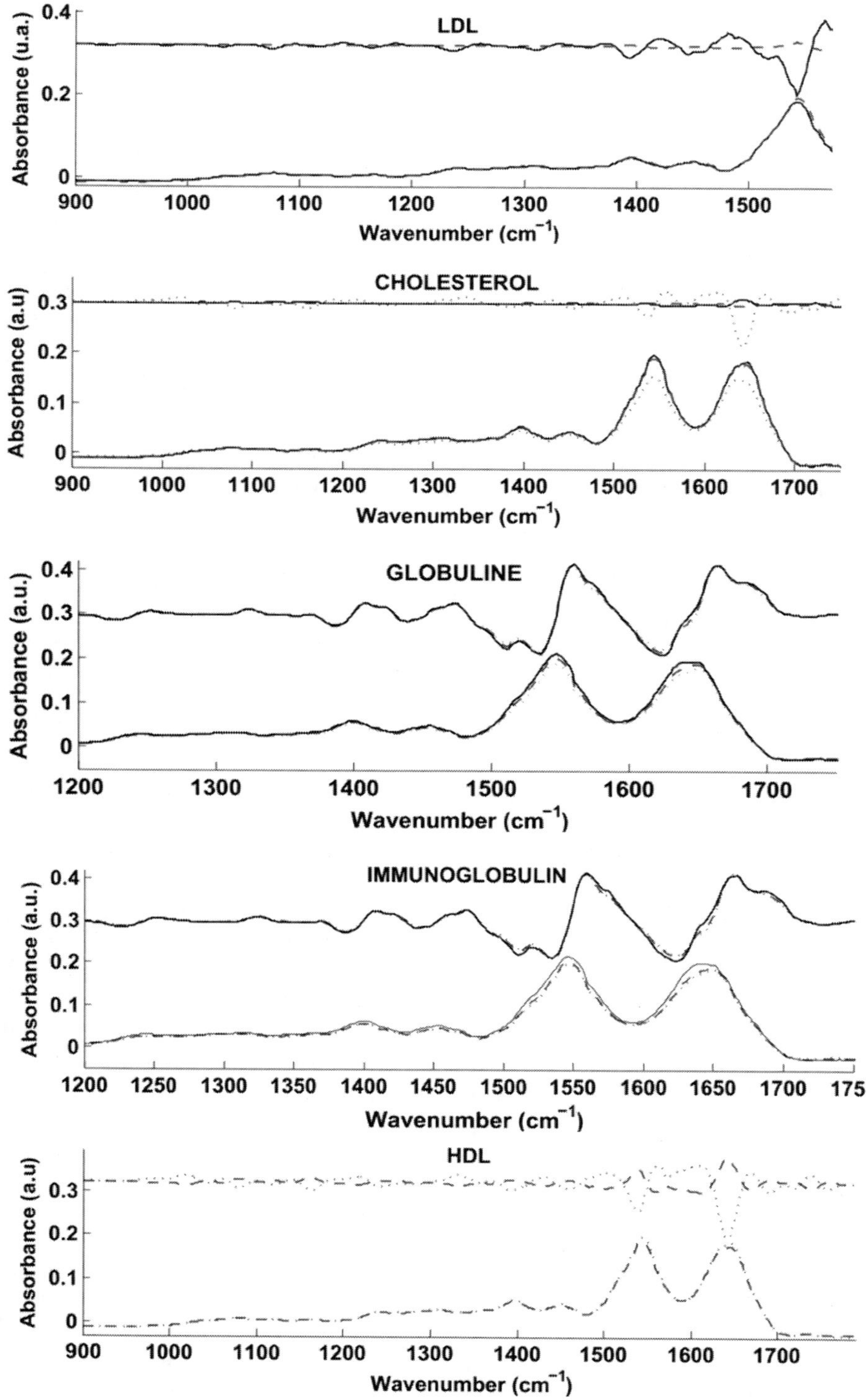

Figure 4. Mean spectra for samples with abnormally low values of the considered analyte (dotted line), abnormally high values of the considered analyte (solid line) or inside the normality values (dashed line) before (bottom) and after (top) preprocessing. Only the regions employed for discrimination model building are shown.

Although this visual differentiation was detected in the raw spectra only for major compounds as albumin, globulin, immunoglobulin and cholesterol, the application of a series of preprocessing treatments clarifies the differentiation among classes of samples. The differentiation was found to be important in the amide I and amide II bands (1500-1700 cm^{-1}) in all the cases. For analytes related to proteins; such as albumin, globulin and immunoglobulin or lipoproteins LDL and HDL, the influence of the aforementioned region in the discrimination between normal and abnormal values of these parameter concentrations should be logical, since the shape and position of these bands contain information about the secondary structure of the proteins present in the sample [42]. However, for the other considered analytes a strong differentiation in this wavenumber interval could be caused by a correlation of the target analyte with the proteins. Nevertheless, when comparing the average spectra of the different classes and the standard spectra measured for creatinine, urea and glucose, it was found strong differences in the absorbance regions of these analytes (see Figure 3). However, for cholesterol (see Figure 4), it is unclear to ensure that the differences in the average spectra are due to correlations of cholesterol content with other components of the sample.

From the aforementioned figures it can be concluded that the preprocessing of sera ATR-FTIR spectra strongly improves the clinical information provided by the spectra and contributes to a clear diagnosis of their content in target analysis.

LDA Classification

LDA treatment reclassified successfully between 71 and 83% of serum samples, depending on the analyte under study (see Table 2). For all the analytes for which three classes were considered, the first function could differentiate around 80-95% of the whole discrimination.

An example of the distribution of samples in the space generated by the discriminant functions is shown in Figure 5, which refers to the classification of samples based on their urea content. The first function discriminated among the three classes of samples whereas the second function differentiated samples with values inside the normal range and samples with abnormal values of urea. Samples with an abnormally high level of urea were reclassified very far for the other classes (see arrows in Figure 5); thus evidencing the manifested discrimination capability of the ATR-IR spectra of serum samples. For analytes with two classes, a Receiver Operator Characteristic (ROC) curve was built in order to evaluate the quality of the classification, providing areas under the curve around 0.87.

Tables 3 and 4 show the confusion matrix obtained in the evaluation of LDA models through cross validation. The percentage of samples well classified in all the cases was found to be between 70 % for creatinine and 79% for LDL. The best predictions were found for classes with a low number of samples such as samples with low values of urea, cholesterol and glucose (less than 100 samples), possibly due to the automatic weight correction applied.

As it was anticipated in Figure 2, it was observed that the proximate Gaussian distribution of analyte values for samples under study splits in two or three Gaussians corresponding to the different predicted classes.

Table 2. Parameters of LDA classification of sera samples from their ATR-FTIR spectra regarding different target analysis

Analyte	Preprocessings used[a]	Range (cm^{-1})	Variance Captured		Area under ROC Curve	% of Samples well reclassified
			F1 (%)	F2 (%)		
Cholesterol	FD/MSC	1776-1684 1591-864	96.36	3.66		83.3
Creatinine	FD	1816-1728 1641-935	84.35	15.74		71.0
Glucose	FD	1170-956 987-896	84.35	15.65		74.7
Urea	FD/SNV	1126-1681 1591-864	83.28	16.42		78.9
HDL	FD/SNV	1753-1664 1492-883			0.886	80.0
LDL	FD	1591-1451 1388-1291 1172-981			0.874	77.0

Notes: Preprocessing methods: FD (First derivate), MSC (Multiplicative scattering correction) and SNV (Standard normal variate). F1 and F2, considered functions and ROC: Receiver Operator Characteristic.

This effect is exemplified in Figure 6 which shows the distribution of the real values (Figure 6a) of urea and the distribution of the different classes provided by cross validation (Figure 6b, 6c and 6d) of the LDA models using the IR data. It can be observed that pathological classes of samples present a wide grey zone of diagnosis, being classified samples with urea values around 128 mg/dL as abnormally low and samples with values of 28 classified as abnormally high. On the other hand, the grey zone for samples with normal values was less wide that the other ones. From samples classified by the model as samples with urea values inside the normal range, only 12% corresponded to false negatives with abnormal values. This behavior was found to be similar for all the analytes, being the percentage of miss-classified samples always less than 18%.

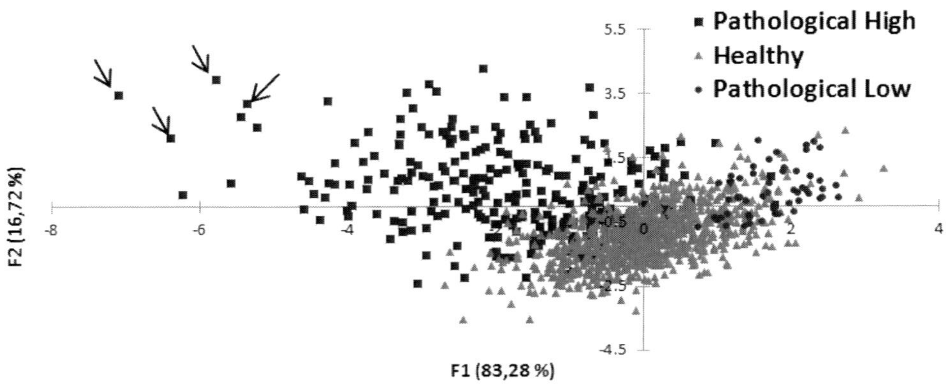

Figure 5. Canonical coordinates for LDA discrimination of sera samples based on urea values. Note: samples indicated with arrows correspond to those with concentrations from 193 till 249 mg/dL.

Table 3. Confusion matrix of Cross Validation for sample discrimination regarding analytes for which two levels were considered

	From/to	Normal	Low	Total	% Correct
HDL	Healthy	425	100	525	80.95%
	Low	60	177	237	74.68%
	Total	485	277	762	
	% Correct	87.63%	63.90%		79.00%
	From/to	High	Normal	Total	% Correct
LDL	High	163	43	206	79.13%
	Healthy	131	388	519	74.76%
	Total	294	431	725	
	% Correct	44.56%	90.02%		76.00%

Results obtained for sample classification according with sera ATR-FTIR spectra are comparable to those obtained by other screening studies for diagnosis based on LDA-FTIR spectra as the diagnosis of cancer in whole blood which raised 90-100% [38] of accuracy with a small number of samples or the detection of β-thalassemia in blood were made through dry films which could discriminate healthy and pathological samples [37]. The diagnostics of diverse illness in gastric tissues afforded between 66 and 90% of well classified samples using ATR the spectra of 100 samples [35, 36].

Table 4. Confusion matrix of Cross Validation for sample discrimination regarding analytes for which three levels were considered

Analyte	From/to	Low	Normal	High	Total	% Correct
	Low	10	0	0	10	100.00%
	Normal	5	534	174	713	74.89%
Cholesterol	High	0	80	263	343	76.68%
	Total	15	614	437	1066	
	% Correct	66.67%	86.97%	60.18%		75.70%
	Low	117	47	1	165	70.91%
	Normal	210	579	33	822	70.44%
Creatinine	High	18	36	146	200	73.00%
	Total	345	662	180	1187	
	% correct	33.91%	87.46%	81.11%		70.94%
	Low	22	2	0	24	91.67%
	Normal	102	580	96	778	74.55%
Glucose	High	22	112	244	378	64.55%
	Total	146	694	340	1180	
	% correct	15.07%	83.57%	71.76%		71.69%
	Low	42	9	0	51	82.35%
	Normal	136	680	31	847	80.28%
Urea	High	16	85	184	285	64.56%
	Total	194	774	215	1183	
	% correct	21.65%	87.86%	85.58%		76.58%

Evaluation of LDA as Screening Tool for the Detection of Samples with Parameter Values Inside the Normal Range

The comparison of LDA results obtained in the described study from those of other studies for the determination of biochemical parameters in serum is difficult because in this last work we have renounced to the quantitative information and we have just tried to detect pathological values of the analytes considered. In addition, this study uses a great number of samples in comparison with the regular size of the calibration sets used for other reported works (100-300 samples). However, taking into account prediction errors acquired by PLS models and errors obtained by LDA classification it can be concluded that LDA cannot compete with PLS quantification in extracting information from sample spectra about their content in glucose, urea, cholesterol, HDL and LDL. On the contrary, this fact is unclear in the case of creatinine because this parameter concerns a minor analyte whose low concentration does not allow its direct quantification by using FTIR spectra [20, 24]. New developments have been applied based on laminar fluid diffusion interface in order to isolate the creatinine for major compounds for its correct determination [13, 43]. However, in the classification study it has been evidenced that LDA can discriminate between samples with normal and abnormally high values of creatinine, being all the samples that contained more than 2.45mg/dL classified as samples with an abnormal high level.

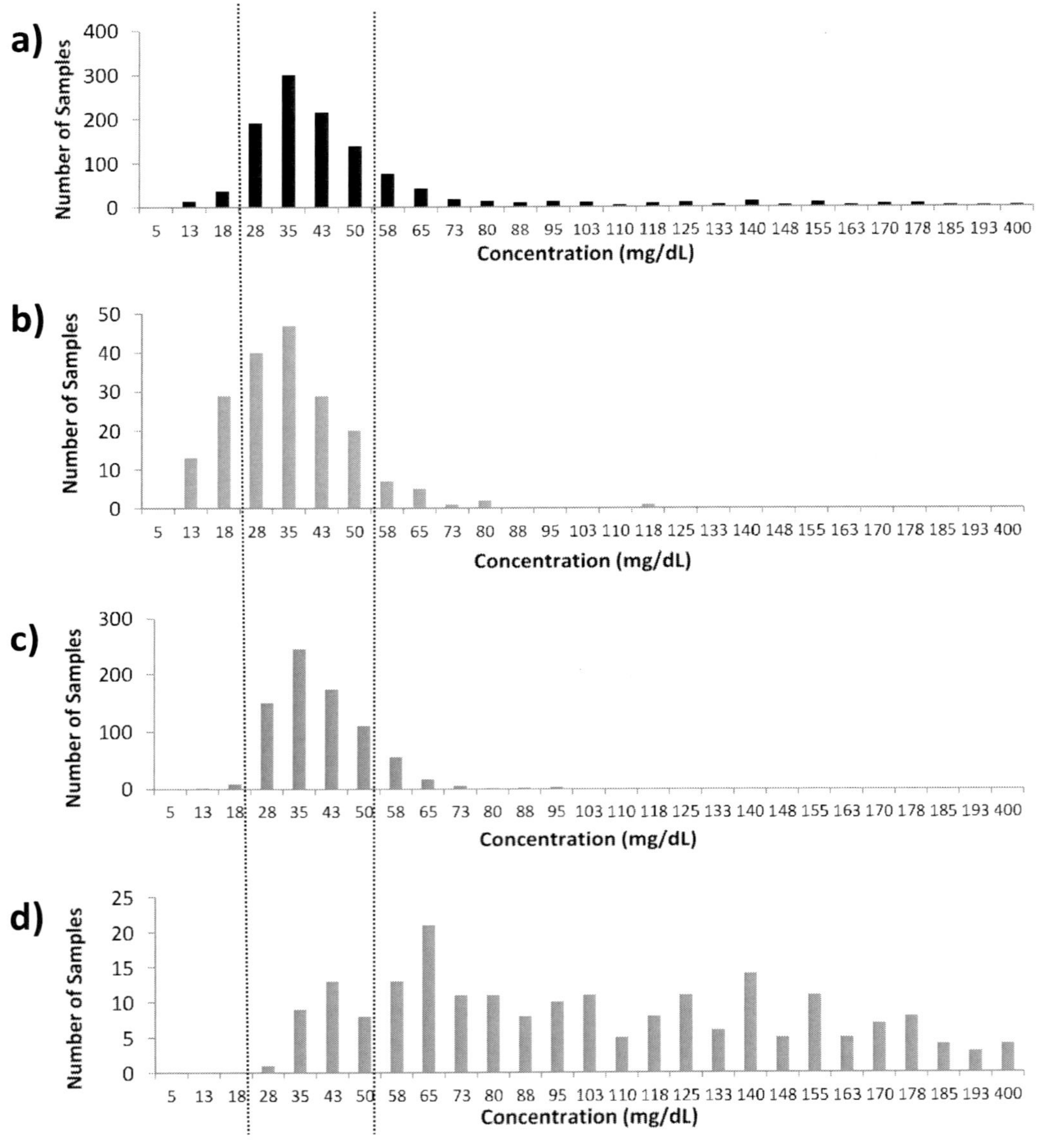

Figure 6. a) Distribution of the urea values of the samples under study, b) distribution of sample values classified as Pathologically Low by cross validation, c) distribution of sample values classified as Normal by cross validation and d) distribution of samples classified as Pathological High by the cross validation. Dashed lines indicate the limits of the classes.

Although the LDA-ATR-FTIR method cannot classify successfully all the samples in their corresponding classes, the classification made could be useful to identify the majority of serum samples with normal values. Table 5 summarizes results obtained for the analysis made with the aim of detecting samples with values inside the normal range and its exclusion of the confirmation analysis.

Table 5. Confirmation analysis saved and errors made considering the cross validation for LDA as a screening tool of abnormal concentration of sera parameters based on ATR -FTIR spectra

Analyte	Total Samples	Classified as Normal	Analysis saved (%)	False negative (%)	Cut-off Values (mg/dL)	False negative by excess		False negative by excess	
						N	X±S (mg/dL)	N’	X±S (mg/dL)
Cholesterol	1066	614	57.60	13.0	80-200	80	221±24	0	
Creatinine	1187	662	55.77	12.5	0.7-1.2	36	1.50±0.44	47	0.62±0.06
Glucose	1180	694	58.81	17.9	70-105	122	116±12	2	60±1
HDL	762	485	63.65	12.3	>40			60	35±4
LDL	725	431	59.45	9.9	<130	43	151±18		
Urea	1183	762	64.41	12.3	19-50	85	59±9	9	16±3

In all the cases the percentage of samples classified as normal and consequently the number of analysis saved was higher than 55 %. However, inside this group of samples classified as samples with normal values, which not need to be analyzed, we found between 10 and 18% of false negatives with concentration of the analytes outside the normal ranges. Nevertheless, the concentrations values of these false negatives were not far from the reference normal values, thus reducing the risk of a bad diagnosis.

PLS -DA of ATR -FTIR Spectra of Serum Samples According to Their Protein Contents

The parameters of the PLS-DA models built for each analyte in serum samples after classification based on the differences of ATR-FTIR spectra between samples with pathologically low and high values and samples with values inside the normal range are reported in Table 6. As the distribution of samples was different for each considered parameter, the number of samples of the calibration and validation sets was different in each case. The classification for the diagnosis of samples with pathological high values of albumin was not tested due to the absence of samples of this type. In Some cases the number of pathological samples was very low and the models were built with only 19 objects as in the class of gamma globulin high. However, for the main part of parameters, discrimination models were built with more than 25 samples, leaving approximately 200-300 samples for validation. In all the cases the number of optimal latent variables was found to be less than three.

The validation of PLS-DA models provided probabilities of samples to be classified in each considered class that can be plotted versus the actual concentration value of the samples (see Figure 7). In all the cases it was found a similar behavior; corresponding the lowest concentration values to samples with a minimum probability to be classified in the non-pathological group by PLS-DA. Around the normal value, which divides the samples between those with and without a normal content of globulin there was a reduced interval for which samples were not clearly classified. This thin diagnostic grey zone was found for all the parameters considered, being that the bottleneck of this classification technique and limiting its application as a screening tool.

In the case of PLS-DA, in order to identify samples with a normal value of each considered analyte and to avoid the need of a confirmation analysis, it must be selected a cut-off value of probability to classify the analytical level of the considered parameters of the sample as healthy, being samples with probabilities under this value considered as pathological ones. On increasing the cut-off value, it was obtained a reduced number of false negatives but also a reduced number of analyses saved. In this study we selected a cut-of value of 0.8, obtaining results summarized in Table 6.

In all cases, and taking into account an independent validation set of more than 200 samples, the number of samples undoubtedly classified by the method was found to be around 60%. Any false negative result was found except for the detection of low immunoglobulin levels, where less than 10 % of pathological samples were classified as healthy. So, it can be concluded that, additionally than to provide quantitative results for many of the parameters under study, the use of PLS-DA offers a good way to reduce the cost and time of diagnosis of different illness associated to proteins.

Table 6. Protein parameters evaluated and results obtained by PLS -DA discrimination of serum samples based on their ATR -FTIR spectra

Analyte (level)	Latent Variables	SET	Healthy Level	Pathological Level	%False Negatives	% Analysis Saved
Albumin (Low)	2	Calibration	32	31	0%	69%
		Validation	222	31	0%	67%
Globulin (High)	2	Calibration	15	19	0%	73%
		Validation	256	19	0%	70%
Globulin (Low)	2	Calibration	13	14	0%	85%
		Validation	278	12	0%	75%
Gamma (High)	1	Calibration	11	8	13%	100%
		Validation	298	4	0%	85%
Gamma (Low)	3	Calibration	14	12	0%	71%
		Validation	184	11	9%	59%

Note: Errors indicated for the calibration set correspond to cross-validation ones.

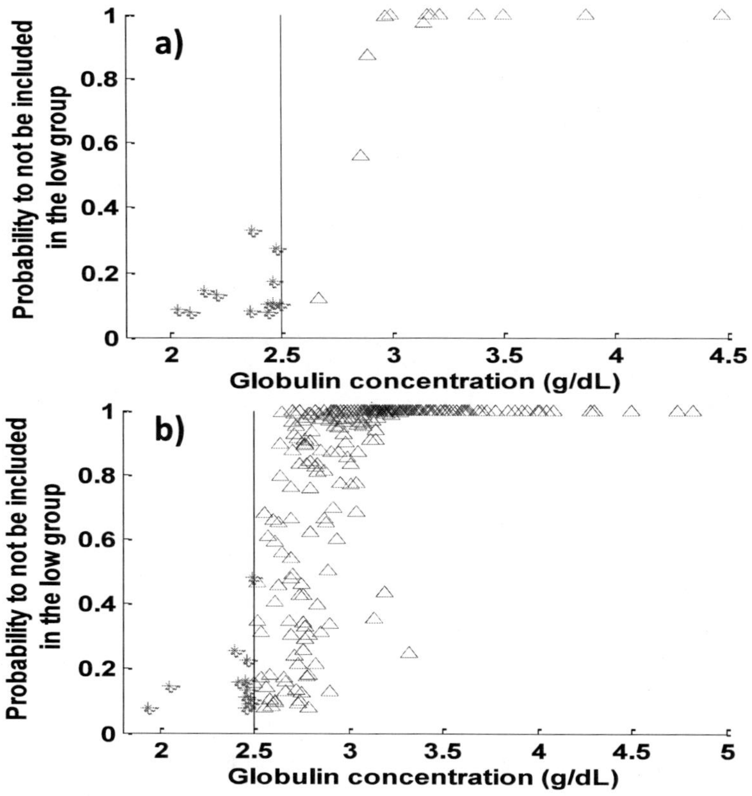

Figure 7. Probabilities to not be included in the concentration group obtained in the classification of serum samples as a function of the globulin concentration. Probabilities obtained by cross validation (a) and by prediction (b) are represented as a function of their globulin concentration. Blue triangles and red crosses represent samples above or below the low normal value (2.5 g/dL), respectively.

CHALLENGES ON THE CHEMOMETRIC TREATMENT OF SERUM SPECTRA

This chapter evaluates the current chemometrical methods in due use for extracting information from the IR spectra about the level of clinical parameters in sera. First it has been performed a revision of the current regression methods and then we have evaluated the use of discriminant chemometric models, LDA and PLS-DA, to evaluate ATR-FTIR spectra of sera in order to detect samples with normal values of different biochemical parameters. Results reported on literature evidences that ATR-FTIR is ready to compete with the reference methods. However, the most difficult task for convincing the clinical community is to perform clinical studies comparable to those performed for validating the reference methods, because the parameters used for evaluate those methods are normally based on univariate determinations. Thereby, the most challenging issue is to make compatible figures of merit from univariate and multivariate methodologies. For this purpose, it is anticipated than chemometric methodologies; such as net analytical signal [44] or science based calibration [45] could be a good starting point in multivariate method evaluation. The implementation of this methodology for the screening and diagnosis situations where an important amount of samples from healthy patients should be analyzed, as in the case of primary-care hospitals, could avoid a large amount of expensive analysis. The developed technique could also be employed to provide point-of-care tools in pharmacies, doctor's practice and in special situations, like in the third world and poor environments, in order to obtain direct, quick and cheap information about the patients.

AKNOWLEDGMENTS

Authors gratefully acknowledge the financial support of the Ministerio de Economía y Competitividad *and FEDER (Projects CTQ2011-25743 and CTQ2012-38635)*, the Generalitat Valenciana (Project *PROMETEO 2010-055)*. DPG acknowledges the "V Segles" grant provided by the University of Valencia to carry out this study.

REFERENCES

[1] Weatherby D, Ferguson S (2004) *Blood Chemistry and CBC Analysis*. Weatherby Associates, LLC.

[2] Wang J, Sowa M, Mantsch HH, Bittner A, Heise HM (1996) Comparison of different infrared measurement techniques in the clinical analysis of biofluids. *Trend Anal. Chem.* 15:286–296.

[3] Wang LQ, Mizaikoff B (2008) Application of multivariate data-analysis techniques to biomedical diagnostics based on mid-infrared spectroscopy. *Anal. Bioanal. Chem.* 391:1641–1654.

[4] Petrich W (2008) From Study Design to Data Analysis. In: Lasch P, Kneipp J (eds) *Biomedical Vibrational Spectroscopy*. John Wiley Sons, Inc., Hoboken, NJ, USA, pp 315–332.

[5] Diem M, Chalmers JM, Griffiths PR (eds) (2008) Vibrational spectroscopy for medical diagnosis. John Wiley Sons, Chichester, England; Hoboken, NJ.

[6] Rohleder D, Kocherscheidt G, Gerber K, Kiefer W, Kohler W, Mocks J, Petrich W (2005) Comparison of mid-infrared and Raman spectroscopy in the quantitative analysis of serum. *J. Biomed. Opt.* 10:031108.

[7] Zhang J, Wang GZ, Jiang N, Yang JW, Gu Y, Yang F (2010) Analysis of urinary calculi composition by infrared spectroscopy: a prospective study of 625 patients in eastern China. *Urol. Res.* 38:111–115.

[8] Schattka B, Alexander M, Ying SL, Man A, Shaw RA (2011) Metabolic Fingerprinting of Biofluids by Infrared Spectroscopy: Modeling and Optimization of Flow Rates for Laminar Fluid Diffusion Interface Sample Preconditioning. *Anal. Chem.* 83:555–562.

[9] Perez-Guaita D, Kokoric V, Wilk A, Garrigues S, Mizaikoff B (2014) Towards the determination of isoprene in human breath using substrate-integrated hollow waveguide mid-infrared sensors. *J. Breath Res.* 8:026003

[10] Sakudo A, Kato YH, Kuratsune H, Ikuta K (2009) Non-invasive prediction of hematocrit levels by portable visible and near-infrared spectrophotometer. *Clin. Chim. Acta* 408:123–127.

[11] Malek K, Wood BR, Bambery KR (2014) FTIR Imaging of Tissues: Techniques and Methods of Analysis. In: Baranska M (ed) *Optical Spectroscopy and Computational Methods in Biology and Medicine*. Springer Netherlands, pp 419–473.

[12] Brandstetter M, Volgger L, Genner A, Jungbauer C, Lendl B (2013) Direct determination of glucose, lactate and triglycerides in blood serum by a tunable quantum cascade laser-based mid-IR sensor. *Appl. Phys. B* 110:233–239.

[13] Shaw RA, Rigatto C, Reslerova M, Ying SL, Man A, Schattka B, Battrell CF, Matthewson J, Mansfield C (2009) Toward point-of-care diagnostic metabolic fingerprinting: quantification of plasma creatinine by infrared spectroscopy of microfluidic-preprocessed samples. *Analyst* 134:1224–1231.

[14] Perez-Guaita D, Sanchez-Illana A, Ventura-Gayete J, Garrigues S, de la Guardia M (2014) Chemometric determination of lipidic parameters in serum using ATR measurements of dry films of solvent extracts. *Analyst* 139:170–178.

[15] Moros J, Garrigues S, de la Guardia M (2010) Vibrational spectroscopy provides a green tool for multi-component analysis. *Trend Anal. Chem.* 29:578–591.

[16] Liu KZ, Shaw RA, Man A, Dembinski TC, Mantsch HH (2002) Reagent-free, simultaneous determination of serum cholesterol in HDL and LDL by infrared spectroscopy. *Clin. Chem.* 48:499–506.

[17] Shen YC, Davies AG, Linfield EH, Taday PF, Arnone DD, Elsey TS (2003) Determination of glucose concentration in whole blood using Fourier-transform infrared spectroscopy. *J. Biol. Phys.* 29:129–133.

[18] Diessel E, Kamphaus P, Grothe K, Kurte R, Damm U, Heise HM (2005) Nanoliter serum sample analysis by mid-infrared spectroscopy for minimally invasive blood-glucose monitoring. *Appl. Spectrosc.* 59:442–451.

[19] Mansfield CD, Man A, Shaw RA (2006) Integration of microfluidics with biomedical infrared spectroscopy for analytical and diagnostic metabolic profiling. *IEE Proceedings-Nanobiotechnology* 153:74–80.

[20] Hosafci G, Klein O, Oremek G, Mantele W (2007) Clinical chemistry without reagents? An infrared spectroscopic technique for determination of clinically relevant constituents of body fluids. *Anal. Bioanal. Chem.* 387:1815–1822.

[21] Vahlsing T, Damm U, Kondepati VR, Leonhardt S, Brendel MD, Wood BR, Heise HM (2010) Transmission infrared spectroscopy of whole blood - complications for quantitative analysis from leucocyte adhesion during continuous monitoring. *J. Biophotonics* 3:567–578.

[22] Perez-Guaita D, Ventura-Gayete J, Pérez-Rambla C, Sancho-Andreu M, Garrigues S, de la Guardia M (2012) Protein determination in serum and whole blood by attenuated total reflectance infrared spectroscopy. *Anal. Bioanal. Chem.* 404:649–56.

[23] Brandstetter M, Sumalowitsch T, Genner A, Posch A, Herwig C, Drolz A, Fuhrmann V, Perkmann T, Lendl B (2013) Reagent-free monitoring of multiple clinically relevant parameters in human blood plasma using a mid-infrared quantum cascade laser based sensor system. *Analyst* 138:4022-4028.

[24] Perez-Guaita D, Ventura-Gayete J, Pérez-Rambla C, Sancho-Andreu M, Garrigues S, de la Guardia M (2013) Evaluation of infrared spectroscopy as a screening tool for serum analysis Impact of the nature of samples included in the calibration set. *Microchem. J.* 106:202–211.

[25] Al-Mbaideen A, Benaissa M (2011) Determination of glucose concentration from NIR spectra using independent component regression. *Chemometr. Intell. Lab.* 105:131–135.

[26] García-García JL, Perez-Guaita D, de la Guardia M, Garrigues S (2014) Determination of biochemical parameters in human serum by near-infrared spectroscopy. *Anal. Meth.*. 6:3982-3989.

[27] Perez-Guaita D, Kuligowski J, Quintás G, Garrigues S, Guardia M de la (2013) Modified locally weighted-Partial least squares regression improving clinical predictions from infrared spectra of human serum samples. Talanta 107:368–375.

[28] Petibois C, Cazorla G, Cassaigne A, Deleris G (2001) Plasma protein contents determined by Fourier-transform infrared spectrometry. *Clin. Chem.* 47:730–738.

[29] Kim YJ, Yoon G (2002) Multicomponent assay for human serum using mid-infrared transmission spectroscopy based on component-optimized spectral region selected by a first loading vector analysis in partial least-squares regression. *Appl. Spectrosc.* 56:625–632.

[30] Kang N, Kasemsumran S, Woo YA, Kim HJ, Ozaki Y (2006) Optimization of informative spectral regions for the quantification of cholesterol, glucose and urea in control serum solutions using searching combination moving window partial least squares regression method with near infrared spectroscopy. *Chemometr. Intell. Lab.* 82:90–96.

[31] Da Costa PA, Poppi RJ (2001) Determination of triglycerides in human plasma using near-infrared spectroscopy and multivariate calibration methods. *Anal. Chim. Acta* 446:39–47.

[32] Kasemsumran S, Du YP, Murayama K, Huehne M, Ozaki Y (2004) Near-infrared spectroscopic determination of human serum albumin, gamma-globulin, and glucose in a control serum solution with searching combination moving window partial least squares. *Anal. Chim. Acta* 512:223–230.

[33] Trygg J, Lundstedt T (2007) Chemometrics Techniques for Metabonomics. In: J.C. Lindon, J.K. Nicholson and E. Holmes (eds) *The Handbook of Metabonomics and Metabolomics*. Elsevier Science B.V., Amsterdam, pp 171–199.

[34] Roggo Y, Chalus P, Maurer L, Lema-Martinez C, Edmond A, Jent N (2007) A review of near infrared spectroscopy and chemometrics in pharmaceutical technologies. *J. Pharm. Biomed. Anal.* 44:683–700.

[35] Fujioka N, Morimoto Y, Arai T, Kikuchi M (2004) Discrimination between normal and malignant human gastric tissues by Fourier transform infrared spectroscopy. *Canc. Detect. Prev.* 28:32–36.

[36] Li QB, Sun XJ, Xu YZ, Yang LM, Zhang YF, Weng SF, Shi JS, Wu JG (2005) Diagnosis of gastric inflammation and malignancy in endoscopic biopsies based on Fourier transform infrared spectroscopy. *Clin. Chem.* 51:346–350.

[37] Liu KZ, Tsang KS, Li CK, Shaw RA, Mantsch HH (2003) Infrared spectroscopic identification of beta-thalassemia. *Clin. Chem.* 49:1125–1132.

[38] Khanmohammadi M, Ansari MA, Garmarudi AB, Hassanzadeh G, Garoosi G (2007) Cancer diagnosis by discrimination between normal and malignant human blood samples using attenuated total reflectance-Fourier transform infrared spectroscopy. *Cancer Invest.* 25:397–404.

[39] Fisher RA (1936) The use of multiple measurements in taxonomic problems. *Annals of Eugenics* 7:179–188.

[40] Martinez AM, Kak AC (2001) PCA versus LDA. *IEEE Pattern Anal.* 23:228–233.

[41] Pérez-Enciso M, Tenenhaus M (2003) Prediction of clinical outcome with microarray data: a partial least squares discriminant analysis (PLS-DA) approach. *Hum. Genet.* 112:581–592.

[42] Barth A (2007) Infrared spectroscopy of proteins. *BBA-Bioenergetics* 1767:1073–1101.

[43] Shaw RA, Low-Ying S, Leroux M, Mantsch HH (2000) Toward reagent-free clinical analysis: Quantitation of urine urea, creatinine, and total protein from the mid-infrared spectra of dried urine films. *Clin. Chem.* 46:1493–1495

[44] Sarraguça MC, Lopes JA (2009) The use of net analyte signal (NAS) in near infrared spectroscopy pharmaceutical applications: Interpretability and figures of merit. *Anal. Chim. Acta* 642:179–185.

[45] Marbach R (2005) A new method for multivariate calibration. *J. Near Infrared Spec.* 13:241-254.

In: Current Applications of Chemometrics
Editor: Mohammadreza Khanmohammadi

ISBN: 978-1-63463-117-4

Chapter 12

MOST COMMON TECHNIQUES OF OUTLIER DETECTION

Amir Bagheri Garmarudi*, Keyvan Ghasemi, Faezeh Mozaffari, Mona Khorani and Mohammadreza Khanmohammadi

Chemistry Department, Faculty of Science,
Imam Khomeini International University, Qazvin, Iran

ABSTRACT

An outlier is an observation that appears to deviate markedly from other observations in the sample as to arouse suspicions that it was generated by a different mechanism. Outlier detection is known as one of the most important tasks in data analysis. The outliers describe the abnormal data behavior, i.e., data which are deviating from the natural data variability. Some other terminologies e.g., abnormalities, discordants, deviants, or anomalies are also commonly used in the data mining and statistics literature. Outliers are categorized in two sectors: some of them are unusual data and should be omitted by one of the outlier detection methods; however there are points which do not belong to bad data category clearly. These data points are due to random variation in measurements or may be scientifically interesting due to something informative inside them. Thus considering a suitable procedure to make decision about them is necessary. Anyway, it is not suggested to eliminate the data point without sufficient consideration and in any case, utilizing outlier detection methodologies would be helpful. This chapter would discuss the most popular methods of outlier detection in chemometrics and analytical chemistry with consideration of their trend and applications.

INTRODUCTION

In most chemical research activities, the data set is created via evaluation of one or more generating approaches or sample sets, which could either reflect the aim parameter in the system or observations collected about entities. When the target media which is considered as

* Corresponding Author: bagheri@sci.ikiu.ac.ir

the generating approach would demonstrate an unusual behavior, outliers are created. In spite of troubles with the outliers in a data set, it should be noticed that they often contain useful information about abnormal characteristics of the data set and entities, which impact the data generation approach. Outlier detection techniques are widely applied in data processing procedures. There are several applications for these techniques in chemical science. Development of novel biodiagnostic methods in medical science is an interesting trend for chemical science researchers. Spectral data collected from a variety of devices such as MRI and MRS may consist of some unusual patterns. Analytical and bio sensors are often used to track various environmental and chemical parameters in many real applications. Unexpected variations in the analysis condition and sudden changes in the underlying patterns may represent outlier outputs. Another field of interest for investigation of outliers is the environmental earth science, in which satellites and remote sensing tools are employed to collect the spatiotemporal data about weather patterns, climate changes, or land cover patterns. Outliers in these types of data sets provide significant insights about hidden environmental trends, which may have caused them [1-3].

Dealing with the spectrochemical data sets, differentiation between noises and outliers is of high importance. As shown in figure 1, data point A is supposed as an outlier in the left side plot. On the other hand, presence of random data points in the right side plot, disables the abalysts to directly decide about the situation of data point A. The initial proposal is to consider both A and B data points as noise. However, more precise evaluations are necessary to make a better decision.

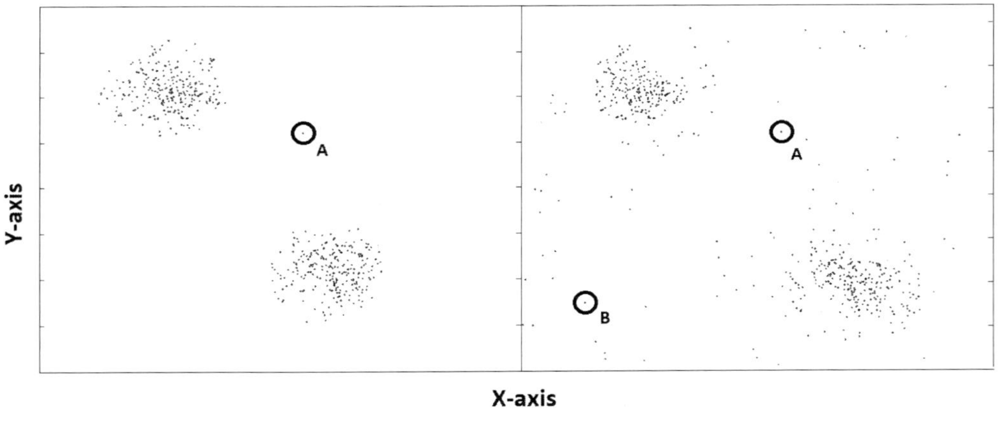

Figure 1. Difference between outlier and noise.

In the next sections, the general classification of outlier detection techniques will be indicated, describing the most common ones.

How Does an Outlier Detection Approach Work

In order the abnormal cases, most outlier detection methods would provide a quantitative output as score which determines the distance between the studied case and the overall dataset. This score would demonstrate the level abnormality for a data point. This score is a useful parameter by which the rank of all data points in terms of their outlier tendency would be determined. In spite of the noticeable capability for this score it cannot provide a concise definition for data points which should be considered as outliers. Another output for the outlier detection methods is a binary label which would indicate whether a data point is an outlier or not. For the techniques which are able to provide the binary labels directly, the outlier score for a data point is converted into binary labels by imposing thresholds on outlier scores, based on their statistical distribution. It is a useful result which is often needed for decision making concerning practical applications of outlier detection. Usually, outlier detection algorithms tend to create a model based on normal patterns in the dataset, and then determine the outlier score of a given data point, according to the deviations from the normal patterns. Normal behavior of a data points depends on the model which may be Gaussian, regression or proximity-based case. Considering the type of data model, different assumptions are made about the "normal" behavior of a given data point. The routine pocedure define the model is to employ an algorithmically strategy. For example, nearest neighbor-based algorithm, makes a model for the outlier tendency of given data points, based on distribution of their k-nearest neighbor distance. Thus, in this case, the assumption is that outliers are located at large distances from most of the data [1,4].

Most common outlier detection strategies are fundamentally classified in 3 categories:

Class A: Unsupervised

Unsupervised techniques determine the outliers with no prior knowledge of the dataset. The simple explanation of such technique is similar to clustering where the dataset is process as a static distribution, allocating the most remote points, which will be then flagged as potential outliers. These techniques assume that errors are separated from the 'normal' data and will thus appear as outliers (similar to left part of Figure 1). It is noticeable that the main cluster of data points may be subdivided if necessary into more clusters to enable classification goals (similar to right part of Figure 1). It is necessary to prepare the whole data before processing and that the data is static. However, in case of large data sets with good coverage, constructed model can compare new data points with the existing ones.

Class B: Supervised

Supervised techniques require pre-labelled data points, in which normal or abnormal cases are determined. The same as supervised classification methods, the area of each class is defined by the borderline of its members. The entire area outside whole classes represents the outlier cases. Classifiers are best suited to static data as the classification needs to be rebuilt from first principles if the data distribution shifts unless the system uses an incremental classifier such as an evolutionary neural network. Supervised outlier detection is useful for on-line classification, where the classifier learns the classification model and then classifies new imported cases. Considering the abovementioned procedure, if a newly imported case lies in a region of normality it is classified as normal, otherwise it is flagged as an outlier. Classification algorithms require good area coverage of both normal and abnormal data points. In the other words, the data set should cover the most possible distribution, enabling the classifier to globalize the model.

Class C: Semi-Supervised

By these techniques the normal class is taught but the algorithm learns to recognize abnormality. Also here it is necessary to define the initial classification but the model is trained only for normal data points. Semi-supervised techniques are suitable for static or dynamic data as they only get trained for a class which provides the model of normality. It is noticeable that these techniques would learn the model incrementally as new data points are added to the initial set, tuning the model to improve the fit as each new cases become available. The final aim is to define a boundary of normality. Most of the approached employed for outlier detection would map the whole data onto mono- or multi-type vectors which will comprise numeric and symbolic features, representing all types of data. The outliers are determined from the distance of vectors using some suitable measurement tools. Different approaches work better for different types of data. It is important to selecting an algorithm which can accurately model the data distribution and accurately highlight outlying points for a clustering, classification or recognition type technique. The desired algorithm should also be scaled to the studied data set. Also selecting a suitable non-trivial neighbourhood of interest for an outlier is an important task. Usually, some boundaries aredefined around normality during processing by the algorithm and autonomously induce a threshold. However, these approaches need the user to define the number of clusters.

Role of Data Model

As mentioned previously, the data model which may be Gaussian, regression or proximity-based would make different assumptions about the "normal" behavior of the data inside it. However, incorrect choice of data model may lead to unreliable results. The whole dataset must fit the generative assumptions of the model and also sufficient number of data points should be available to train the model and its parameters appropriately. In case of multi-dimensional data, it is difficult to define data location correctly. Thus, reliable model construction is to be performed after precise evaluation of modeling characteristics of the desired domain. Type of the data set and its size, availability of relevant typical outliers and interpretability in a model are the most important factors, affecting the decision about choosing an outlier detection approach. The analyst must realize the reason for which a particular data point is called as an outlier in terms of its relative properties with respect to the remaining members of the data set.

Statistical Models Based on Probability

Probability related statistical models tend to model the data in the form of probability distribution to train the model parameters. Thus, the most effective task is to choose the data distribution with which the modeling is performed [5]. Gaussian mixture models are generative and organize the dataset in the form of a process consisting of Gaussian clusters. Usually, expectation-maximization algorithm is utilized to train the parameters of distributions on the data set. The main result of this technique is membership probability of a given data point to different clusters, as well as the density-based fit to the modeled distribution. Thus outliers will be modeled due to their very low fit values.

Advantages	**Disadvantage**
easily applied to any data type an appropriate generative model is available for each mixture component for a mixture of different types of attributes, a product of the attribute specific generative components may be used provides a generic framework, which is relatively easy to apply	try to fit the data to a particular kind of distribution, which may often not be appropriate for the underlying data as the number of model parameters increases, over-fitting becomes more common some models are hard to interpret, especially when the parameters of the model cannot be intuitively presented to an analyst in terms of underlying attribute

Extreme Value Analysis

This is known as the most basic type of outlier detection, especially for 1-dimensional datastes. For a 1-D dataset, outliers are the data points which their values are either too large or too small along the data series. The main step in outlier detection is to determine the statistical tails of the underlying distribution. If there is a normal distribution along the data set, then analysis would be easy, because most statistical tests can be interpreted directly in terms of probabilities of significance.

Advantages	**Disadvantage**
value analysis is directly helpful many variables are often tracked as statistical aggregates, in which extreme value analysis provides useful insights about outliers can also be extended to multivariate data	applicable only where outliers are known to be present at the boundaries of the data have often not found much utility in the literature for generic outlier analysis unable to discover outlier in the sparse interior regions of a data set

It is important to understand the difference between extreme value statistics and traditional definition of outliers which consider them as detected objects by their generative probabilities rather than the extremity in their values [6]. Confusions between extreme value analysis and outlier analysis are common for chemometricians, especially in the context of multivariate data analysis. Chemometric application of extreme value analysis is naturally designed for univariate data. Nonetheless, it is also possible to be globalized for multivariate data sets, by determining the points at the multidimensional borders of the data. In case of multi-dimensional data sets, during the outlier detection procedure, a vector of outlier scores may be obtained of extreme values which enables the conversion of multiple outlier scores into a single value, and also generate a binary label output.

Linear Models

Linear models tend to reflect the data into lower dimensional subspaces by linear correlations. General steps are:

- performing a least squares to determine the optimal lower dimensional subspace
- determination of distances of the data points from drown plane in the data space
- extreme values analysis on obtained deviations to determine the outliers.

Dimension reduction would also be conducted by principal component analysis (PCA), which can be derived through multivariate regression analysis, by determining the plane which optimizes the least squares error of representation in terms of the normal distance to the plane. PCA provides the subspaces, such that by projecting the data into these subspaces, the aggregate least square errors of the residuals are minimized. The absolute sizes of these residuals can be analyzed in order to determine the outliers. Data points, which have large residuals, are more likely to be outliers, because their behavior does not conform to the natural subspace patterns in the data set.

PCA Supported Approach

PCA is an unsupervised dimension reduction method which can retain those characteristics of the data set that contribute most to its variance by few principal components which are in lower orders while contain the most important informative part of the data. Eigenvalue decomposition in the covariance matrix of the data is the main part of PCA. Only few components would remain in the data space (high eigenvalues) to represent the overall data. Output of PCA is sensitive to outliers and the direction of first few principal components will be influenced seriously if the dataset contain some outliers. Figure 2 demonstrates how the existence of an outlier would influence the PC direction. Graph A is a data set free of outlier while graph B contains 1 outlier.

Leave one out is an appropriate strategy for detection of probable outliers. In this procedure, the PCA will be modeled, omitting one of the data points. If an extremely small angle would be formed between the old first principal direction and the new one, then the left out data point is not an outlier. Else, it can be considered as outlier. The only concern for leave one out procedure is when the analyst deals with an extremely large sized data set because the computation in estimating the PC directions will be heavy due to necessity for considering all the probabilities.

Duplicate sampling would be employed to overcome this problem. In order to identify an outlier via leave one out strategy, the target data point is duplicated instead of being removing it. If it is an outlier, the deviation of PC direction would become more severe and if it is a normal data point the observed change would be slight. Figure 3 demonstrates this strategy schematically.

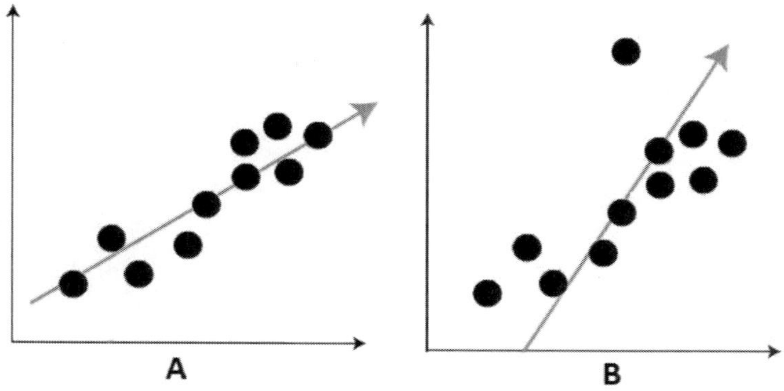

Figrue 2. Role of outliers in PCA output.

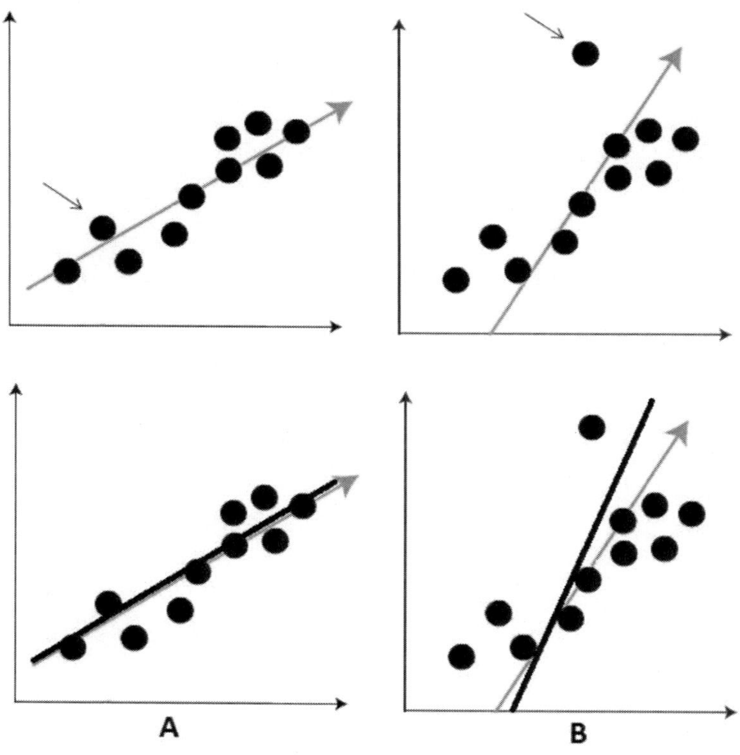

Figure 3. Role of duplicate sampling in PC direction for normal data point (A) and outlier (B).

In order to avoid the heavy computation and calculations procedures, two strategies can be proposed to accelerate the estimation of PC directions. The first one is fast updating for the covariance matrix. The other one is to solve the eigenvalue equation via the power method. During the PCA computation, it is unnecessary re-compute the covariance matrix in the leave one out procedure, completely. The difference of covariance matrix can be easily adjusted while only one data point is duplicated. Thus PCA could be a powerful and easy to perform approach by which the outlier data points are detected [7-9].

Cook's Distance

Cook's distance is useful for identification of outliers in the observations for predictor variables so it is employed to estimate the influence of a data point when we are performing least squares regression analysis [10]. It is named after the American statistician R. Dennis Cook, who introduced the concept in 1977. [11] In a regression model, high effective points cause high changes in the model equation, this high impact value is outlier and we can remove it based on its cook's distance.

In this method, first a data point with high residual in the regression model is removed, then the rate of changes of regression line is surveyed by Cook's distance .[12]

$$Hii = X(X^TX)^{-1}X^T \quad \text{COOK'distance} = \frac{e_i^2}{pMSE}\left[\frac{hii}{(1-hii)^2}\right]$$

Hii is the ith diagonal element of the hat matrix which shows the residual (i.e., the difference between the observed value and the value fitted by the proposed model). MSE is the mean square error of the regression model; P is the number of fitted parameters in the model. In case of significant variations, evaluated data point is considered as an outlier (Figure 4) [12-14].

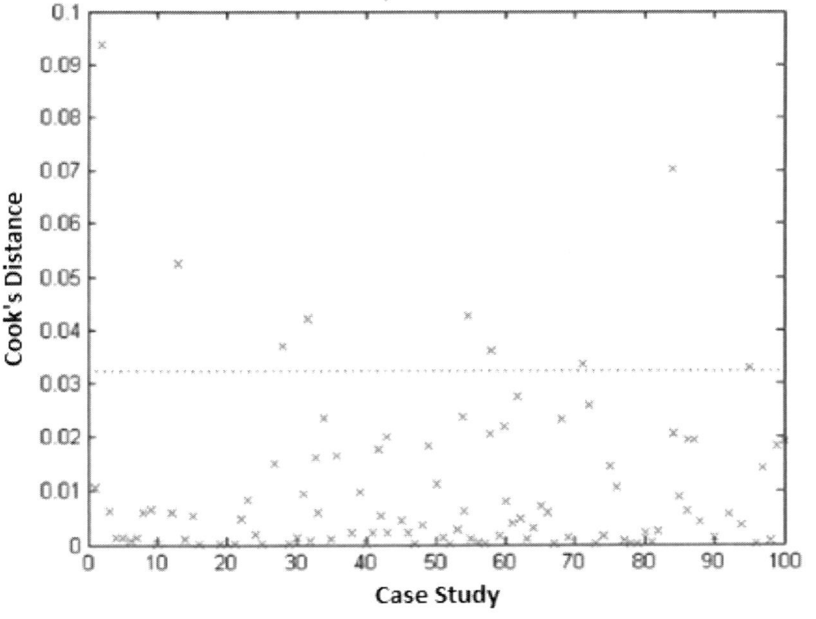

Figure 4. Typical Case order plot of Cook's distance.

Welsch and Kuh improved cook's distance by scale selection [15]. Modifications for cook's distance technique were reported [16] by different approaches e.g., evaluation of local Influence [17] extended cook's distance[18] forward plot based on modified cook's distance [19], asymptotic distribution of Cook's distance in logistic regression [20] and development of scaled Cook's distances[21].

Cook's distance as a basic method has been employed to identify outliers in many research fields e.g., in biological data of child growth, biological data of animals agricultural datasets, survey dataset of Mental Health Organizations and logistic regression models.

Advantages	**Disadvantage**
Powerful in detection of highly influential observations	Unable to detect several outliers with same value

K-Nearest Neighbors

This method is a nonparametric one introduced by Fix and Hodges 1951, which does not need to estimate parameters such as in the regression [22]. In this method there is a criterion (k) to compare objects distance with each other in the data space. In this way a cell is created with x center and let the radius of the cell partly extent which includes K^{th} data. These samples are the nearest neighbors of x. If the density of the training points around x is high, cells would become small and therefore a better result is obtained and if the density of points around x is low, cell will be large. Actually the density distribution for each point of x is calculated by its equations [23].

The method states that those objects with the largest distance to their k^{th} nearest neighbors are likely to be outliers respective to the data set, because it can be assumed that those objects have many scattered neighborhoods in compare with the average of all of objects. As this effectively provides a simple ranking over all the objects in the data set according to the distance to their k^{th} nearest neighbors, the user can specify a number of n objects to be the top-n outliers in the data set.

Employing k-nearest neighbor algorithm for outlier detection needs to estimate optimal distance based on measuring of distance. Some of the most common approaches for this aim are Euclidean distance and Mahalanobis distance [22,23]. Euclidean distance is usually used to determine the distance when there is j objects in data space ;in Cartesian coordinates, if $p = (p_1, p_2,..., p_n)$ and $q = (q_1, q_2,..., q_n)$ are two points in Euclidean n-space, then the distance from p to q, or from q to p is given by [24]:

$$\text{Euclidean distance} = \sqrt{\sum_{i=1}^{n} (q_1 - p_1)^2 + (q_2 - p_2)^2 + \cdots + (q_n - p_n)^2}$$

$$\text{Mahalanobis distance} = \sqrt{(x - m)^T C^{(-1)} (x - m)}$$

where x is data vector, m is vector of mean values of independent variables, C^{-1} is inverse covariance matrix of independent variables and T is indicate vector should be transposed.

How many neighbors is the best amount? This is a challenge of this approach. [25]. We can utilize k and realize that unknown object belongs to which class. Actually in this method the data have normal distribution and we can classify them. This method involves two important factors:

- Selection of the function and their weights
- Choosing the best number of neighbors

In order to investigate the quality of assessment, root mean squared error, correlation coefficient, mean size and the error estimates are determined. One of the challenges of this algorithm is its running time. In order to overcome this defect, we can reduce data space by sampling of initial data. [23]. Another possibility is to is to split the data set into clusters, then detecting the outliers in each cluster.

Over the years, several reported papers have been published, dealing with developments for this method e.g., proposing mutual k nearest neighbor graph for outlier detection, introducing an optimized k-nearest neighbor algorithm to identify outliers, using new algorithm basis of k-nearest neighbor for rapid outlier detection and presenting approximated k-nearest neighbors search algorithm [26-28].

Recently various techniques of k-nearest neighbors have been demonstrated as useful approaches for detection of outliers in different domains. Some of the most important examples are:

- direct application in mammographic images [29]
- in combination with model graph for statistical data analysis of players performance [30]
- optimized algorithm in network connections [31]
- reverse algorithm for credit card fraud detection [32]
- novel algorithm b in real-life NBA database [24]

Advantages	**Disadvantages**
Its simplicity Yields competitive results when cleverly combined with prior knowledge, it has significantly advanced the state-of-the-art Ability of it to scale to large datasets Analytically tractable simple implementation Uses local information, which can yield highly adaptive behavior	better performance is observed on most of the large data sets. Computationally intensive recall Highly susceptible to the curse of dimensionality The number of model parameters grow exponentially with number of locations spatial dependency also needs to be captured

Leverage Value

In statistics, leverage is a manner used in connection with regression analysis introduced in 1978 by D.C. Hoaglin and R. Welsch [32]. In particular, it is used in analytical chemometric data processing to identify those observations that are far away from corresponding average predictor values. Leverage points do not necessarily have a large effect on the outcome of fitting regression models [33].

Leverage points are those observations, if any, made outlying values of the independent variables such that the lack of neighboring observations means that the fitted regression model will pass close to that particular observation [33, 34]. Standardized coefficients or refer to how many standard deviations a dependent variable will change, per standard deviation increase in the predictor variable. Standardization of the coefficient is usually done to answer the question which of the independent variables has a greater effect on the dependent variable in a multiple regression analysis, when the variables are measured in different units of measurement [35]. If the leverage of points is high, our estimate of the $\boldsymbol{\beta}$ coefficient will be wrong. Leverage is a tool to survey extreme variables.

Actually this method is useful for outlier detection by survey of points with large residuals or high leverages [34]. Each sample's leverage is considered to detect outliers in the space of x matrix. Leverage of a sample is defined as its distance from the main samples distance in the x matrix space (Figure 5).

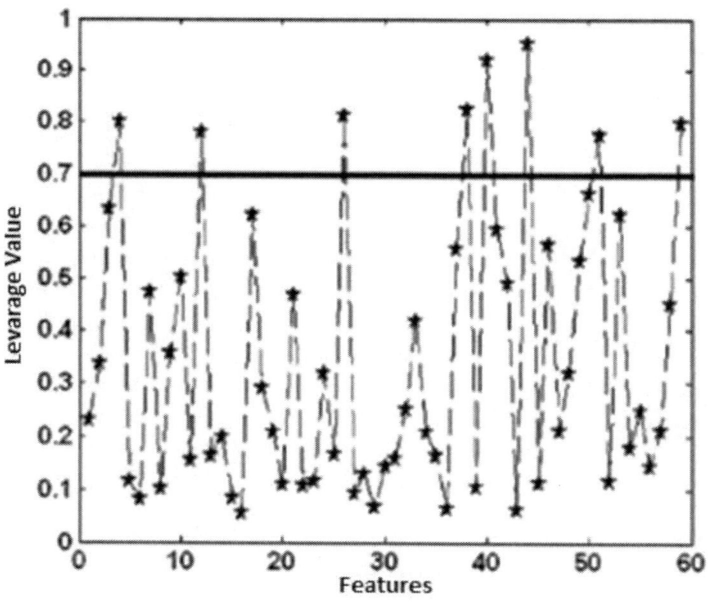

Figure 5. Typical plot of leverage based outlier detection (Threshold is 0.7 here).

This value shows that if a unique sample is an outlier, how is it effective on the estimate of parameters?

For given information in the x matrix with M × N dimension, the leverage at the Ith sample is given by the diagonal elements (pii) of the leverage matrix (P), calculated according to following equation [34, 36]:

P=X 〖〖(XX〗 ^T)〗 ^(-1) X^T

Samples with h greater than h_{limit} from the set of processed points are eliminated and model will be repeated again. For a spectroscopic data set:

M = number of samples
N = number of wavelength

In the next step, by forming a matrix, its diagonal elements are compared with the h_{limit} and if they were given greater distractions, they would be eliminated. The diagonal elements of P matrix are considered to realize whether the leverage is high or not and this points (points with high leverage) are very effective on the estimate of parameters .Thereby much attention must be paid to them [36, 37].

Since application of k-clustring over modified hat matrix to identify influential sets [38], there are several developments in this technique e.g., proposing elemental sets [39], using the off diagonal elements of hat matrix and the modified hat matrix [40], and recently presented discriminative hat matrix for detection of outliers [41].

Simplicity and efficiency of this method has resulted in various domains to attend it. Leverage based procedures have shown very good results for detecting outliers for example in ecosystem dataset [43, 44], spectroscopy [45, 46] , statistic dataset [47] or fuel data set[48,49].

Advantages	**Disadvantages**
Understanding leverage is essential in regression because leverage exposes the potential role of individual data points combines distance and leverage to identify unusually influential observations	High leverage observations might don't be influential observation. Influential observations might don't not be outliers.

Mahalanobis Distance

The Mahalanobis distance is a descriptive statistic topic that provides a relative measure of a data point's distance (residual) from a desired point. It is a unit less measure introduced by P. C. Mahalanobis in 1936[50]. Mahalanobis distance is defined versus Euclidean distance.

In this distance it is assumed that there is a normal distribution curve for data in it (it means these data are predictable). In mathematics, the Euclidean distance between two points is a typical distance which is obtained from Pythagorean theorem. The distance between two points (p and q) is the linear segment that connects them to each other. In Cartesian coordinates,if $p=(p_1,p_2,\dots,p_n)$ and $q=(q_1,q_2,\dots,q_n)$ are two points in a dimensional Euclidean space, then the distance between them is defined as: [51,52]

$$D(p,q) = \sqrt{(p_1-q_1)^2 + (p_2-q_2)^2 + \dots + (p_n-q_n)^2} = \sqrt{\sum_{i=1}^{n}(p_i-q_i)^2}$$

This distance is calculated so quickly and easily, but includes some problems:

- In terms of geometric aspects, all variables are measured in the same units while the real variables may be in different scales and natures, such as temperature, pressure, etc. which are not comparable.
- Euclidean distance does not take into account the correlated variables. As an example, assume a dataset that contains five variables which one of them is exactly replicates of another variable, then these two variables must have correlation, but Euclidean distance does not give us any information about this correlation. Consequence of this matter is that Euclidean distance only determines the distance between two points the same as measured distances by a ruler. The identity matrix can be entered in the Euclidean distance which does not affect the calculation of the distance. Mahalanobis distance is the same as Euclidean distance with this difference that inverse of covariance matrix is used instead of identity matrix in its equation. This matrix poses the weight matrix's properties, i.e., all entries are positive [50,52,53].

Mahalanobis distance's features (This items enter into the calculation) are: [54]

- the variance is different in each direction
- covariance between variables is computed.

A function to evaluate that the data is outlier or not is defined now. The realted equation defines the maximum and minimum values which are inside the normal data. Thus the equation relies that largest amount is

$$\frac{n-1}{p}$$

n: number of observations
p: number of dimensions

and if the function output for a data point is out of this value, it will be an outlier. In order to perform an easier comparison, the function result is scaled and weighted locally between 0 and 1 [53].

Distance is calcualted by:

$$D^2 = (x - m)^T C^{-1} (x - m)$$

where
D^2 is Mahalanobis distance

D is matrix with m*n dimensions (m is number of samples and n is number of classes)
x is the data vector
m is the vector of mean values of independent variables
C^{-1} is the inverse covariance matricx of independent variables
T is the indicate vector that should be transposed

In recent years, the Mahalanobis distances based approaches have been utilized along with robust estimators in order to provide appropriate measuring of the mean and covariance matrix for outlier detection. Some of the most interesting examples are minimum covariance determinant (MCD) estimator [54,55], minimum volume ellipsoid (MVE) [56], fast MCD [57,58], hybrid method [59], modified Stahel-Donoho estimators [60], resampling by half-means (RHM) and smallest half volume (SHV) [62] which have been in improvement of this technique. However there had been a growing attentions towards this technique and recently locally centered squared Mahalanobis distance was introduced for outlier detection in complex multivariate data sets [53].

Several reports have illustrated the practical application of Mahalanobis distance based appraoches for outlier detection. It has been successfully applied for the linear and logistic regression [65]. Mahalanobis distance using modified Stahel-Donoho estimators has been used in domain industrial information [61], locally centered squared Mahalanobis distance has been examined in treatment of biological data [53]. The technique in combination with MCD has been employed in medical imaging [57], and in combination with fast MCD in satellite observations [60]. RHM and SHV have been coupled with it for dataset of blast furnace. [66]

Advantages	**Disadvantages**
considers the correlations between the variables that is very important in pattern recognition	if the number of observation of each class be less than dimension number, a singular covariance matrix will be obtained
The region of constant Mahalanobis distance around the mean forms an ellipse, ellipsoid or hyper ellipsoid for multidimensional datasets	equal adding up of the variance normalized squared distances of the feature which makes troubles for noisy data sets
It is useful for time domain data and observations	Difficult to detect individual outliers
Possible to use robust and independent choices for the centroid and covariance matrix values	If the clusters are not well defined , its measure would not be well defined
provides an alternative to regression techniques when there is no obvious value to be predicted	

Regression

The concept of regression comes from genetics and was popularized by Sir Francis Galton during the late 19th century with the publication of Regression towards mediocrity in hereditary stature [67-70].

Usually the main part of our information about two or several participating variables is achievable in terms of the relationship between them. Depending on the variables nature, checking can be performed by regression analysis or their correlation with each other should perform to assess the association between two or more variables. It is very important to know whether the distribution of dependent variables is normal or not and also if variance for different values of the independent variables is constant [71].

Several important parts of regression analysis are

- investigation of the linearity assumption
- normality
- immutability of variance
- independence of observations

While a model is fitted to data, the residuals play a very important role. By examining the distribution of residuals and their relationship with other variables can be determined the exactness of regression assumptions [72].

Generally, the difference between the observed value of dependent variables and the value which is predicted by the regression line is residual. If we survey in 2 variables state, first all the data points are drawn, then the line is crossed while pass from all points (by scatter plot) [70]. In the multivariate regression, the equation which predicts the response variable as a linear function of p variable is estimated by data. Regarding the statistical aspect, we can write multiple regression equation as follows [71]:

$$Y = \beta_0 + \beta_1 X_1 + \beta_2 X_2 + \cdots + \beta_p X_p + e$$

Y: dependent variable
X: independent variable
e: error term (a vector of random variables)

The estimate of β vector is selected such that the sum of squares becomes minimum. Probable existence of outlier would cause some trouble due to its influence on slope and intercept of the regression line.

Outliers pull the regression line towards themselves [70,71]. During the regression line fitting, the data points which are far from the estimated line and demonstrate high residual value are known as outlier data. There are also some data points for which the value of independent variable in far from mean value after regression. These data points are known as leverage. Actually high leverage up warnings that their related data points can impact on the regression line (Figure 6) [70].

Nowadays there are several methods to deal with outlying data in regression models. The most commonly used are least absolute values (LAV) regression [72], ordinary least squares (OLS) regression, least median of squares (LMS) regression, the least trimmed squares (LTS) regression [73].

Regression models have been widely applied in different fields but about the removal outliers in regression models can be point to use least median square regression in calibration data of electrochemistry for outlier detection[74], statistic information[75], LMS regression and forward search have been applied in physical chemistry data [76], difference algorithms have been compared in air pollution dataset [77] ans several multiple outlier detection procedures have been used in children's performance management [71].

Advantages	Disadvantages
Capable of detecting multiple outliers suitable for the cases in which most of the data is distributed along linear correlation planes can also be used in a limited way over discrete attribute values, when the number of possible values of an attribute is not too large In case of almost linear relationship between independent and dependent variables, shows optimal results	difficult to formulate an exact procedure for the multiple outlier case in multivariate regression may work poorly, if the underlying data is clustered arbitrarily only helpful when the relationship between variables is linear severely depends on the assumption about the error distribution Partitioning the independent and dependent variables is mandatory

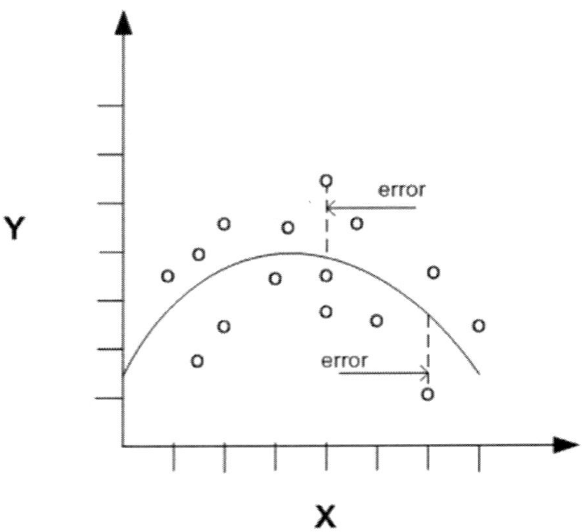

Figure 6. Regression strategy for outlier detection.

M-Estimators

M-estimators use maximum likelihood formulations by deriving optimal weighting for the data set in non-normal conditions. It was introduced by Huber [78, 79] as a generalization of the familiar least squares criterion [80]. The maximum likelihood estimator is robust and can calculate the distribution center and for data that are far from center takes less weight .In statistics maximum likelihood estimators are a large range of estimators that obtain as the minimum sum of functions of the data. In fact it is verfied maximum likelihood estimators that use a weight function to down weight, extreme values [79, 81].

The least square estimator, is the estimator of the type of maximum likelihood estimators [81], and M estimator is the same as the least square estimator which in the robust case define a new case to it [78,79] i.e., least-squares estimators are M-estimators. It is obtained as the minima of sums of functions of the data. We can use M-estimates to find a balance between efficiency and the resistance of the unusual observations [78,79] which is an important class of robust estimates [82].

When the model is Gaussian one, the mean is not very efficient, and in order to make a comparison between robustness and efficiency we can use M- estimator. M- estimator is more mathematically complex, but that's what computers are good to the center of the distribution and give less weight to values that are further away from center. Different M-Estimators give different weights for deviating values e.g., MM estimation which is a sub-method of estimation MM estimation is a special type of M-estimation which performs the best overall against a comprehensive set of outlier conditions [83,84]while high breakdown estimation and efficient estimation are combined. It is known as the first estimate with a high breakdown point and high efficiency under normal error [85]. Maximum likelihood estimation to the minimization of $\sum_{i=1}^{n} \rho(x_i, \theta)$, where ρ is a function with certain properties (see below). The solution is [86]:

$$\hat{\theta}=\arg\min_{\theta}\left(\sum_{i=1}^{\theta} \rho(x_i, \theta)\right)$$

The function ρ, can be chosen in such a way to provide the estimator desirable properties when the data are aright from the assumed distribution, and 'not bad' behavior when the data are generated from a model that is, in some understand, close to the assumed distribution. [87,88]. Hampel in 1986 introduced ψ function with three part M-estimators [89] which was later modified by bi-weight estimators that biweight is one member of the family of m-estimators used to estimate locations. [90]. Later new ρ-function was proposed with its corresponding ψ and w-functions, thus giving development to a new weighted least square method [91]. Modified robust M-estimate has been also applied to find the outliers in leverage points [91]. Redescending M-estimators are Ψ-type M-estimators which have ψ functions that are non-decreasing near the origin, but decreasing toward 0 far from the origin. Due to properties of the ψ function, these kinds of estimators are very efficient, have a high breakdown point and, unlike other outlier rejection techniques, they do not suffer from a masking effect. They are efficient because they completely reject gross outliers, and do not completely ignore moderately large outliers (like median). Recently a new descending M-estimator, called Alamgir redescending M- estimator has been introduced that is based on a modified tangent hyperbolic (tanh) type weight function [92,93].

Several reports of successful applications of this method to remove outliers can be found in the literature. Re-descending M-estimator has been used in statistics data for price growth studies [80], a in medical data dealing with cardiovascular diseases [91]. The other examples are modified M-estimator used in biological data [89], or a new robust algorithm used in real image data [95]. Also M-estimator has been compared with other methods in satellite images processing for outlier detection [96].

Advantages	**Disadvantages**
robust techniques to estimate location and scale in the presence of outliers Combining robustness with efficiency under the regression model with normal errors robust against any type of outliers	not very resistant to leverage points does not have rigorous statistical basis unable to identify multiple leverage points

REFERENCES

[1] Aggarwal C.C., 2013, Outlier Analysis, Wiley, New York

[2] Aggarwal C.C., Yu P.S. Outlier Detection in High Dimensional Data, *ACM SIGMOD Conference*, 2001.

[3] Hazel G.G. *GeoRS*, 2000, 38(3), 1199.

[4] Aggarwal C.C. On the Effects of Dimensionality Reduction on High Dimensional Similarity Search, *ACM PODS Conference*, 2001.

[5] Desforges M., Jacob P., Cooper J., *Proceedings of Institute of Mechanical Engineers*, 1998, 212, 687.

[6] Ruts I., Rousseeuw P., Computational Statistics and Data Analysis, 1996, 23, 153.

[7] Chandola V., Banerjee A., Kumar V., Anomaly detection: a survey. ACM Computing Surveys, 2009.

[8] Erdogmus D., Rao Y., Peddaneni H., Hegde A., Principe J.C., *Jornal of Applied Signal Process.* 2004, 13, 2034.

[9] Yeh Y., Lee Z.Y., Lee Y.J., New Advances in Intelligent Decision Technologies: *Studies in Computational Intelligence*, 2009, 199, 449.

[10] Mendenhall W., Sincich T., 1996, A Second Course in Statistics: Regression Analysis (5th ed.). Upper Saddle River, NJ: Prentice-Hall.

[11] Cook R. D., *Technometricso,* 1977, 19, 1.

[12] Jagadeeswari T., Harini N., Satya Kumar C., *International Journal Of Engineering And Computer Science* , 2013, 2 , 2045-2049.

[13] Cousineau D., Chartier S., *International Journal of Psychological Research*, 2010, 3, 1.

[14] Barrett B. E., Gray J. B., *Computational Statistics & Data Analysis* , 1997, 26, 39-52.

[15] Welsch R. E.and Kuh E., *Linear regression diagnostics. Technical report* , 1977, 923-77, *Solan School of Management, Massachusetts Institute of Technology.*

[16] Atkinson A.C., Journal of the Royal Statistical Society, 1982, *Series B, Methodological,* 44, 1-36.

[17] Cook R. D., Roy J., *Statist. Soc*, 1986, B, 48, 133-69.

[18] Korn E. L., and Graubard B. I., *The American Statistician*, 1990, 44, 270-276.

[19] Atkinson A. C., and Riani M., 2000, *Robust Diagnostic Regression Analysis,* New York: Springer-Verlag.

[20] Martin N. and Pardo L., *Journal of Applied Statistics*, 2009, 36 , 10, 1119-1146.

[21] Zhu H., Ibrahim J. G. and Cho H., *the annals of statistics*, 2012, 40, 2, 785–811.

[22] Fix E., Hodges J. L., Technical Report 4, 1951, USAF *School of Aviation Medicine, Randolph* Field, Texas.

[23] Tran T. N., Wehrens R., Lutgarde M. C., *Computational Statistics & Data Analysis,* 2006, 51, 513 – 525.

[24] Zhao J., Long C., Xiong S., Liu C., Yuan Z., *Elektronika ir Elektrotechnika,* 2014, 20, 1 , 1392-1215.

[25] Deza M. M., Deza E., Encyclopedia of Distances, 2009, Springer, Library of Congress Control Number: 2009921824.

[26] Kim S., Wook Cho N., Kang B., Kang S. H., *Expert Systems with Applications* , 2011, 68, 9587–9596.

[27] Weinberger K. Q., Saul L. K., *Journal of Machine Learning Research*, 2009, 10, 207-244.
[28] Duda R.O., Hart P. E., *Pattern classification and scene analysis*, 1973, Wiley, New York.
[29] Chen Y., Miao D., Zhang H., *Expert Systems with Applications*, 2010, 37, 8745–8749.
[30] Hautamaki V., Karkkainen I. and Franti P., *Proceedings of the 17th International Conference on Pattern Recognition (*ICPR'04).1051-4651/04 $ 20.00 IEE.
[31] Prerau M. J., Eskin E., http://www.music.columbia.edu/ ~mike/publications /thesis.pdf
[32] Ratnam Ganji V.and et al. / *International Journal on Computer Science and Engineering* (IJCSE), 2012, 4, 06, 1035-1039.
[33] Hoaglin D.C. and Welsch R., *American Statistician.JSTOR*, 1978, 32, 1 , 17-22.
[34] Everitt B. S., 2002, Cambridge Dictionary of Statistics.CPU.*ISBN* 0-521-81099-X.
[35] Habshah M., Norazan M. R., Rahmatullah Imon A. H. M., *Journal of Applied Statistics*, 2009, 36, 5, 507–520.
[36] link: Glossary of social science terms
[37] Barrett B.E., Gray J.B., *Computational Statistics & Data Analysis*, 1997, 26, 39-52.
[38] Rousseeuw P. J. and Hubert M., Robust statistics for outlier detection, John Wiley & Sons, Inc. *WIREs Data Mining Knowl Discov*, 2011, 1, 73–79.
[39] Gray J. B., Ling R. F., *thechnometrics,* 1984, 26, 305-330.
[40] Hawkins D. M., Bradu D., Kass G. V., *Technometrics* , 1984, 26, 3, 197-208.
[41] Seung Gu K., Joon Keun Y., *Journal of the KSQC*, 1989, 17, 1, 107-115.
[42] Dufrenois F. and Noyer, neural networks(IJCNN) the 2011 *international joint Conference* on, 777-784.
[43] Hoaglin D. C. and Welsch R. E., *The American Statistician*, 1978, 32, 1, 17-22.
[44] Salehi E., Abdi J., Aliei M. H., *Journal of Saudi Chemical Society*, 2014 , Available online 13 March 2014.
[45] Yuping W., Dissertation, *Optimization methods for quantitative measurement of glucose based on near-infrared spectroscopy,* 2008, 237, 3347257.
[46] Khanmohammadi M., Bagheri Garmarudi A., Ghasemi K., and et al., *Microchemical Journal,* 2009, 91, 47–52.
[47] Ampanthong P., Suwattee P., *Proceedings of the International MultiConference of Engineers and Computer Scientists* 2009 Vol I, IMECS 2009, March 18 - 20, 2009, Hong Kong.
[48] Khanmohammadi M., Bagheri Garmarudi A., De la Guardia M., *Talanta,* 2013, 104, 128-134.
[49] Khanmohammadi M., Bagheri Garmarudi A., Ghasemi K. De la Guardia M., Fuel, 2013, 111, 96-102.
[50] Mahalanobis, Prasanta Chandra. *Proceedings of the National Institute of Sciences of India,* 1936, 2, 1, 49–55.
[51] Roy D. M., Delphine J. R., and Désiré L. M., *Chemometrics and Intelligent Laboratory Systems*, 2000, 50, 1–18.
[52] Gnanadesikan R., and Kettenring J. R., *Biometrics,* 1952, 28, 81–124.
[53] Todeschini R. and et al., *Analytica Chimica Acta*, 2013, 787, 1– 9.
[54] http://blogs.sas.com/content/iml/2012/02/15/what-is-mahalanobis-distance/
[55] Deza E. & Deza M. M., Encyclopedia of Distances, 2009, page 94, Springer.

[56] De Maesschalck R., Jouan-Rimbaud D., Massart D. L., *Chemometrics and Intelligent Laboratory Systems* , 2000, 50, 1–18.
[57] Fritsch V. and et al., Medical *Image Analysis*, 2012, 16, 1359–1370.
[58] Rousseeuw P. J., van Zomeren B. C., *J. Am. Stat. Assoc.*, 1990, 85, 633–651.
[59] Bartkowaik A., *biocybernetrics and biomedical engineering* , 2005, 25, 1 , 7-21.
[60] Giménez E. and et al., *International Journal of Applied Earth Observation and Geoinformation,* 2012, 16, 94–100.
[61] Franklin S., Thomas S., Brodeur M., http://www.amstat.org/meetings/ices/2000 /proceedings/S33.pdf
[62] Egan W. J., Morgan S. L., *Analytical Chemistry*, 1998, 70, 2372–2379.
[63] Kapoor S., Khanna S., Bhatia R., (IJCSIS) *International Journal of Computer Science and Information Security,* 2010, 7, 2 , 267-272.
[64] Warren R., Smith R. F., Cybenko A. K., Use of Mahalanobis Distance for Detecting Outliers and Outlier Clusters in Markedly Non-Normal Data: A Vehicular Traffic Example Interim rept. Mar 2009-Jun 2011.
[65] Tiwari K. and et al., NESUG 2007 Statistics and Data Analysis.
[66] Zeng J., Gao Ch., *Journal of Process Control*, 2009, 19, 1519–1528.
[67] Bulmer M., Galton F., Pioneer of Heredity and Biometry. *Johns Hopkins University* Press., 2003.
[68] Galton F., *The Journal of the Anthropological Institute of Great Britain and Ireland*, 1886, 15, 246–263.
[69] Clauser B. E., Journal of Educational and Behavioral Statistics, 2007, 32, 4 , 440-444.
[70] Cao D. S., Liang Y. Z., Xu Q. S., Li H. D., Chen X., *J. Comput. Chem.,* 2010, 31, 3, 592-602.
[71] http://www.bisrg.uwaterloo.ca/archive/RR-06-07.pdf
[72] http://www.sagepub.com/upm-data/17839_Chapter_4.pdf
[73] Rousseeuw P.J., *J. Amer. Statist. Assoc.*, 1984, 79, 871–880.
[74] Cruz Ortiz M., et al., *Talanta*, 2006, 70 , 499–512.
[75] http://statistika.vse.cz/konference/amse/PDF/Blatna.pdf
[76] http://www.iasri.res.in/seminar/AS-299
[77] McCann L., Thesis (Ph. D.) Massachusetts Institute of Technology, Sloan School of Management, Operations Research Center, 2006.Includes bibliographical references (p. 191-196).
[78] Huber P. J., Ann Math Statist, 1964, 35, 1, 73–101.
[79] Yu C., Yao W., and Bai X., http://www-personal.ksu.edu/ ~wxyao/material/submitted /reviewrobreg.pdf
[80] Ullah I., Qadir M. F., Ali A., *Pak. j. stat. oper. res.,* 2006, II, 2 , 135-144.
[81] Franklin S., Thomas S., Brodeur M., Statistics Canada, 2000.
[82] Bellio R., Ventura L., 2005, An Introduction to Robust Estimation with R Functions
[83] Özlem Gürünlü Alma, *Int. J. Contemp. Math. Sciences,* 2011, 6 , 9, 409 – 421,
[84] Yohai V. J., *The Annals of Statistics*, 1987, 15, 642-656.
[85] Stromberg A. J., *Journal of the American Statistical Association*, 1993, 88, 421.
[86] Maumet C. , Maurel P., Ferré J. C. , Barillot C., *Magnetic Resonance Imaging* , 2014, 32, 497–504.
[87] Andrews D. F., *Technometrics*, 1974, 16, 523-531.

[88] Hampel F. R., Ronchetti E. M., Rousseeuw P. J., and Stahel W. A., 1986, Robust Statistics: The Approach Based on Influence Functions, John Wiley & Sons, New York.
[89] Mosteller F., and Tukey J. W., Data Analysis and Regression, 1977 Addison- Wesley Publishing Company, Inc. Philippines.
[90] Ali A., Qadir M. F., PJSOR, 2005, 1, 49-64.
[91] Koch K. R., AVN, 2007, 10, 330-336.
[92] Alamgir, A. Amjad and et al., *Research Journal of Recent Sciences*, 2013, 2, 8, 79-91.
[93] Rousseeuw P., LeRoy A., Robust regression and outlier detection, *Probability and mathematical statistics.* Wiley-Interscience; 2003.
[94] Neugebauer, Shawn Patrick, Robust Analysis of M-Estimators of Nonlinear Models 1996, Master's Thesis
[95] Huang J. F., Lai S. H., Cheng C. M., *Journal of information science and engineering*, 2007, 23, 1213-1225.
[96] Knight N. L. and Wang J., *Journal of Navigation*, 2009, 62, 04, 699-709.

INDEX

#

A

B

C

L

M

N

O

P

Q

R

S

T

U

V

W